# METHODS IN MOLECULAR BIOLOGY



*Series Editor*
**John M. Walker**
**School of Life and Medical Sciences**
**University of Hertfordshire**
**Hatfield, Hertfordshire, UK**

For further volumes:
http://www.springer.com/series/7651

For over 35 years, biological scientists have come to rely on the research protocols and methodologies in the critically acclaimed *Methods in Molecular Biology* series. The series was the first to introduce the step-by-step protocols approach that has become the standard in all biomedical protocol publishing. Each protocol is provided in readily-reproducible step-by-step fashion, opening with an introductory overview, a list of the materials and reagents needed to complete the experiment, and followed by a detailed procedure that is supported with a helpful notes section offering tips and tricks of the trade as well as troubleshooting advice. These hallmark features were introduced by series editor Dr. John Walker and constitute the key ingredient in each and every volume of the *Methods in Molecular Biology* series. Tested and trusted, comprehensive and reliable, all protocols from the series are indexed in PubMed.

# Animal Models of Diabetes

## Methods and Protocols

Edited by

**Aileen J. F. King**

*Department of Diabetes, School of Life Course Sciences, King's College London, London, UK*

Humana Press

*Editor*
Aileen J. F. King
Department of Diabetes
School of Life Course Sciences
King's College London
London, UK

ISSN 1064-3745 ISSN 1940-6029 (electronic)
Methods in Molecular Biology
ISBN 978-1-0716-0387-1 ISBN 978-1-0716-0385-7 (eBook)
https://doi.org/10.1007/978-1-0716-0385-7



This Humana imprint is published by the registered company Springer Science+Business Media, LLC, part of Springer Nature.
The registered company address is: 1 New York Plaza, New York, NY 10004, U.S.A.

# Preface

Mechanisms of glucose homeostasis are well conserved between species, and therefore there is much to be learned through the large variety of animal models of diabetes currently available. Normal blood glucose homeostasis is dependent on the actions of insulin, which is secreted by the pancreatic beta cells in the islets of Langerhans. Diabetes is a collection of diseases characterized by the development of hyperglycemia due to inadequate supply and/or action of insulin. Some animal models develop diabetes spontaneously with a pathogenesis similar to that seen in humans. Indeed, several models of diabetes that are now commonly used were discovered rather than designed. Other models have been genetically manipulated or selectively bred, or chemicals or dietary modifications have been used to induce a phenotype of relevance to studies of diabetes. Moreover, some models do not develop overt diabetes, but nonetheless play key roles in understanding fundamental aspects of glucose homeostasis and islet biology. In common with many animal models of disease, no single model perfectly recapitulates all aspects of the disease: the choice of animal model will be determined by the experimental question, and it is recommended that any key findings should be validated in an additional model.

The first chapter of this book outlines some important considerations when working with mouse models of diabetes, highlighting factors that new investigators to the field may be unaware of and which could potentially affect experimental outcome. Subsequent chapters describe the main characteristics of some of the most commonly used animal models in diabetes research, ranging from mice to larger mammals such as pigs and primates. Many of these animals develop signs of diabetes that are well documented in human patients (such as hyperglycemia, impaired insulin secretion, impaired insulin action). These models are useful in studying the pathogenesis of diabetes and for testing novel therapies. This book also includes chapters that describe animal models that may not develop diabetes but are useful for studying specific aspects of beta cell biology such as proliferation. The last chapters focus on how to assess blood glucose homeostasis, insulin action, and islet function in vivo and ex vivo. Indeed, the endpoints of many studies include ex vivo analysis of tissues (most commonly pancreas and insulin-sensitive tissues). It is anticipated that this book will help diabetes researchers design and carry out in vivo studies, which best suit their experimental question.

*London, UK* *Aileen J. F. King*

# Contents

# Contributors


TANJA ARNDT • *Institute of Clinical Biochemistry, Hannover Medical School, Hannover, Germany*

AKIRA ASAI • *Department of Endocrinology, Diabetes and Metabolism, Graduate School of Medicine, Nippon Medical School, Tokyo, Japan; Food and Health Science Research Unit, Graduate School of Agricultural Science, Tohoku University, Sendai, Japan*

PATRICIO ATANES • *Department of Diabetes, School of Life Course Sciences, King's College London, London, UK*

GAVIN A. BEWICK • *Department of Diabetes, School of Life Course Sciences, Faculty of Life Science and Medicine, King's College London, London, UK*

JOANNE BOLDISON • *Division of Infection and Immunity, Cardiff University School of Medicine, Cardiff, UK*

JAMES E. BOWE • *Department of Diabetes, School of Life Course Sciences, King's College London, London, UK*

CHUNGUANG CHEN • *Paul Langerhans Institute Dresden (PLID) of the Helmholtz Zentrum München at the University Clinic Carl Gustav Carus of Technische Universität Dresden, Helmholtz Zentrum München, Neuherberg, Germany; Institute of Physiology, Faculty of Medicine, Technische Universität Dresden, Dresden, Germany; German Center for Diabetes Research (DZD), München-Neuherberg, Germany*

DAWEI CHEN • *Division of Infection and Immunity, Cardiff University School of Medicine, Cardiff, UK*

GUSTAF CHRISTOFFERSSON • *Department of Medical Cell Biology, Uppsala University, Uppsala, Sweden*

CHRISTIAN M. COHRS • *Paul Langerhans Institute Dresden (PLID) of the Helmholtz Zentrum München at the University Clinic Carl Gustav Carus of Technische Universität Dresden, Helmholtz Zentrum München, Neuherberg, Germany; Institute of Physiology, Faculty of Medicine, Technische Universität Dresden, Dresden, Germany; German Center for Diabetes Research (DZD), München-Neuherberg, Germany*

CRISTINA COSENTINO • *Université Libre de Bruxelles, Bruxelles, Belgium*

IRENE CÓZAR-CASTELLANO • *Instituto de Biología y Genética Molecular (Universidad de Valladolid-CSIC), Valladolid, Spain*

LYDIA F. DANIELS GATWARD • *Department of Diabetes, School of Life Course Sciences, King's College London, London, UK*

LENA ELIASSON • *Islet Cell Exocytosis, Lund University Diabetes Centre, Department of Clinical Sciences Malmö, Lund University, Malmö, Sweden; Clinical Research Centre, Skåne University Hospital, Lund and Malmö, Sweden*

MATTHIAS ELSNER • *Institute of Clinical Biochemistry, Hannover Medical School, Hannover, Germany*

JONATHAN LOU S. ESGUERRA • *Islet Cell Exocytosis, Lund University Diabetes Centre, Department of Clinical Sciences Malmö, Lund University, Malmö, Sweden; Clinical Research Centre, Skåne University Hospital, Lund and Malmö, Sweden*

ELISABET ESTIL·LES • *CIBER of Diabetes and Associated Metabolic Diseases (CIBERDEM), Barcelona, Spain; Bellvitge Biomedical Research Institute (IDIBELL), Barcelona, Spain*

MALIN FLODSTRÖM-TULLBERG • *The Center for Infectious Medicine (CIM), Department of Medicine Huddinge, Karolinska Institutet and Karolinska University Hospital Huddinge, Stockholm, Sweden*
THOMAS G. HILL • *Department of Diabetes, School of Life Course Sciences, King's College London, London, UK*
MARIANA IGOILLO-ESTEVE • *Université Libre de Bruxelles, Bruxelles, Belgium*
ANNE JÖRNS • *Institute of Clinical Biochemistry, Hannover Medical School, Hannover, Germany*
DIMITRI KAKABADSE • *Division of Infection and Immunity, Cardiff University School of Medicine, Cardiff, UK*
MATILDA R. KENNARD • *Department of Diabetes, School of Life Course Sciences, King's College London, London, UK*
AILEEN J. F. KING • *Department of Diabetes, School of Life Course Sciences, King's College London, London, UK*
SIGURD LENZEN • *Institute of Experimental Diabetes Research, Hannover Medical School, Hannover, Germany; Institute of Clinical Biochemistry, Hannover Medical School, Hannover, Germany*
BARBARA LUDWIG • *Department of Medicine III, University Hospital Carl Gustav Carus, Dresden, Germany; Paul Langerhans Institute Dresden (PLID) of the Helmholtz Center Munich at the University Hospital Carl Gustav Carus and Faculty of Medicine of the TU Dresden, Dresden, Germany; DFG-Center for Regenerative Therapies Dresden, Technische Universität Dresden, Dresden, Germany; Department of Endocrinology and Diabetology, University Hospital Zurich, Zurich, Switzerland*
STEFAN LUDWIG • *Department of Visceral, Thoracic and Vascular Surgery, University Hospital Carl Gustav Carus, Dresden, Germany*
THOMAS A. LUTZ • *Institute of Veterinary Physiology, Vetsuisse Faculty University of Zurich, Zurich, Switzerland*
EDUARD MONTANYA • *CIBER of Diabetes and Associated Metabolic Diseases (CIBERDEM), Barcelona, Spain; Bellvitge Biomedical Research Institute (IDIBELL), Barcelona, Spain; Hospital Universitari de Bellvitge, Barcelona, Spain; University of Barcelona, Barcelona, Spain*
MOTOTSUGU NAGAO • *Islet Cell Exocytosis, Lund University Diabetes Centre, Department of Clinical Sciences Malmö, Lund University, Malmö, Sweden; Clinical Research Centre, Skåne University Hospital, Lund and Malmö, Sweden; Department of Endocrinology, Diabetes and Metabolism, Graduate School of Medicine, Nippon Medical School, Tokyo, Japan*
MANASI NANDI • *School of Cancer & Pharmaceutical Sciences, King's College London, London, UK*
JAMES C. NEEDELL • *Innate Biotechnologies LLC, Denver, CO, USA*
NIKOLAY NINOV • *DFG-Center for Regenerative Therapies Dresden, Cluster of Excellence, Technische Universität Dresden, Dresden, Germany; Paul Langerhans Institute Dresden of the Helmholtz Zentrum München at the University Hospital and Faculty of Medicine Carl Gustav Carus of Technische Universität Dresden, Dresden, Germany; German Center for Diabetes Research (DZD e.V.), Neuherberg, Germany*
SHINICHI OIKAWA • *Department of Endocrinology, Diabetes and Metabolism, Graduate School of Medicine, Nippon Medical School, Tokyo, Japan; Diabetes and Lifestyle-Related Disease Center, Japan Anti-Tuberculosis Association, Fukujuji Hospital, Tokyo, Japan*

Oladapo E. Olaniru • *Department of Diabetes, School of Life Course Sciences, King's College London, London, UK*

Julia K. Panzer • *Paul Langerhans Institute Dresden (PLID) of the Helmholtz Zentrum München at the University Clinic Carl Gustav Carus of Technische Universität Dresden, Helmholtz Zentrum München, Neuherberg, Germany; Institute of Physiology, Faculty of Medicine, Technische Universität Dresden, Dresden, Germany; German Center for Diabetes Research (DZD), München-Neuherberg, Germany*

Patricia Fonseca Pedro • *Department of Diabetes, School of Life Course Sciences, Faculty of Life Science and Medicine, King's College London, London, UK*

Germán Perdomo • *Departmento de Ciencias de la Salud, Facultad de Ciencias de la Salud, Universidad de Burgos, Burgos, Spain*

Shanta J. Persaud • *Department of Diabetes, School of Life Course Sciences, King's College London, London, UK*

Inmaculada Ruz-Maldonado • *Department of Diabetes, School of Life Course Sciences, King's College London, London, UK*

Paraskevi Salpea • *Université Libre de Bruxelles, Bruxelles, Belgium*

Uwe Schönmann • *German Primate Center, Leibniz-Institute for Primate Research, Göttingen, Germany*

Sumeet Pal Singh • *DFG-Center for Regenerative Therapies Dresden, Cluster of Excellence, Technische Universität Dresden, Dresden, Germany*

Lorna I. F. Smith • *Department of Diabetes, School of Life Course Sciences, King's College London, London, UK*

Stephan Speier • *Paul Langerhans Institute Dresden (PLID) of the Helmholtz Zentrum München at the University Clinic Carl Gustav Carus of Technische Universität Dresden, Helmholtz Zentrum München, Neuherberg, Germany; Institute of Physiology, Faculty of Medicine, Technische Universität Dresden, Dresden, Germany; German Center for Diabetes Research (DZD), München-Neuherberg, Germany*

Hitoshi Sugihara • *Department of Endocrinology, Diabetes and Metabolism, Graduate School of Medicine, Nippon Medical School, Tokyo, Japan*

Noèlia Téllez • *CIBER of Diabetes and Associated Metabolic Diseases (CIBERDEM), Barcelona, Spain; Bellvitge Biomedical Research Institute, IDIBELL, Barcelona, Spain; University of Barcelona, Barcelona, Spain*

Terri C. Thayer • *Division of Infection and Immunity, Cardiff University School of Medicine, Cardiff, UK*

Anastasia Tsakmaki • *Department of Diabetes, School of Life Course Sciences, Faculty of Life Science and Medicine, King's College London, London, UK*

Dirk Wedekind • *Institute for Laboratory Animal Science, Hannover Medical School, Hannover, Germany*

Li Wen • *Section of Endocrinology, Yale School of Medicine, New Haven, CT, USA*

Anna Wendt • *Islet Cell Exocytosis, Lund University Diabetes Centre, Department of Clinical Sciences Malmö, Lund University, Malmö, Sweden; Clinical Research Centre, Skåne University Hospital, Lund and Malmö, Sweden*

Eckhard Wolf • *Department of Molecular Animal Breeding and Biotechnology, Ludwig Maximilian University, Munich, Germany*

F. Susan Wong • *Division of Infection and Immunity, Cardiff University School of Medicine, Cardiff, UK*

Danny Zipris • *Innate Biotechnologies LLC, Denver, CO, USA*

# Chapter 1

# Practical Considerations when Using Mouse Models of Diabetes

Aileen J. F. King, Lydia F. Daniels Gatward, and Matilda R. Kennard

## Abstract

Mouse models of diabetes are important tools used in preclinical diabetes research. However, when working with these models, it is important to consider factors that could influence experimental outcome. This is particularly important given the wide variety of models available, each with specific characteristics that could be influenced by extrinsic or intrinsic factors. Blood glucose concentrations, a commonly used and valid endpoint in these models, are particularly susceptible to manipulation by these factors. These include potential effects of intrinsic factors such as strain, sex, and age and extrinsic factors such as husbandry practices and experimental protocols. These variables should therefore be taken into consideration when the model is chosen and the experiments are designed. This chapter outlines common variables that can impact the phenotype of a model, as well as describes the methods used for assessing onset of diabetes and monitoring diabetic mice.

**Key words** Mouse models of diabetes, Blood glucose, Husbandry

## 1 Introduction

Animal models are an important tool in preclinical diabetes research, and many different models are currently available [1–3]. The model of choice will depend on the scientific question being asked and the endpoints of interest, but the researcher should also consider variables that could impact the outcome and/or interpretation of the experiments. For example, in models developing hyperglycemia, a reduction in blood glucose concentration is a commonly used and valid endpoint. However, the definition of normoglycemia in mice is poorly defined [4] and may differ between different strains and different studies, especially if they are defining onset of hyperglycemia as opposed to defining "cure" (return to normoglycemia after a state of diabetes). Normoglycemia is often defined as <11.1 mM (200 mg/dL), whereas the definition of hyperglycemia tends to be more variable (>13.9 mM (250 mg/dL), >16.7 mM (300 mg/dL), >20 mM (360 mg/

Aileen J. F. King (ed.), *Animal Models of Diabetes: Methods and Protocols*, Methods in Molecular Biology, vol. 2128,
https://doi.org/10.1007/978-1-0716-0385-7_1, 

dL)). In models of type 1 diabetes, where hyperglycemia is caused by insulin deficiency, any reduction in blood glucose should be accompanied by an increase in body weight to ensure that blood glucose reductions are not due to lack of food intake. The ethical endpoints should also be carefully considered and comply with local ethical regulations [5]. This chapter will outline specific variables that can be important to consider when working with diabetic mouse models ranging from husbandry considerations to the potential impacts of strain and sex. Methods for assessing onset of diabetes and monitoring diabetic animals will also be described.

## 2 Materials

### 2.1 Measuring Glucosuria

1. Reagent strips for urinalysis, such as Diastix strips.

### 2.2 Measuring Blood Glucose

1. 27G or 30G needles (*see* **Note 1**).
2. Blood glucose meter and strips (*see* **Note 2**).

### 2.3 Administration of Insulin

1. Long-acting insulin (*see* **Note 3**).
2. Insulin needle and syringe.

## 3 Methods

### 3.1 Choosing an Appropriate Model

There are a wide variety of models of diabetes available, and each has specific characteristics that may be relevant to studying type 1 diabetes, type 2 diabetes, or monogenic diabetes [1–3]. The animal model of choice will depend on the experimental question and may or may not involve animals that develop overt diabetes as part of their phenotype. An important consideration is the background strain and gender, since both can markedly alter the phenotype of a model. Indeed, several models ranging from high-fat-feeding models to genetic models show a more severe phenotype in male mice [4, 6–8]. Male mice are also reported to be more susceptible to streptozotocin [9]. An exception to this is the non-obese diabetic (NOD) mouse, which has higher incidences of diabetes development in females [10]. The reasons for these sex differences are not well documented, although it has been suggested that estrogen has a protective effect in female mice [9, 11, 12]. The translational implications are not known, although a higher incidence of type 1 diabetes has been reported in postpubertal males in humans [13] and the incidence of type 2 diabetes in women increases after the menopause [14].

Strain should also be carefully considered as it is well documented that background strain and even substrains can affect the severity of a diabetic phenotype and the susceptibility to interventions such as a high-fat diet [15–17]. In line with this, it should be noted that different models and strains also have varying susceptibilities to developing diabetic complications [18–21].

Adult mice are typically used for studying diabetes; however, it is important to note that the age of diabetes development is dependent upon the model, and this has implications for experimental design. For example, diabetes develops as early as 3–4 weeks in the Akita mouse but between 4–8-weeks in the $Lepr^{db/db}$ mouse. In NOD mice, onset can be less predictable. Females start to develop diabetes at around 10–14 weeks, but this can vary depending on the colony [2]. Furthermore, while in many diabetic models (including the Akita and NOD mice) glycemic control worsens with age, diabetic phenotype is improved in the $Lep^{ob/ob}$ mouse after 6 months as a result of compensatory islet hypertrophy and increased insulin secretion [22]. These changes in phenotype can cause problems in long-term studies as, in cases where diabetes phenotype worsens, it is important that the health of the mouse does not deteriorate to a critical level. On the other hand, in the case of the $Lep^{ob/ob}$ mouse, the spontaneous improvement in glycemia could make interpretation of the data difficult. Finally, off-target effects should be considered and controlled for wherever possible. Off-target effects specific to animal models of diabetes include nephrotoxicity and neurotoxicity when using alloxan and streptozotocin [23, 24]. Obese models may have aberrations in fertility and thermogenesis [22], whereas hypothalamic expression of transgenes under the insulin promoter can lead to difficulties in interpreting data [25]. Any intervention that causes stress to the animal (such as surgery) may also impact blood glucose concentrations and therefore could affect interpretation of data. The researcher should therefore conduct a thorough literature review prior to starting their study to ensure the most appropriate model, strain, age, and sex are selected.

#### 3.1.1 General Husbandry and Handling Considerations when Working with Diabetic Mice

A key characteristic of diabetes is hyperglycemia which leads to polydipsia and polyuria. Therefore, cages may need to be changed more frequently, and water availability should be closely monitored. In general, mice should have ad libitum access to water and standard chow. However, some models involve a change of diet (e.g., in the case of high-fat feeding). With a change of diet, the feeding habits of the mice should be initially monitored as there may be variability in consumption between mice [26]. Furthermore, some procedures may involve fasting the mice. Although overnight fasting is commonly used in humans, mice consume the majority of their food at night, and consequently an overnight fast can cause substantial weight loss, nearly deplete liver glycogen stores, and cause metabolic stress [27–29]. Consequently, overnight fasts should be avoided on both ethical and scientific grounds.

Aspects of husbandry can also affect the incidence of diabetes. For example, in NOD mice, the incidence of diabetes may decrease if the mice are not kept in specific pathogen-free (SPF) conditions.

Mice should be housed in groups where possible, preferably with siblings to reduce fighting, particularly in males. However, if there is prolonged aggressive behavior, single housing may be preferable [30]. Stress can impact blood glucose concentrations due to the effects of adrenalin and glucocorticoids; therefore, in addition to the negative implications on mouse welfare, stress can potentially affect results if an endpoint of blood glucose concentrations is being used. Changing cages, which is required for husbandry, can also cause stress responses which may be due to loss of familiar smells, such as urine and pheromones, which can alter perception of the mouse's society and hierarchy [31]. Indeed, cage changing has been associated with increased fighting in males which is reduced when nesting material from the dirty cage is reused [31]. Corticosterone levels are also increased following cage change suggesting that blood glucose concentrations could be affected [31]. To reduce stress, it is recommended that enrichment is used in cages such as nesting materials and shelters. Finally, handling may cause stress, but this is often overlooked due to its necessity. Standard handling involves picking up the animal by the base of its tail, but potential alternatives such as cupping and tunnelling have been reported to reduce blood glucose following glucose injection as well as reduce anxiety-like behaviors such as urination and defecation [32, 33].

## 3.2 Assessing Onset of Diabetes

The onset of diabetes can be rapid and predictable if using a chemically induced model such as streptozotocin but is less predictable in models of spontaneous development of diabetes, such as NOD mice. In the latter case, it may be preferable to initially measure glucose in the urine to detect onset of diabetes. This procedure is noninvasive and therefore appropriate to use repeatedly over many days or weeks. However, this only detects whether the animal has hyperglycemia, and blood glucose concentrations should then be measured to determine exact levels of glycemia.

### 3.2.1 Measuring Glucosuria in Mice

1. Scruff the mouse; this will often lead to spontaneous urination. If not, carefully press on the bladder (*see* **Note 4**).
2. Place a small drop of urine on a Diastix test strip.
3. Compare the color to the reference color chart.

### 3.2.2 Measuring Blood Glucose Concentrations in Mice

1. While allowing the mouse to move freely, hold the end of the tail.
2. Using as small a needle as possible, prick the very end of the tail (Fig. 1; *see* **Note 5**).

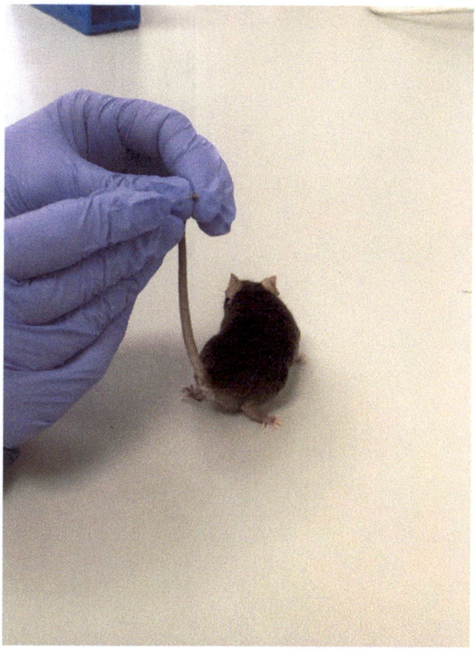

**Fig. 1** A small needle is used to make a small prick at the very end of the tail. With most glucose meters requiring blood droplets around 1 μL, a 30G is sufficient

3. Form a droplet of blood by running your thumb and index finger up the tail while applying pressure, starting from the base and ending at the tip.
4. Apply the droplet of blood to the strip placed inside the glucose meter (Fig. 2).

### *3.3 Monitoring Diabetes*

Diabetes should be carefully monitored for both experimental and ethical reasons particularly in models of type 1 diabetes where mice may experience significant weight loss. Although weight loss is symptomatic of type 1 diabetes and thus arguably adds to the validity of a model, it is advisable that animals losing more than 20% body weight should be culled. However, it is important that local guidelines should be followed. If local guidelines allow and it fits the experimental design, excessive weight loss can be prevented by administering long-acting insulin. In mouse models of type 1 diabetes, improved glucose homeostasis should generally lead to both a decrease in blood glucose and an increase in weight. If the mice exhibit continued weight loss despite decreased blood glucose, it is possibly due to ill health and/or lack of feeding rather than improved blood glucose homeostasis. The overall condition of the mice should also be monitored for signs of pain or distress (e.g., hunched posture, piloerection). This is especially important when diabetes has been chemically induced (e.g., using as streptozotocin

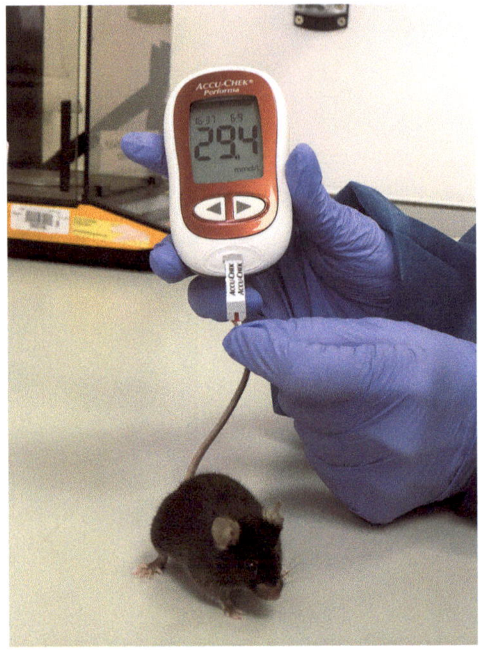


**Fig. 2** The blood droplet is applied to the strip placed in the blood glucose meter which gives a reading within a few seconds

or alloxan) since these can have off-target toxic effects [34, 35]. It is recommended that animals showing poor condition are humanely killed for both ethical and scientific reasons as meaningful and valid data is unlikely to be collected from mice with poor health.

To monitor glycemic control, blood glucose should be measured as described in Subheading 3.2.2. The frequency of measuring blood glucose will depend on the study and will typically range from once a day to once a week. Daily testing over prolonged periods (weeks to months) should be avoided to prevent excessive tail damage. We typically test daily for the first week of a study and thereafter 2–3 times a week.

#### *3.3.1 Considerations when Monitoring Diabetes in Mice*

1. Blood glucose should be measured at the same time on each day due to blood glucose dynamicity, and this should be reported. In particular, it should be noted that blood glucose concentrations may be substantially higher during the night as this is typically when rodents feed [36]. Generally, most labs measure non-fasted blood glucose concentrations in the morning (7–9 am).
2. Stress will increase blood glucose concentrations. As mentioned, stress can be caused even by simple husbandry procedures such as changing cage. Therefore, mice should be handled carefully and be habituated to their surroundings or experimental protocol as much as possible.

### 3.4 Administration of Insulin

While administration of insulin could potentially impact the experimental outcome, it can be warranted in some cases. For example, it could be used to prevent excessive weight loss prior to starting an experiment. In some studies, it may be acceptable to administer insulin without impacting the endpoints of interest. Regardless of this, use of insulin should be discussed with a veterinary surgeon as overdose can be fatal.

1. Scruff skin between the shoulder blades of the mouse.
2. Inject insulin subcutaneously at a suitable dose and frequency as discussed with a veterinary surgeon.

### 3.5 Reporting Experiments Using Animal Models of Diabetes

As discussed above, many factors can affect blood glucose concentrations and/or development of diabetes in rodents. These factors include husbandry, sex, strain, handling, and time of day of blood glucose measurements. It is therefore very important that manuscripts contain as much detail as possible to allow readers to fully understand under which conditions the experiments were carried out. Indeed, use of the ARRIVE guidelines is fully recommended [37].

## 4 Notes

1. With most modern blood glucose meters requiring less than 1 μL of blood, a 30G needle is sufficient to produce a sufficiently sized droplet of blood.
2. Many researchers use blood glucose meters designed for human use as meters and strips designed and marketed for animal use are, in general, more expensive. One difference between rodents and humans is hematocrit levels [38], and thus meters that are designed for rodent use have reported to be more accurate than consumer meters [39]. Another consideration, especially for the use of models of type 1 diabetes, is the upper limit of the meter. Many meters only capture blood glucose concentrations of up to 27 or 33 mM, whereas others can read concentrations up to 50 mM. This may be important if the animals are regularly showing very high blood glucose concentrations. The amount of blood required for the meter should also be considered; some meters require larger samples than others, with small samples being favorable.
3. A long-acting insulin can be useful as it can be injected just once or twice a day to prevent excessive hyperglycemia and weight loss. A suitable insulin and dose should be chosen with advice from a veterinary surgeon as overdose can be lethal.
4. Care should be taken in male mice that semen is not inadvertently measured.

5. The end of the tail can be pricked while still allowing the mouse to roam in order to reduce stress. The end of the tail has minimal nerve endings and thus will be most comfortable for the mouse. A local anesthetic is generally not required, but this can be discussed with a veterinary surgeon if appropriate.

## References


1. King A, Bowe J (2016) Animal models for diabetes: understanding the pathogenesis and finding new treatments. Biochem Pharmacol 99:1–10. https://doi.org/10.1016/j.bcp.2015.08.108
2. King AJF (2012) The use of animal models in diabetes research. Br J Pharmacol 166(3):877–894. https://doi.org/10.1111/j.1476-5381.2012.01911.x
3. Bowe JE, Franklin ZJ, Hauge-Evans AC, King AJ, Persaud SJ, Jones PM (2014) Metabolic phenotyping guidelines: assessing glucose homeostasis in rodent models. J Endocrinol 222(3):G13–G25. https://doi.org/10.1530/joe-14-0182
4. Leiter EH (2009) Selecting the "right" mouse model for metabolic syndrome and type 2 diabetes research. Methods Mol Biol 560:1–17. https://doi.org/10.1007/978-1-59745-448-3_1
5. Morton DB (2000) A systematic approach for establishing humane endpoints. ILAR J 41(2):80–86. https://doi.org/10.1093/ilar.41.2.80
6. Leiter EH, Schile A (2013) Genetic and pharmacologic models for type 1 diabetes. Curr Protoc Mouse Biol 3(1):9–19. https://doi.org/10.1002/9780470942390.mo120154
7. Yoshioka M, Kayo T, Ikeda T, Koizumi A (1997) A novel locus, Mody4, distal to D7Mit189 on chromosome 7 determines early-onset NIDDM in nonobese C57BL/6 (Akita) mutant mice. Diabetes 46(5):887–894. https://doi.org/10.2337/diab.46.5.887
8. Herbach N, Rathkolb B, Kemter E, Pichl L, Klaften M, de Angelis MH, Halban PA, Wolf E, Aigner B, Wanke R (2007) Dominant-negative effects of a novel mutated Ins2 allele causes early-onset diabetes and severe beta-cell loss in Munich Ins2C95S mutant mice. Diabetes 56(5):1268–1276. https://doi.org/10.2337/db06-0658
9. Le May C, Chu K, Hu M, Ortega CS, Simpson ER, Korach KS, Tsai MJ, Mauvais-Jarvis F (2006) Estrogens protect pancreatic beta-cells from apoptosis and prevent insulin-deficient diabetes mellitus in mice. Proc Natl Acad Sci U S A 103(24):9232–9237. https://doi.org/10.1073/pnas.0602956103
10. Pozzilli P, Signore A, Williams AJ, Beales PE (1993) NOD mouse colonies around the world--recent facts and figures. Immunol Today 14(5):193–196. https://doi.org/10.1016/0167-5699(93)90160-m
11. Tiano JP, Mauvais-Jarvis F (2012) Importance of oestrogen receptors to preserve functional beta-cell mass in diabetes. Nat Rev Endocrinol 8(6):342–351. https://doi.org/10.1038/nrendo.2011.242
12. Xu B, Allard C, Alvarez-Mercado AI, Fuselier T, Kim JH, Coons LA, Hewitt SC, Urano F, Korach KS, Levin ER, Arvan P, Floyd ZE, Mauvais-Jarvis F (2018) Estrogens promote misfolded proinsulin degradation to protect insulin production and delay diabetes. Cell Rep 24(1):181–196. https://doi.org/10.1016/j.celrep.2018.06.019
13. Gale EA, Gillespie KM (2001) Diabetes and gender. Diabetologia 44(1):3–15. https://doi.org/10.1007/s001250051573
14. Mauvais-Jarvis F (2015) Sex differences in metabolic homeostasis, diabetes, and obesity. Biol Sex Differ 6:14. https://doi.org/10.1186/s13293-015-0033-y
15. Hull RL, Willard JR, Struck MD, Barrow BM, Brar GS, Andrikopoulos S, Zraika S (2017) High fat feeding unmasks variable insulin responses in male C57BL/6 mouse substrains. J Endocrinol 233(1):53–64. https://doi.org/10.1530/joe-16-0377
16. Andrikopoulos S, Massa CM, Aston-Mourney-K, Funkat A, Fam BC, Hull RL, Kahn SE, Proietto J (2005) Differential effect of inbred mouse strain (C57BL/6, DBA/2, 129T2) on insulin secretory function in response to a high fat diet. J Endocrinol 187(1):45–53. https://doi.org/10.1677/joe.1.06333
17. Montgomery MK, Hallahan NL, Brown SH, Liu M, Mitchell TW, Cooney GJ, Turner N (2013) Mouse strain-dependent variation in obesity and glucose homeostasis in response to high-fat feeding. Diabetologia 56(5):1129–1139. https://doi.org/10.1007/s00125-013-2846-8

18. Bugger H, Abel ED (2009) Rodent models of diabetic cardiomyopathy. Dis Model Mech 2 (9–10):454–466. https://doi.org/10.1242/dmm.001941
19. Azushima K, Gurley SB, Coffman TM (2018) Modelling diabetic nephropathy in mice. Nat Rev Nephrol 14(1):48–56. https://doi.org/10.1038/nrneph.2017.142
20. Mi X-S, Yuan T-F, Ding Y, Zhong J-X, So K-F (2014) Choosing preclinical study models of diabetic retinopathy: key problems for consideration. Drug Des Devel Ther 8:2311–2319. https://doi.org/10.2147/DDDT.S72797
21. O'Brien PD, Sakowski SA, Feldman EL (2014) Mouse models of diabetic neuropathy. ILAR J 54(3):259–272. https://doi.org/10.1093/ilar/ilt052
22. Lindstrom P (2007) The physiology of obese-hyperglycemic mice [ob/ob mice]. ScientificWorldJournal 7:666–685. https://doi.org/10.1100/tsw.2007.117
23. Andersson DA, Filipovic MR, Gentry C, Eberhardt M, Vastani N, Leffler A, Reeh P, Bevan S (2015) Streptozotocin stimulates the ion channel TRPA1 directly: involvement of peroxynitrite. J Biol Chem 290 (24):15185–15196. https://doi.org/10.1074/jbc.M115.644476
24. Evan AP, Mong SA, Connors BA, Aronoff GR, Luft FC (1984) The effect of alloxan, and alloxan-induced diabetes on the kidney. Anat Rec 208(1):33–47. https://doi.org/10.1002/ar.1092080105
25. Wicksteed B, Brissova M, Yan W, Opland DM, Plank JL, Reinert RB, Dickson LM, Tamarina NA, Philipson LH, Shostak A, Bernal-Mizrachi E, Elghazi L, Roe MW, Labosky PA, Myers MG Jr, Gannon M, Powers AC, Dempsey PJ (2010) Conditional gene targeting in mouse pancreatic ss-cells: analysis of ectopic Cre transgene expression in the brain. Diabetes 59(12):3090–3098. https://doi.org/10.2337/db10-0624
26. De Francesco PN, Cornejo MP, Barrile F, Garcia Romero G, Valdivia S, Andreoli MF, Perello M (2019) Inter-individual Variability for high fat diet consumption in inbred C57BL/6 Mice. Front Nutr 6:67. https://doi.org/10.3389/fnut.2019.00067
27. Andrikopoulos S, Blair AR, Deluca N, Fam BC, Proietto J (2008) Evaluating the glucose tolerance test in mice. Am J Physiol Endocrinol Metab 295(6):E1323–E1332. https://doi.org/10.1152/ajpendo.90617.2008
28. McGuinness OP, Ayala JE, Laughlin MR, Wasserman DH (2009) NIH experiment in centralized mouse phenotyping: the Vanderbilt experience and recommendations for evaluating glucose homeostasis in the mouse. Am J Physiol Endocrinol Metab 297(4): E849–E855. https://doi.org/10.1152/ajpendo.90996.2008
29. Ayala JE, Samuel VT, Morton GJ, Obici S, Croniger CM, Shulman GI, Wasserman DH, McGuinness OP (2010) Standard operating procedures for describing and performing metabolic tests of glucose homeostasis in mice. Dis Model Mech 3(9–10):525–534. https://doi.org/10.1242/dmm.006239
30. Kappel S, Hawkins P, Mendl MT (2017) To group or not to group? Good practice for housing male laboratory mice. Animals 7(12). https://doi.org/10.3390/ani7120088
31. Rasmussen S, Miller MM, Filipski SB, Tolwani RJ (2011) Cage change influences serum corticosterone and anxiety-like behaviors in the mouse. J Am Assoc Lab Animal Sci 50 (4):479–483
32. Ghosal S, Nunley A, Mahbod P, Lewis AG, Smith EP, Tong J, D'Alessio DA, Herman JP (2015) Mouse handling limits the impact of stress on metabolic endpoints. Physiol Behav 150:31–37. https://doi.org/10.1016/j.physbeh.2015.06.021
33. Hurst JL, West RS (2010) Taming anxiety in laboratory mice. Nat Methods 7(10):825–826. https://doi.org/10.1038/nmeth.1500
34. Goyal SN, Reddy NM, Patil KR, Nakhate KT, Ojha S, Patil CR, Agrawal YO (2016) Challenges and issues with streptozotocin-induced diabetes - a clinically relevant animal model to understand the diabetes pathogenesis and evaluate therapeutics. Chem Biol Interact 244:49–63. https://doi.org/10.1016/j.cbi.2015.11.032
35. Lenzen S (2008) The mechanisms of alloxan- and streptozotocin-induced diabetes. Diabetologia 51(2):216–226. https://doi.org/10.1007/s00125-007-0886-7
36. King AJ, Austin AL, Nandi M, Bowe JE (2017) Diabetes in rats is cured by islet transplantation...But only during daytime. Cell Transplant 26(1):171–172. https://doi.org/10.3727/096368916X692258
37. Kilkenny C, Browne W, Cuthill IC, Emerson M, Altman DG (2010) Animal research: reporting in vivo experiments: the ARRIVE guidelines. Br J Pharmacol 160 (7):1577–1579. https://doi.org/10.1111/j.1476-5381.2010.00872.x
38. Weitgasser RDAWG (1999) Measurement of glucose concentrations in rats: differences

between glucose meter and plasma laboratory results. Diabetologia 42(2):256–256. https://doi.org/10.1007/s001250051147

39. Peterson RG, Brockway R (2012) Assessment of Nova Biomedical StatStrip® glucose meters and test strips in rodent glucose studies. FASEB J 26(1_supplement):1127.1111. https://doi.org/10.1096/fasebj.26.1_supplement.1127.11

# Chapter 2

# An Overview of Rodent Models of Obesity and Type 2 Diabetes

**Thomas A. Lutz**

## Abstract

Many animal models that are currently used in appetite and obesity research share at least some main features of human obesity and its comorbidities. Hence, even though no animal model replicates all aspects of "common" human obesity, animal models are imperative in studying the control of energy balance and reasons for its imbalance that may eventually lead to overt obesity. The most frequently used animal models are small rodents that may be based on mutations or manipulations of individual or several genes and on the exposure to obesogenic diets or other manipulations that predispose the animals to gaining or maintaining excessive weight. Characteristics include hyperphagia or changes in energy metabolism and at least in some models the frequent comorbidities of obesity, like hyperglycemia, insulin resistance, or diabetes-like syndromes. Some of the most frequently used animal models of obesity research involve animals with monogenic mutations of the leptin pathway which in fact are useful to study specific mechanistic aspects of eating controls, but typically do not recapitulate "common" obesity in the human population. Hence, this review will mention advantages and disadvantages of respective animal models in order to build a basis for the most appropriate use in biomedical research.

**Key words** Monogenetic models, Polygenetic models, Surgical models, Diabetes mellitus

## 1 Introduction

Selected animal models are discussed in this chapter; the selection was based on their frequency of use in obesity research and their value in understanding the controls of eating and body weight but of course also reflects a subjective opinion about which animal models may be interesting. The different phenotypes express differences in respect to hyperphagia or alterations in energy metabolism and in respect to obesity-associated comorbidities like hyperglycemia, insulin resistance, or diabetes-like syndromes [1, 2].

Most animal models of obesity in biomedical research are small rodents, but it should be mentioned that most mammals, when kept in captivity or as companion animals [3], will develop obesity. In fact, large animal models may bear some advantages over rodents

Aileen J. F. King (ed.), *Animal Models of Diabetes: Methods and Protocols*, Methods in Molecular Biology, vol. 2128, https://doi.org/10.1007/978-1-0716-0385-7_2, 

(e.g., larger body size, large sample volumes, similar pancreas size and structure to humans, pharmacokinetics more similar to humans) [2], but their maintenance and housing are more challenging and, of course, more costly. The most commonly used animal models of obesity are probably the leptin-deficient ob/ob mouse, the leptin-receptor-deficient db/db mouse and its rat counterparts like the Zucker or ZDF rats, and a large number of models of diet-induced obesity (DIO).

## 2 Monogenic Mutations in the Leptin Pathway

Some of the most dramatic phenotypes of obesity are seen in animals with a defect in the leptin signaling pathway in the hypothalamus. The models are characterized by either a lack of leptin production or leptin receptor mutations with extreme leptin resistance. Mutations are spontaneous (e.g., $Lep^{ob}/Lep^{ob}$ mouse, $Lep^{db}/Lep^{db}$ mouse) or genetically engineered. Animals with mutations that lie downstream of the leptin-sensing neurons in the hypothalamus develop similar phenotypes. Many of these animals also develop a diabetic phenotype. It is important to note that the cause of obesity in these animals in general reflects only a negligible proportion of human obesity in the general population.

### 2.1 ob/ob Mouse ($Lep^{ob}/Lep^{ob}$ Mouse, the "Obese" Mouse)

- Spontaneous mutation that leads to the markedly obese phenotype [4, 5]; obesity due to a single-base spontaneous mutation of the ob gene that codes for leptin.
- The mutation prevents the secretion of bioactive leptin.
- Early-onset obesity with hyperphagia, reduced energy expenditure, and hypothermia.
- Further characteristics include high glucocorticoid levels, insulin resistance associated with hyperglycemia and hyperinsulinemia, hypothyroidism, and growth hormone deficiency which leads to a decrease in linear growth.
- ob/ob mice are infertile; hence, breeding has to be done with heterozygous animals.
- Phenotypic defects in $Lep^{ob}/Lep^{ob}$ mice including obesity, symptoms of the metabolic syndrome, and reproductive function are normalized by exogenous leptin [5–11].

Leptin deficiency is observed in rare cases of human obesity [12]; hence, this animal model does not represent most cases of human obesity. Appropriate controls may include leptin-treated $Lep^{ob}/Lep^{ob}$ mice to test whether leptin substitution restores the parameter of interest to be tested.

### 2.2 db/db Mouse (The "Diabetic" Mouse)

- $Lep^{db}/Lep^{db}$ mouse has a spontaneously mutated leptin receptor.
- Phenotypically similar to the $Lep^{ob}/Lep^{ob}$ mouse but with more marked hyperglycemia on some background strains than in the $Lep^{ob}/Lep^{ob}$ mouse [2].
- Hyperphagia and reduced energy expenditure lead to marked early-onset obesity.
- Mice are hypothermic, have decreased linear growth due to growth hormone deficiency, and are also infertile [5, 9, 13].
- Major difference from the $Lep^{ob}/Lep^{ob}$ mouse is the marked resistance to leptin and that leptin levels are markedly elevated.
- $Lep^{db}/Lep^{db}$ mice are insulin resistant and develop diabetes, but mice do not develop the full phenotype of type 2 diabetes.

Mutations of the leptin receptor gene are found in some human families, but the mutation is very rare. Hence, the obese and diabetic phenotype may be a useful feature of these animals, but direct links to human pathophysiology have to be made with caution.

### 2.3 Leptin-Receptor-Deficient Rats: Zucker Rat, ZDF Rat, and Koletsky Rat

- Obese Zucker (fa/fa or "fatty" rat) and Koletsky rats carry mutated forms of the leptin receptor and hence are similar to the $Lep^{db}/Lep^{db}$ mouse.
- Hyperphagia and reduced energy expenditure lead to morbid obesity [6, 14].
- Rats exhibit impaired glucose tolerance, a growth deficit possibly related to a lower activity of the growth hormone axis and hypothyroidism, and reduced fertility.
- There is an important difference between Koletsky rats (with a nonsense and null mutation of the receptor, hence undetectable levels of leptin receptor mRNA expression [13, 15–19]) and the fa/fa mutation of Zucker fatty rats (with a processing defect of the leptin receptor, resulting in intracellular trapping of the receptor).
- Koletsky rats are more hypertensive and have a more severe phenotype of insulin resistance than Zucker fatty rats; Koletsky rats may no longer be available via commercial suppliers.
- Zucker diabetic fatty (ZDF) rats are a substrain of obese Zucker fatty rats and were created by selective breeding of obese and hyperglycemic Zucker rats and exhibit early dysregulation of glucose metabolism [20].

Mutations of the leptin receptor gene are found in some human families, but the mutation is very rare. Hence, the obese and diabetic phenotype may be a useful feature of these animals, but direct links to human pathophysiology have to be made with caution.

### 2.4 s/s Mouse

- Genetically engineered mouse model, similar to the $Lep^{db}/Lep^{db}$ mouse [21, 22].
- Specific disruption of the signaling pathway that mediates leptin's effects on energy homeostasis; disruption of pSTAT3 signaling pathway that mediates leptin's action at the long form of the leptin receptor [23].
- Homozygous (s/s) mice are hyperphagic and obese, but they are fertile indicating that different leptin actions use different signaling pathways.

### 2.5 POMC Knockout Mouse

- Genetically engineered mice that lack proopiomelanocortin (POMC; $POMC^{-/-}$) and that have a deficit in a number of biologically active neuropeptides, in particular α-melanocyte-stimulating hormone (αMSH) which is a potent eating-inhibitory neuropeptide and which mediates parts of leptin's effects on energy balance [24].
- $POMC^{-/-}$ mice overeat and develop marked obesity [25, 26].
- Heterozygous mutants have an intermediate phenotype.
- Mutations in the gene may also cause obesity in other species [27, 28].

POMC deficiency has also been reported in rare cases of human obesity [12]; mutations in this pathway are relatively more frequent than other monogenic causes of human obesity.

### 2.6 MC4R Knockout Models

- αMSH and its functional antagonist agouti-related protein (AgRP) which are released from neurons in the hypothalamic arcuate nucleus impact on energy homeostasis via melanocortin (MC) receptors.
- The MC4 receptor subtype mediates the eating-inhibitory effect of αMSH and the eating-stimulatory effect of AgRP.
- MC4 receptor knockout produces hyperphagia and morbid obesity [29].
- $MC4^{-/-}$ mice are hyperinsulinemic, hyperglycemic, and hyperleptinemic.
- $MC4^{-/-}$ mice do not respond to leptin, AgRP, or αMSH.
- The MC4 knockout rat [30] has many characteristics in common with MC4 KO mice (e.g., increased body weight, food intake and body length, and lower spontaneous activity); however, $MC4^{-/-}$ mice have increased expression of NPY but not of POMC in parts of the hypothalamus, but adult MC4 knockout rats have unchanged NPY levels but increased POMC expression.

Similar mutations of the MC4 receptor are often stated to be the most frequent (mono)genetic cause of obesity in humans.

## 3 Monogenic Models of Obesity and Diabetes that Are Independent of the Leptin Pathway

### 3.1 Otsuka Long-Evans Tokushima Fatty Rat (OLETF)

- Otsuka Long-Evans Tokushima Fatty (OLETF) rats are deficient of the receptor for the satiating hormone cholecystokinin (CCK), i.e., these rats can be considered CCK-1 receptor knockouts.
- Compared to their lean counterparts, the Long-Evans Tokushima Otsuka (LETO) rats, OLETF rats overeat and develop obesity which is relatively mild [31–33].
- Hyperphagia of OLETF rats is due to increases in the size of meals which is not compensated by reduced meal numbers.
- The increase in meal size in OLETF rats is a phenomenon observed early in life, i.e., already in rat pups 2 days of age which reflects the early maturation of the hindbrain systems of satiation which occurs earlier than in the centers for metabolic control in the hypothalamus [34].
- At least later in life, overexpression of the eating-stimulatory peptide neuropeptide Y (NPY) in the hypothalamus may contribute to overeating in OLETF rats.
- The relative importance of the latter effect in rats versus mice may explain the paradoxical finding that CCK1 receptor knockout mouse does not develop obesity on a normal diet [35].
- Subsequent to obesity, OLETF rats develop diabetes from about 4–5 months of age and develop polyuria and polydipsia.

Human CCK receptor allele variants have been associated to larger meal sizes and proneness to develop obesity [36].

### 3.2 Human Islet Amyloid Polypeptide (hIAPP) Overexpressing Mice and Rats

- Several mouse and rat models that are transgenic for the human form of islet amyloid polypeptide (IAPP, or amylin) have been created [37–40].
- Human, but not rodent, IAPP forms intra- and extracellular aggregates and fibrils that contribute to progressive pancreatic beta-cell destruction and loss.
- Insulin sensitivity typically unaltered.
- Progression of beta-cell destruction and of clinical signs depends on the animals' genetic background.

hIAPP mice and rats are the only available rodent models that mimic the beta-cell pathology in type 2 diabetic patients. Some of the rat and mouse strains may not be easily available from commercial breeders.

## 4 Polygenic Obesity Models

### 4.1 Diet-Induced Obese (DIO) Versus Diet-Resistant (DR) Rodents

- The outbred Sprague Dawley (SD) rats is the best characterized DIO model
- On exposure to a high-energy diet (HE; 31% fat), rats with the DIO phenotype become obese, while DR rats maintain lower body weight and a rather lean phenotype, similar to chow-fed controls.
- DIO rats remain obese even if returned to a low-fat diet.
- Selective breeding of DIO and DR rats exacerbates the phenotypic differences.
- The DIO versus DR phenotype shares many characteristics with the common form of human obesity [41–44].
- The phenotype seems to be inherited in a polygenic fashion, but the involved genes are largely unknown.
- Important metabolic characteristics of the DIO rat (e.g., leptin resistance, insulin resistance, hypertension, and hypertriglyceridemia) develop before the onset of overt obesity [43, 45–48].
- The extent of weight gain and obesity depends on the type of diet [42].

DIO rats and DR controls are probably the best available rodent model to mimic most types of human obesity.

### 4.2 ZDSD Rat

- The ZDSD rat is derived from crossbreeding selectively bred DIO obese Charles River Sprague Dawley rats with $ZDF^{fa/+}$ lean rats [49, 50].
- The existing literature is inconsistent whether $ZDF^{fa/+}$ or $ZDF^{+/+}$ rats have been used for the original crossing; in either case, the ZDSD rat is a model with a functional leptin signaling pathway.
- Rats are characterized by an age-dependent development of hyperglycemia.
- Progression of diabetes mellitus is slower than in the ZDF rats and resembles better the time course in humans.
- Initial increase in baseline insulin followed by progressive loss of insulin secretion.

ZDSD rats recapitulate the progressive beta-cell pathology in human type 2 diabetes. These rats have not yet been used extensively in biomedical research so that less data are available compared to some other rodent models described in this chapter.

### 4.3 Cafeteria-Diet-Induced and High-Fat-Diet-Induced Obesity

- Palatable ("cafeteria") diets often lead to obesity in rats [51–53].
- This type of obesity is due to hyperphagia, including an increase in the average meal size and meal frequency.
- Simple high-fat (HF) diets also often result in the development of obesity.
- Increased caloric density of HF diets may contribute to the effect, but an important factor most likely is reduced leptin (and insulin) sensitivity [54–58].
- HF diets affect critical intracellular signaling pathways in hypothalamic target neurons, e.g., POMC neurons.
- Saturated fat (e.g., palmitic acid) may be more deleterious than unsaturated fat and contribute to the inflammatory responses in the hypothalamus [59–62].

Use of these diets is often combined with the use of genetically obese models, as discussed earlier in this chapter.

### 4.4 Maternal Overfeeding and Exposure to High-Fat Diets

- The metabolic situation in the intrauterine and perinatal period of life may have long-lasting effects for the control of energy balance [63–67].
- Feeding pregnant dams a high-fat (HF) diet affects the offspring's metabolic health [67–71].
- Offspring of DIO dams are heavier and more obese than offspring of DR animals.
- The effect is accentuated in DIO dams exposed to HF diet.

If well-matched control groups are used, the model may be useful to study developmental prenatal or early postnatal effects and their relevance for metabolic disease. It is important to note that the developmental stages differ across mammalian species; in other words, at the time of birth, mouse and rat pups are developed much less than other mammalian species, including humans.

### 4.5 Early Postnatal Overfeeding-Induced Obesity; Rearing in Small Litters

- High food intake in the immediate postnatal period leads to lasting increases in body weight and obesity with ensuing insulin resistance, hyperinsulinemia, and glucose intolerance.
- Manipulation of the litter size is the best established model of early postnatal overnutrition in rats and mice [72–74].
- Pups reared in small litters ingest more than pups reared in large litters.
- The effect does not depend on changes in maternal behavior and generally leads to increased susceptibility to become obese as adults [75].
- The effect of early postnatal hyperphagia on obesity is potentiated when offered a HF diet directly after weaning.

- Rearing obesity-prone rats or mice in large litters, on the other hand, provides some protective effect against the development of obesity [48, 66].

## 5 Polygenic Diabetes Model

### 5.1 Goto-Kakizaki (GK) Rat

- Nonobese rat diabetes mellitus model with specific defects in insulin secretion [76]
- Spontaneous development of a type 2 diabetes-like phenotype without insulin resistance
- Defective insulin secretion and reduction in pancreatic beta-cell mass

Well-defined rat model that is used to study the pathophysiology of pancreatic beta-cells in the development of diabetes mellitus. (see Chap. 3)

## 6 Genetically Engineered Mutants in Obesity Research

### 6.1 Enhanced Expression of Glucose Transporter Subtype 4 (GLUT4) and Gluconeogenic Enzymes

- Mice overexpressing GLUT4 in white and brown adipose tissue develop early-onset obesity [77].
- Increased glucose uptake and hence increased supply of substrate promote adipogenesis.
- Mice mimic the occurrence of increased expression of GLUT4 in adipose tissue that is present in certain forms of obesity.
- Similar phenotype in mice overexpressing phosphoenolpyruvate carboxykinase (PEPCK) to increase the substrate for synthesizing triglycerides; obesity also develops when feeding low-fat diets [78].

These and similar animal models allow the study of specific aspects of adipogenesis.

### 6.2 Neuronal Insulin Receptor Knockout Mice (NIRKO Mice)

- Neuron-specific insulin receptor knockout mice (NIRKO mice) are moderately obese [79].
- Obesity is largely due to a moderate increase in food intake [80] and is paralleled by hypertriglyceridemia.
- Mice are characterized by elevations of systemic plasma insulin levels and insulin resistance.
- Mice have a defect in the insulin-mediated control of hepatic glucose production [80].

These mice are a good example of animal models that allow assessment of specific receptor populations. However, species-specific receptor distribution must be considered if data are to be translated from one to another species.

## 7 Surgical or Chemical Models of Obesity

### *7.1 Lesion of the Ventromedial Hypothalamus (VMH Lesion) or the Hypothalamic Paraventricular Nucleus (PVN Lesion)*

- Specific brain lesions result in obesity in rodents.
- Classical lesion encompasses the mediobasal hypothalamus, including the ventromedial and arcuate nuclei.
- Obesity is due to hyperphagia and probably reduced energy expenditure and reduced ambulatory activity [81–83].
- Extensive lesions of the PVN also induce obesity which is mainly linked to increased eating and unchanged energy expenditure.
- Similar to VMH lesions, PVN-lesioned rats often develop insulin resistance and hyperinsulinemia [6, 84, 85].

These models are classical models in obesity research and helped to define critical brain regions, but the models lost most of their relevance due to the unspecific destruction of brain structures and the availability of more selective modifications leading to obesity.

### *7.2 Ovariectomy (OVX)*

- OVX leads to a rapid increase in body weight and adiposity.
- OVX animals lack the cyclic and tonic inhibitory effects of estradiol on eating.
- The obesity in OVX animals is due to a transient increase in eating which is in part due to reduced action of physiological satiation signals, e.g., cholecystokinin [86–89].

Ovariectomy (OVX) in female rats and mice is used as a model for the increased obesity in menopausal women [90]. Species-specific differences in the animals' estrus cycle need to be considered when hormone replacement studies are planned.

## Acknowledgment

I gratefully acknowledge the financial support from many funding sources that helped me perform my research with some of the animal models mentioned here, in particular the Swiss National Science Foundation, the National Institutes of Health, the EU Frame Program 7, and the University of Zurich.

### References

1. Lutz TA, Woods SC (2012) Overview of animal models of obesity. Curr Protoc Pharmacol Chapter 5:Unit5 61. https://doi.org/10.1002/0471141755.ph0561s58
2. Kleinert M, Clemmensen C, Hofmann SM, Moore MC, Renner S, Woods SC, Huypens P, Beckers J, de Angelis MH, Schurmann A, Bakhti M, Klingenspor M, Heiman M, Cherrington AD, Ristow M, Lickert H, Wolf E, Havel PJ, Muller TD, Tschop MH (2018) Animal models of obesity and diabetes mellitus. Nat Rev Endocrinol. https://doi.org/10.1038/nrendo.2017.161
3. Osto M, Lutz TA (2015) Translational value of animal models of obesity-Focus on dogs and cats. Eur J Pharmacol 759:240–252. https://doi.org/10.1016/j.ejphar.2015.03.036

4. Mayer J, Bates MW, Dickie MM (1951) Hereditary diabetes in genetically obese mice. Science 113(2948):746–747
5. Coleman DL (1978) Obese and diabetes: two mutant genes causing diabetes-obesity syndromes in mice. Diabetologia 14(3):141–148
6. Bray GA, York DA (1979) Hypothalamic and genetic obesity in experimental animals: an autonomic and endocrine hypothesis. Physiol Rev 59(3):719–809
7. Zhang Y, Proenca R, Maffei M, Barone M, Leopold L, Friedman JM (1994) Positional cloning of the mouse obese gene and its human homologue. Nature 372 (6505):425–432. https://doi.org/10.1038/372425a0
8. Campfield LA, Smith FJ, Guisez Y, Devos R, Burn P (1995) Recombinant mouse OB protein: evidence for a peripheral signal linking adiposity and central neural networks. Science 269(5223):546–549
9. Halaas JL, Gajiwala KS, Maffei M, Cohen SL, Chait BT, Rabinowitz D, Lallone RL, Burley SK, Friedman JM (1995) Weight-reducing effects of the plasma protein encoded by the obese gene. Science 269(5223):543–546
10. Pelleymounter MA, Cullen MJ, Baker MB, Hecht R, Winters D, Boone T, Collins F (1995) Effects of the obese gene product on body weight regulation in ob/ob mice. Science 269(5223):540–543
11. Friedman JM (1998) Leptin, leptin receptors, and the control of body weight. Nutr Rev 56 (2 Pt 2):s38–s46. discussion s54–75
12. O'Rahilly S (2009) Human genetics illuminates the paths to metabolic disease. Nature 462(7271):307–314. https://doi.org/10.1038/nature08532
13. Chua SC Jr, Chung WK, Wu-Peng XS, Zhang Y, Liu SM, Tartaglia L, Leibel RL (1996) Phenotypes of mouse diabetes and rat fatty due to mutations in the OB (leptin) receptor. Science 271(5251):994–996
14. Bray GA (1977) The Zucker-fatty rat: a review. Fed Proc 36(2):148–153
15. Takaya K, Ogawa Y, Hiraoka J, Hosoda K, Yamori Y, Nakao K, Koletsky RJ (1996) Nonsense mutation of leptin receptor in the obese spontaneously hypertensive Koletsky rat. Nat Genet 14(2):130–131. https://doi.org/10.1038/ng1096-130
16. Friedman JM (1997) Leptin, leptin receptors and the control of body weight. Eur J Med Res 2(1):7–13
17. Wu-Peng XS, Chua SC Jr, Okada N, Liu SM, Nicolson M, Leibel RL (1997) Phenotype of the obese Koletsky (f) rat due to Tyr763Stop mutation in the extracellular domain of the leptin receptor (Lepr): evidence for deficient plasma-to-CSF transport of leptin in both the Zucker and Koletsky obese rat. Diabetes 46 (3):513–518
18. Crouse JA, Elliott GE, Burgess TL, Chiu L, Bennett L, Moore J, Nicolson M, Pacifici RE (1998) Altered cell surface expression and signaling of leptin receptors containing the fatty mutation. J Biol Chem 273(29):18365–18373
19. da Silva BA, Bjorbaek C, Uotani S, Flier JS (1998) Functional properties of leptin receptor isoforms containing the gln-->pro extracellular domain mutation of the fatty rat. Endocrinology 139(9):3681–3690
20. Zierath JR, Ryder JW, Doebber T, Woods J, Wu M, Ventre J, Li Z, McCrary C, Berger J, Zhang B, Moller DE (1998) Role of skeletal muscle in thiazolidinedione insulin sensitizer (PPARgamma agonist) action. Endocrinology 139(12):5034–5041
21. Bates SH, Kulkarni RN, Seifert M, Myers MG Jr (2005) Roles for leptin receptor/STAT3-dependent and -independent signals in the regulation of glucose homeostasis. Cell Metab 1 (3):169–178. https://doi.org/10.1016/j.cmet.2005.02.001
22. Bates SH, Stearns WH, Dundon TA, Schubert M, Tso AW, Wang Y, Banks AS, Lavery HJ, Haq AK, Maratos-Flier E, Neel BG, Schwartz MW, Myers MG Jr (2003) STAT3 signalling is required for leptin regulation of energy balance but not reproduction. Nature 421(6925):856–859. https://doi.org/10.1038/nature01388. nature01388 [pii]
23. Allison MB, Myers MG Jr (2014) 20 years of leptin: connecting leptin signaling to biological function. J Endocrinol 223(1):T25–T35. https://doi.org/10.1530/joe-14-0404
24. Elias CF, Aschkenasi C, Lee C, Kelly J, Ahima RS, Bjorbaek C, Flier JS, Saper CB, Elmquist JK (1999) Leptin differentially regulates NPY and POMC neurons projecting to the lateral hypothalamic area. Neuron 23(4):775–786
25. Yaswen L, Diehl N, Brennan MB, Hochgeschwender U (1999) Obesity in the mouse model of pro-opiomelanocortin deficiency responds to peripheral melanocortin. Nat Med 5 (9):1066–1070. https://doi.org/10.1038/12506
26. Challis BG, Coll AP, Yeo GS, Pinnock SB, Dickson SL, Thresher RR, Dixon J, Zahn D, Rochford JJ, White A, Oliver RL, Millington G, Aparicio SA, Colledge WH, Russ AP, Carlton MB, O'Rahilly S (2004) Mice lacking pro-opiomelanocortin are sensitive to high-fat feeding but respond normally to the acute anorectic effects of peptide-YY

(3-36). Proc Natl Acad Sci U S A 101 (13):4695–4700. https://doi.org/10.1073/pnas.0306931101

27. Mankowska M, Krzeminska P, Graczyk M, Switonski M (2017) Confirmation that a deletion in the POMC gene is associated with body weight of Labrador Retriever dogs. Res Vet Sci 112:116–118. https://doi.org/10.1016/j.rvsc.2017.02.014
28. Raffan E, Dennis RJ, O'Donovan CJ, Becker JM, Scott RA, Smith SP, Withers DJ, Wood CJ, Conci E, Clements DN, Summers KM, German AJ, Mellersh CS, Arendt ML, Iyemere VP, Withers E, Soder J, Wernersson S, Andersson G, Lindblad-Toh K, Yeo GS, O'Rahilly S (2016) A deletion in the canine pomc gene is associated with weight and appetite in obesity-prone labrador retriever dogs. Cell Metab 23(5):893–900. https://doi.org/10.1016/j.cmet.2016.04.012
29. Huszar D, Lynch CA, Fairchild-Huntress V, Dunmore JH, Fang Q, Berkemeier LR, Gu W, Kesterson RA, Boston BA, Cone RD, Smith FJ, Campfield LA, Burn P, Lee F (1997) Targeted disruption of the melanocortin-4 receptor results in obesity in mice. Cell 88 (1):131–141
30. Mul JD, van Boxtel R, Bergen DJ, Brans MA, Brakkee JH, Toonen PW, Garner KM, Adan RA, Cuppen E (2011) Melanocortin receptor 4 deficiency affects body weight regulation, grooming behavior, and substrate preference in the rat. Obesity (Silver Spring). https://doi.org/10.1038/oby.2011.81
31. Kawano K, Hirashima T, Mori S, Saitoh Y, Kurosumi M, Natori T (1992) Spontaneous long-term hyperglycemic rat with diabetic complications. Otsuka Long-Evans Tokushima Fatty (OLETF) strain. Diabetes 41 (11):1422–1428
32. Moran TH (2008) Unraveling the obesity of OLETF rats. Physiol Behav 94(1):71–78. https://doi.org/10.1016/j.physbeh.2007.11.035. S0031-9384(07)00454-4 [pii]
33. Moran TH, Bi S (2006) Hyperphagia and obesity in OLETF rats lacking CCK-1 receptors. Philos Trans R Soc Lond B Biol Sci 361 (1471):1211–1218. https://doi.org/10.1098/rstb.2006.1857. 565077G7K5T6246K [pii]
34. Smith GP (2006) Ontogeny of ingestive behavior. Dev Psychobiol 48(5):345–359. https://doi.org/10.1002/dev.20145
35. Lo CM, King A, Samuelson LC, Kindel TL, Rider T, Jandacek RJ, Raybould HE, Woods SC, Tso P (2010) Cholecystokinin knockout mice are resistant to high-fat diet-induced obesity. Gastroenterology. https://doi.org/10.1053/j.gastro.2010.01.044. S0016-5085(10)00150-2 [pii]
36. de Krom M, van der Schouw YT, Hendriks J, Ophoff RA, van Gils CH, Stolk RP, Grobbee DE, Adan R (2007) Common genetic variations in CCK, leptin, and leptin receptor genes are associated with specific human eating patterns. Diabetes 56(1):276–280. https://doi.org/10.2337/db06-0473. 56/1/276 [pii]
37. Butler AE, Jang J, Gurlo T, Carty MD, Soeller WC, Butler PC (2004) Diabetes due to a progressive defect in beta-cell mass in rats transgenic for human islet amyloid polypeptide (HIP Rat): a new model for type 2 diabetes. Diabetes 53(6):1509–1516
38. Hay DL, Chen S, Lutz TA, Parkes DG, Roth JD (2015) Amylin: pharmacology, physiology, and clinical potential. Pharmacol Rev 67 (3):564–600. https://doi.org/10.1124/pr.115.010629
39. Matveyenko AV, Butler PC (2006) Islet amyloid polypeptide (IAPP) transgenic rodents as models for type 2 diabetes. ILAR J 47 (3):225–233
40. Westermark P, Andersson A, Westermark GT (2011) Islet amyloid polypeptide, islet amyloid, and diabetes mellitus. Physiol Rev 91 (3):795–826. https://doi.org/10.1152/physrev.00042.2009
41. Levin BE, Dunn-Meynell AA (2000) Defense of body weight against chronic caloric restriction in obesity-prone and -resistant rats. Am J Physiol Regul Integr Comp Physiol 278(1):R231–R237
42. Levin BE, Dunn-Meynell AA (2002) Defense of body weight depends on dietary composition and palatability in rats with diet-induced obesity. Am J Physiol Regul Integr Comp Physiol 282(1):R46–R54
43. Levin BE, Dunn-Meynell AA, Balkan B, Keesey RE (1997) Selective breeding for diet-induced obesity and resistance in Sprague-Dawley rats. Am J Physiol 273(2 Pt 2):R725–R730
44. Levin BE, Triscari J, Sullivan AC (1986) The effect of diet and chronic obesity on brain catecholamine turnover in the rat. Pharmacol Biochem Behav 24(2):299–304
45. Levin BE, Dunn-Meynell AA, McMinn JE, Alperovich M, Cunningham-Bussel A, Chua SC Jr (2003) A new obesity-prone, glucose-intolerant rat strain (F.DIO). Am J Physiol Regul Integr Comp Physiol 285(5):R1184–R1191. https://doi.org/10.1152/ajpregu.00267.2003
46. Levin BE, Dunn-Meynell AA, Ricci MR, Cummings DE (2003) Abnormalities of leptin and

ghrelin regulation in obesity-prone juvenile rats. Am J Physiol Endocrinol Metab 285(5): E949–E957. https://doi.org/10.1152/ajpendo.00186.2003. 00186.2003 [pii]

47. Bouret SG, Gorski JN, Patterson CM, Chen S, Levin BE, Simerly RB (2008) Hypothalamic neural projections are permanently disrupted in diet-induced obese rats. Cell Metab 7 (2):179–185. https://doi.org/10.1016/j.cmet.2007.12.001. S1550-4131(07)00374-9 [pii]
48. Gorski JN, Dunn-Meynell AA, Levin BE (2007) Maternal obesity increases hypothalamic leptin receptor expression and sensitivity in juvenile obesity-prone rats. Am J Physiol Regul Integr Comp Physiol 292(5): R1782–R1791. https://doi.org/10.1152/ajpregu.00749.2006
49. Peterson RG, Jackson CV, Zimmerman K, de Winter W, Huebert N, Hansen MK (2015) Characterization of the ZDSD rat: a translational model for the study of metabolic syndrome and type 2 diabetes. J Diabetes Res 2015:487816. https://doi.org/10.1155/2015/487816
50. Reinwald S, Peterson RG, Allen MR, Burr DB (2009) Skeletal changes associated with the onset of type 2 diabetes in the ZDF and ZDSD rodent models. Am J Physiol Endocrinol Metab 296(4):E765–E774. https://doi.org/10.1152/ajpendo.90937.2008
51. Rothwell NJ, Stock MJ (1979) Combined effects of cafeteria and tube-feeding on energy balance in the rat. Proc Nutr Soc 38(1):5A
52. Rogers PJ, Blundell JE (1984) Meal patterns and food selection during the development of obesity in rats fed a cafeteria diet. Neurosci Biobehav Rev 8(4):441–453
53. Perez C, Fanizza LJ, Sclafani A (1999) Flavor preferences conditioned by intragastric nutrient infusions in rats fed chow or a cafeteria diet. Appetite 32(1):155–170. https://doi.org/10.1006/appe.1998.0182
54. Banks WA, Coon AB, Robinson SM, Moinuddin A, Shultz JM, Nakaoke R, Morley JE (2004) Triglycerides induce leptin resistance at the blood-brain barrier. Diabetes 53 (5):1253–1260
55. Benoit SC, Kemp CJ, Elias CF, Abplanalp W, Herman JP, Migrenne S, Lefevre AL, Cruciani-Guglielmacci C, Magnan C, Yu F, Niswender K, Irani BG, Holland WL, Clegg DJ (2009) Palmitic acid mediates hypothalamic insulin resistance by altering PKC-theta subcellular localization in rodents. J Clin Invest 119 (9):2577–2589. https://doi.org/10.1172/jci36714
56. Clegg DJ, Gotoh K, Kemp C, Wortman MD, Benoit SC, Brown LM, D'Alessio D, Tso P, Seeley RJ, Woods SC (2011) Consumption of a high-fat diet induces central insulin resistance independent of adiposity. Physiol Behav 103 (1):10–16. https://doi.org/10.1016/j.physbeh.2011.01.010
57. Hariri N, Thibault L (2010) High-fat diet-induced obesity in animal models. Nutr Res Rev 23(2):270–299. https://doi.org/10.1017/s0954422410000168
58. Woods SC, D'Alessio DA, Tso P, Rushing PA, Clegg DJ, Benoit SC, Gotoh K, Liu M, Seeley RJ (2004) Consumption of a high-fat diet alters the homeostatic regulation of energy balance. Physiol Behav 83(4):573–578. https://doi.org/10.1016/j.physbeh.2004.07.026. S0031-9384(04)00411-1 [pii]
59. Koch CE, Lowe C, Pretz D, Steger J, Williams LM, Tups A (2014) High-fat diet induces leptin resistance in leptin-deficient mice. J Neuroendocrinol 26(2):58–67. https://doi.org/10.1111/jne.12131
60. Garcia-Caceres C, Yi CX, Tschop MH (2013) Hypothalamic astrocytes in obesity. Endocrinol Metab Clin North Am 42(1):57–66. https://doi.org/10.1016/j.ecl.2012.11.003
61. Thaler JP, Guyenet SJ, Dorfman MD, Wisse BE, Schwartz MW (2013) Hypothalamic inflammation: marker or mechanism of obesity pathogenesis? Diabetes 62(8):2629–2634. https://doi.org/10.2337/db12-1605
62. Thaler JP, Yi CX, Schur EA, Guyenet SJ, Hwang BH, Dietrich MO, Zhao X, Sarruf DA, Izgur V, Maravilla KR, Nguyen HT, Fischer JD, Matsen ME, Wisse BE, Morton GJ, Horvath TL, Baskin DG, Tschop MH, Schwartz MW (2012) Obesity is associated with hypothalamic injury in rodents and humans. J Clin Invest 122(1):153–162. https://doi.org/10.1172/jci59660
63. Tamashiro KL, Moran TH (2010) Perinatal environment and its influences on metabolic programming of offspring. Physiol Behav 100 (5):560–566. https://doi.org/10.1016/j.physbeh.2010.04.008
64. Bouret SG (2009) Early life origins of obesity: role of hypothalamic programming. J Pediatr Gastroenterol Nutr 48(Suppl 1):S31–S38. https://doi.org/10.1097/MPG.0b013e3181977375. 00005176-200903001-00006 [pii]
65. Le Foll C, Irani BG, Magnan C, Dunn-Meynell A, Levin BE (2009) Effects of maternal genotype and diet on offspring glucose and fatty acid-sensing ventromedial hypothalamic nucleus neurons. Am J Physiol Regul Integr Comp Physiol 297(5):R1351–R1357.

https://doi.org/10.1152/ajpregu.00370.2009
66. Levin BE, Dunn-Meynell AA (2002) Maternal obesity alters adiposity and monoamine function in genetically predisposed offspring. Am J Physiol Regul Integr Comp Physiol 283(5): R1087–R1093. https://doi.org/10.1152/ajpregu.00402.2002
67. Levin BE, Govek E (1998) Gestational obesity accentuates obesity in obesity-prone progeny. Am J Physiol 275(4 Pt 2):R1374–R1379
68. Sullivan EL, Grayson B, Takahashi D, Robertson N, Maier A, Bethea CL, Smith MS, Coleman K, Grove KL (2010) Chronic consumption of a high-fat diet during pregnancy causes perturbations in the serotonergic system and increased anxiety-like behavior in nonhuman primate offspring. J Neurosci 30 (10):3826–3830. https://doi.org/10.1523/JNEUROSCI.5560-09.2010. 30/10/3826 [pii]
69. Sullivan EL, Smith MS, Grove KL (2010) Perinatal exposure to high-fat diet programs energy balance, metabolism and behavior in adulthood. Neuroendocrinology. https://doi.org/10.1159/000322038. 000322038 [pii]
70. Tamashiro KL, Terrillion CE, Hyun J, Koenig JI, Moran TH (2009) Prenatal stress or high-fat diet increases susceptibility to diet-induced obesity in rat offspring. Diabetes 58 (5):1116–1125. https://doi.org/10.2337/db08-1129
71. West DB, Diaz J, Woods SC (1982) Infant gastrostomy and chronic formula infusion as a technique to overfeed and accelerate weight gain of neonatal rats. J Nutr 112 (7):1339–1343
72. Faust IM, Johnson PR, Hirsch J (1980) Long-term effects of early nutritional experience on the development of obesity in the rat. J Nutr 110(10):2027–2034
73. Schmidt I, Fritz A, Scholch C, Schneider D, Simon E, Plagemann A (2001) The effect of leptin treatment on the development of obesity in overfed suckling Wistar rats. Int J Obes Relat Metab Disord 25(8):1168–1174. https://doi.org/10.1038/sj.ijo.0801669
74. Morris MJ, Velkoska E, Cole TJ (2005) Central and peripheral contributions to obesity-associated hypertension: impact of early overnourishment. Exp Physiol 90(5):697–702. https://doi.org/10.1113/expphysiol.2005.030783
75. West DB, Diaz J, Roddy S, Woods SC (1987) Long-term effects on adiposity after preweaning nutritional manipulations in the gastrostomy-reared rat. J Nutr 117 (7):1259–1264
76. Akash MS, Rehman K, Chen S (2013) Goto-Kakizaki rats: its suitability as non-obese diabetic animal model for spontaneous type 2 diabetes mellitus. Curr Diabetes Rev 9 (5):387–396
77. Shepherd PR, Gnudi L, Tozzo E, Yang H, Leach F, Kahn BB (1993) Adipose cell hyperplasia and enhanced glucose disposal in transgenic mice overexpressing GLUT4 selectively in adipose tissue. J Biol Chem 268 (30):22243–22246
78. Thorburn AW, Baldwin ME, Rosella G, Zajac JD, Fabris S, Song S, Proietto J (1999) Features of syndrome X develop in transgenic rats expressing a non-insulin responsive phosphoenolpyruvate carboxykinase gene. Diabetologia 42(4):419–426. https://doi.org/10.1007/s001250051174
79. Schwartz MW, Woods SC, Porte D Jr, Seeley RJ, Baskin DG (2000) Central nervous system control of food intake. Nature 404 (6778):661–671
80. Bruning JC, Gautam D, Burks DJ, Gillette J, Schubert M, Orban PC, Klein R, Krone W, Muller-Wieland D, Kahn CR (2000) Role of brain insulin receptor in control of body weight and reproduction. Science 289 (5487):2122–2125
81. Penicaud L, Larue-Achagiotis C, Le Magnen J (1983) Endocrine basis for weight gain after fasting or VMH lesion in rats. Am J Physiol 245(3):E246–E252
82. King BM (2006) The rise, fall, and resurrection of the ventromedial hypothalamus in the regulation of feeding behavior and body weight. Physiol Behav 87(2):221–244. https://doi.org/10.1016/j.physbeh.2005.10.007. S0031-9384(05)00482-8 [pii]
83. King BM (1991) Ventromedial hypothalamic obesity: a reexamination of the irritative hypothesis. Neurosci Biobehav Rev 15 (3):341–347
84. Sims JS, Lorden JF (1986) Effect of paraventricular nucleus lesions on body weight, food intake and insulin levels. Behav Brain Res 22 (3):265–281
85. Tokunaga K, Matsuzawa Y, Fujioka S, Kobatake T, Keno Y, Odaka H, Matsuo T, Tarui S (1991) PVN-lesioned obese rats maintain ambulatory activity and its circadian rhythm. Brain Res Bull 26(3):393–396
86. Asarian L, Geary N (2007) Estradiol enhances cholecystokinin-dependent lipid-induced satiation and activates estrogen receptor-alpha-expressing cells in the nucleus tractus solitarius of ovariectomized rats. Endocrinology 148 (12):5656–5666. https://doi.org/10.1210/en.2007-0341. en.2007-0341 [pii]

87. Asarian L, Geary N (2002) Cyclic estradiol treatment normalizes body weight and restores physiological patterns of spontaneous feeding and sexual receptivity in ovariectomized rats. Horm Behav 42(4):461–471. S0018506X02918350 [pii]
88. Thammacharoen S, Lutz TA, Geary N, Asarian L (2008) Hindbrain administration of estradiol inhibits feeding and activates estrogen receptor-alpha-expressing cells in the nucleus tractus solitarius of ovariectomized rats. Endocrinology 149(4):1609–1617. https://doi.org/10.1210/en.2007-0340
89. Asarian L, Geary N (2006) Modulation of appetite by gonadal steroid hormones. Philos Trans R Soc Lond B Biol Sci 361 (1471):1251–1263. https://doi.org/10.1098/rstb.2006.1860
90. Asarian L, Geary N (2013) Sex differences in the physiology of eating. Am J Physiol Regul Integr Comp Physiol 305(11):R1215–R1267. https://doi.org/10.1152/ajpregu.00446.2012

# Chapter 3

# Selectively Bred Diabetes Models: GK Rats, NSY Mice, and ON Mice

Mototsugu Nagao, Jonathan Lou S. Esguerra, Anna Wendt, Akira Asai, Hitoshi Sugihara, Shinichi Oikawa, and Lena Eliasson

## Abstract

The polygenic background of selectively bred diabetes models mimics the etiology of type 2 diabetes. So far, three different rodent models (Goto-Kakizaki rats, Nagoya-Shibata-Yasuda mice, and Oikawa-Nagao mice) have been established in the diabetes research field by continuous selective breeding for glucose tolerance from outbred rodent stocks. The origin of hyperglycemia in these rodents is mainly insulin secretion deficiency from the pancreatic β-cells and mild insulin resistance in insulin target organs. In this chapter, we summarize backgrounds and phenotypes of these rodent models to highlight their importance in diabetes research. Then, we introduce experimental methodologies to evaluate β-cell exocytosis as a putative common defect observed in these rodent models.

**Key words** Goto-Kakizaki rats, Nagoya-Shibata-Yasuda mice, Oikawa-Nagao mice, Islets, β-Cells, Insulin secretion, Exocytosis, Capacitance measurement

## 1 Introduction

Selective breeding is an approach to develop animal and plant strains carrying particular phenotypic traits. During the selection process, genetic variants regulating the desirable traits can be accumulated into the later generation. Therefore, the polygenic background of selectively bred diabetes models mimics the etiology of type 2 diabetes (T2D) in humans. They are also useful to investigate genotype-phenotype interactions in complex traits involving multiple genes, e.g., glucose tolerance and insulin secretion. So far, mainly three models have been established by repetitive selective breeding for glucose tolerance: Goto-Kakizaki (GK) rats, Nagoya-Shibata-Yasuda (NSY) mice, and Oikawa-Nagao (ON) mice. GK rats and NSY mice were developed by the selection for spontaneous glucose intolerance in outbred rodent stocks, whereas ON mice (Diabetes-Prone and Diabetes-Resistant lines) were developed by the selections for both inferior and superior glucose tolerance after

Aileen J. F. King (ed.), *Animal Models of Diabetes: Methods and Protocols*, Methods in Molecular Biology, vol. 2128, https://doi.org/10.1007/978-1-0716-0385-7_3, 

high-fat diet (HFD) feeding in hybrid mice of three different inbred strains. Here, we summarize backgrounds and phenotypes, specifically glucose tolerance and insulin secretion, of these rodent models. Thereafter, we compare the phenotypic features to highlight the advantages in investigating the selectively bred rodent models in diabetes research. Finally, we present protocols to investigate β-cell defects in late stages of the insulin secretion process.

## 1.1 Goto-Kakizaki (GK) Rat

### 1.1.1 History

The selective breeding for establishing GK rat was launched by Yoshio Goto and colleagues in 1973. The original aim of the selective breeding study was to test if diabetes is a disease of inheritance or not. In the initial oral glucose tolerance tests (OGTT; glucose 2 g/kg) on 130 male and 81 female Wistar rats, nine male and nine female rats with higher blood glucose levels during the OGTT were selected as a breeding stock [1]. Selection and breeding of rats showing higher blood glucose levels during an OGTT were continued over multiple generations. The sum of blood glucose levels during the OGTT ($SUM_{BG}$; the sum of blood glucose measured at fasting and at time points of 30, 60, 90, and 120 min) was increased gradually as the generations proceeded. The distribution of $SUM_{BG}$ was completely separated from that of control Wistar rats at the ninth generation and finally reached a plateau at the 15th generation [2]. The spontaneously diabetic rats were designated as Goto-Kakizaki (GK) rats by the founders. From the end of the 1980s, breeding pairs of GK rats from the original colony (the Sendai colony; GK/Sen) were distributed worldwide, and local colonies were established in Sweden (the Stockholm colony; GK/Sto [3]), France (the Paris colony; GK/Par [4]), the UK (the Cardiff and London colony; GK/Card [5, 6] and GK/Lon [6], respectively), etc. In 2002, Luthman H established a colony of GK rats at Lund University (Malmö, Sweden), called as GK/KyoSwe [7]. The colony was separated from GK/Sto which originated in 1989 from breeding pairs of GK/Sen rats sent from Japan [3] and renamed here to the Lund University colony (GK/LU). Nowadays, GK rats are available not only from various local colonies but also from some commercial breeders in Japan, the USA, and Europe. Most of the colonies are maintained as inbred lines from the ninth generation through sister-brother mating [8, 9]. Curiously, the founder Goto Y [10] described in Japanese literature that the brother-sister mating was introduced already at the fourth generation.

### 1.1.2 Phenotype

The GK rat develops non-insulin-dependent diabetes, and the glucose intolerance is suggested to be primarily caused by impaired insulin secretion due to defective β-cell function [11, 12]. To demonstrate such phenotypes of the GK rat, we performed measurements of GK/LU rats. Non-fasting blood glucose levels were 20–25 mM in 10 weeks old GK/LU rats as compared to 6–8 mM

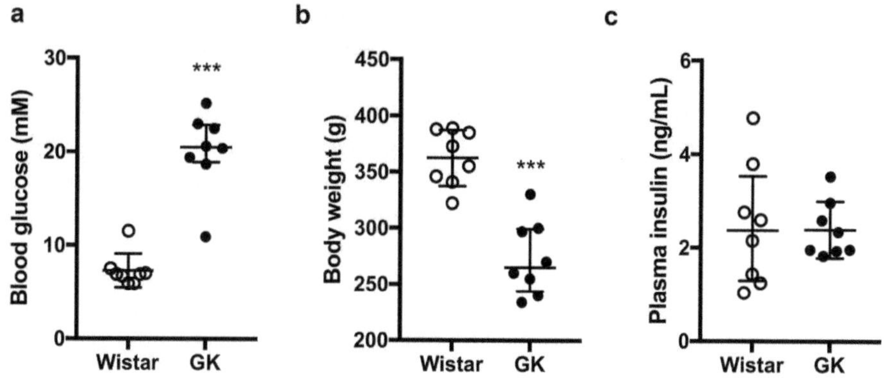


**Fig. 1** Representative metabolic phenotypes of Goto-Kakizaki (GK) and Wistar rats. Blood glucose levels (**a**), body weights (**b**), and plasma insulin levels (**c**) of male Goto-Kakizaki (GK/LU) and control Wistar rats. The evaluations were performed on 11–14 week-old mice under non-fasting condition ($n = 8$ for each group). The lines display the median with interquartile range of the plots. Wistar rats, open circles. GK rats, closed circles. $***P < 0.001$ vs. control Wistar rats using Student's *t*-test

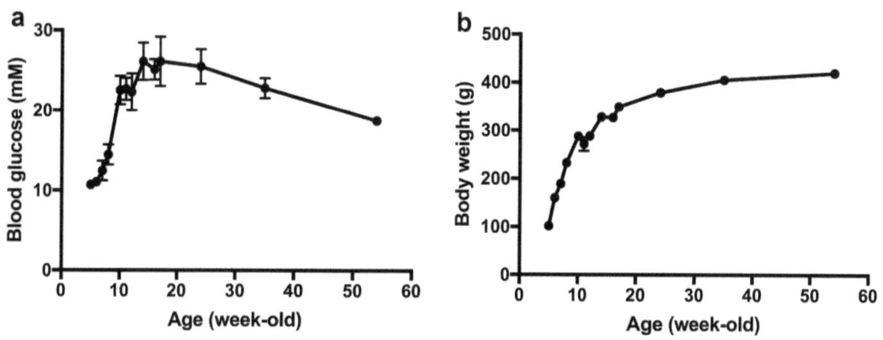


**Fig. 2** Longitudinal observations for blood glucose levels and body weights of Goto-Kakizaki (GK) rats. Blood glucose levels (**a**) and body weights (**b**) of male Goto-Kakizaki (GK/LU) rats at 5, 6, 7, 8, 10, 11, 12, 14, 16, 17, 24, 35, and 54 weeks of age ($n = 3$–13 for each point). The data are expressed as mean ± SEM

in control Wistar rats (Fig. 1a). Body weight was ~20% lower in GK/LU rats than Wistar rats of the same age (Fig. 1b), whereas there was no significant difference in plasma insulin levels (Fig. 1c). The blood glucose levels of GK/LU rats were relatively higher than those in the other colonies [9, 13]. Such differences in glycemic status and other characteristics (e.g., islet morphology and function) among colonies may be influenced by local breeding environment and/or genetic variation [8, 13]. (For more information about the characteristics of each colony, see reviews by Portha et al. [13] and Östenson et al. [9, 14].)

To follow the development of hyperglycemia in GK/LU rats, we measured blood glucose and body weight at different ages (Fig. 2a, b). Young GK/LU rats (5–8 weeks of age) had lower non-fasting blood glucose levels (~10 mM) as compared to matured GK/LU rats (over 10 weeks of age). The blood glucose level of GK/LU rats increased rapidly up to 10 weeks of age and reached a maximum at ~15 weeks of age followed by a small decline from ~25 weeks of age (Fig. 2a). The acute development of hyperglycemia after weaning in GK/LU rats follows that in the other colonies and has been explained primarily by the acquired hepatic insulin resistance [12, 15] and secondarily by the insufficient β-cell compensation for the insulin resistance [16–18].

Young GK/LU rats had normal-shaped islets as compared to control Wistar rats (Fig. 3a). However, the proportion of normal-shaped islets decreases with age [12, 19, 20]. Indeed, the majority of islets in adult GK/LU rats showed an irregular shape, so-called starfish-shaped islets, being characterized by decreased β-cell number, increased amount of fibrosis tissue, and random distribution of α- and δ-cells throughout the core [12, 19, 21]. The development of fibrosis in the islets is suggested to be the result of chronic inflammation [22]. We measured a decreased insulin content in the islets of GK/LU rats (Fig. 3b), which is in accordance with a decreased β-cell population in islets and/or degranulation in the β-cells [12, 19, 20]. However, this is not always true as there are examples where neither β-cell mass nor islet insulin content was decreased even in adult GK rats [13, 14].

### *1.1.3 β-Cell Function*

Although non-fasting insulin levels of GK/LU rats were like those of control Wistar rats (Fig. 1c), impaired insulin secretion during an OGTT has been demonstrated repeatedly in the GK rat colonies [4, 12, 23]. Moreover, also in our hands, we measured severely reduced glucose-induced insulin secretion from isolated GK/LU rat islets (Fig. 3c). Note that the insulin secretion was reduced to a larger extent than the insulin content (Fig. 3b, c), which is likely to be a combination of reduced insulin content and impaired secretion capacity of each β-cell. Interestingly, the functional deterioration of insulin secretion was observed even in the islets of young GK/LU rats (Fig. 3c) with a normal round shape (Fig. 3a), suggesting that the β-cell function is impaired prior to the morphological change in the islets.

Glucose is a main stimulator of insulin secretion from β-cells. Uptake and metabolism of glucose lead to the generation of ATP, which closes the ATP-dependent $K^+$ channel leading to membrane depolarization and opening of voltage-activated $Na^+$- and $Ca^{2+}$-channels. The influx of $Ca^{2+}$ initiates exocytosis of insulin granules, and insulin is released. This is known as the β-cell stimulus-secretion coupling. Several defects have been reported in the stimulus-secretion coupling in GK rat β-cells. A lower expression level of the glucose transporter 2 (GLUT2) has been demonstrated [24, 25], which most likely reduces the uptake of glucose into the cell. Glycolysis has been suggested to be intact [3, 6, 26, 27], but the mitochondrial glycerol phosphate shuttle has consistently been reported to be impaired in GK rat β-cells [28–32]. These defects

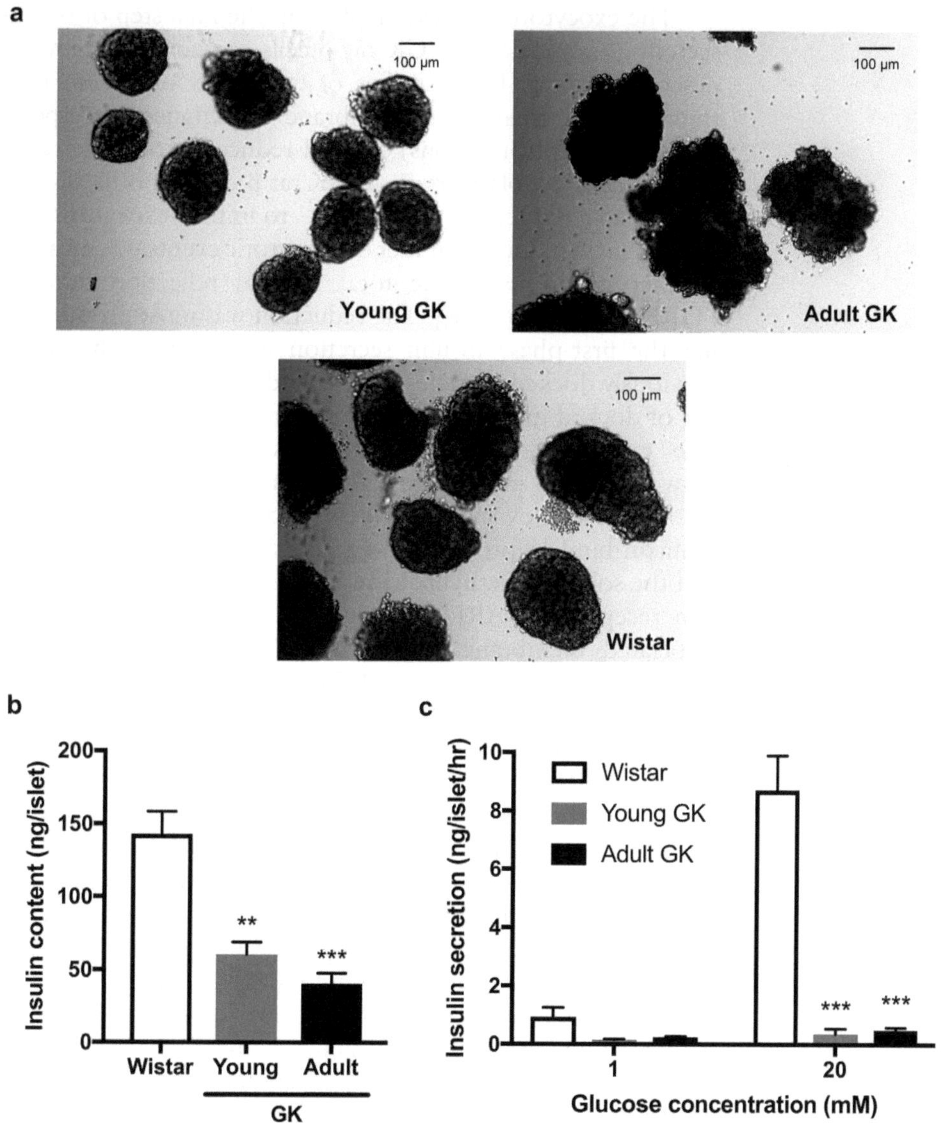


**Fig. 3** Features of pancreatic islets in Goto-Kakizaki (GK) and Wistar rats. Appearances of pancreatic islets (**a**) in 5 week-old Goto-Kakizaki (GK/LU) rats (young GK), 12 week-old Goto-Kakizaki (GK/LU) rats (adult GK), and 12 week-old Wistar rats as indicated. The images were taken using a confocal microscope (×10). Insulin content (**b**) and glucose-stimulated insulin secretion (**c**) in/from the pancreatic islets from young and adult GK/LU rats ($n = 4$ and 5, respectively) and control Wistar rats ($n = 6$). The data are expressed as mean ± SEM. $^{**}P < 0.01$ and $^{***}P < 0.001$ vs. control Wistar rats using Student's *t*-test

may account for the reported abnormalities in glucose-stimulated ATP production [27, 33], downstream closure of the ATP-dependent $K^+$ channel [34], and elevation of cytoplasmic $Ca^{2+}$ [35, 36]. The functions of the ATP-dependent $K^+$ channels and the voltage-activated $Ca^{2+}$ channels seem to be intact as suggested by a series of patch clamp experiments on GK rat β-cells [27, 34, 37].

The exocytotic process involved in the final step of the insulin secretion is impaired in GK rat β-cells as exemplified by studies described below. First, studies of exocytosis of insulin granules (measured as changes in membrane capacitance in response to membrane depolarizations) showed reduced exocytotic responses to the initial depolarizations in GK rat β-cells. Specifically, GK rat β-cells had a reduced efficacy of $Ca^{2+}$ to initiate exocytosis [37]. In agreement, a reduced number of exocytotic events was measured by another group using the total internal reflection fluorescence (TIRF) microscopy [38]. The reduction was most prominent during the first-phase insulin secretion representing the release of previously docked granules. These studies suggest a reduced number of docked granules at the plasma membrane in GK rat β-cells. The exocytotic machinery mediates the docking and fusion between insulin granules and the plasma membrane. This process involves several exocytotic proteins including the docking protein, syntaxin-binding protein 1 (STXBP1; also known as MUNC-18a), and the soluble *N*-ethylmaleimide-sensitive factor attachment protein receptor (SNARE) proteins, syntaxin 1A (STX1A), vesicle-associated membrane protein 2 (VAMP2), and synaptosomal-associated protein 25 (SNAP25) [39]. The GK rat islets have marked reductions of the exocytotic proteins (e.g., STX1A, VAMP2, SNAP25, and STXBP1) [40–43]. Moreover, the clustering of SNAP25 and STX1A, which is important for functional and rapid exocytosis, is severely impaired in GK rat β-cells and could be overcome by insulin treatment [42]. STXBP1 is important for granular docking and priming [39], which is why reduced expression of this protein contributes to the observed exocytotic phenotype of the GK rat β-cells with reduced first-phase exocytosis and decreased number of docked granules.

One of the main second messengers to amplify insulin secretion is cAMP [44]. Several impairments in mechanisms to potentiate cAMP-dependent amplifying pathway have been reported in GK rat β-cells [45–47]. Moreover, one of the main genetic variants in the GK rat was associated with reduced cAMP levels in the β-cells [48]. Hence, it is of interest that cAMP inducers, i.e., forskolin, acetylcholine, and glucagon-like polypeptide-1 (GLP-1), can restore the impaired glucose-induced insulin secretion observed in GK rat islets [16, 49, 50].

### 1.1.4 Pathogenesis

Quantitative trait locus (QTL) linkage analyses have identified the genetic loci responsible for quantitative traits in GK rats, e.g., fasting glycemia [51, 52], glucose tolerance [51–53], and insulin secretion [51, 54] (for more details, see a review by Bihoreau et al. [8]). Most of the identified QTLs are located at different genomic regions [8]. Moreover, recent full genome sequencing of GK rats in the Oxford colony (GK/Ox) revealed 3,584,504 single-nucleotide polymorphisms (SNPs), 1,147,996 indels, and 58,877 structural variants compared to the control inbred rat reference genome [55].

Several studies have been performed to identify the genomic regions in the GK rat corresponding to T2D loci in humans [48, 54, 56]. In a congenic strain of the GK rat, one locus was coupled to defective β-cell function [56]. The region in focus, *Niddm1i,* is present on chromosome 1 in the GK rat genome. Interestingly, different segments within *Niddm1i* are suggested to be associated with distinct defects in glucose metabolism and insulin exocytosis. The defect in insulin exocytosis, including also deteriorated granular docking, is coupled to a 1.4 Mb genetic segment of *Niddm1i* [48]. The segment contains five known protein-coding genes: *Pdcd4*, *Lysmd3*, *Shoc2*, *Adra2a*, and *ENSRNOG00000036577*. Among them, we demonstrated overexpression of *Adra2a* leading to reduced granule docking at the plasma membrane and decreased exocytosis [48]. Interestingly, the congenic strain had not developed diabetes, suggesting that the defective β-cell exocytosis is not the result of hyperglycemia [48, 56]. In contrast, phlorizin treatment can partially restore SNARE protein expression and improve glucose-stimulated insulin secretion in GK rat islets [40], suggesting that the defective exocytosis, at least when due to reduced expression of exocytotic proteins in the β-cells, is secondary to glucotoxicity.

We have shown that GK rat islets have a perturbed microRNA (miRNA) network and that the upregulated miRNAs may target proteins involved in β-cell exocytosis, as in the case of miR-335 targeting *Stxbp1* [57]. Indeed, overexpression of miR-335 directly impaired insulin secretion through defective priming of insulin granules [58]. Aside from targeting exocytotic genes, we have also shown that other upregulated miRNAs in GK rat islets, i.e., miR-130a/b and miR-152, negatively regulate pyruvate dehydrogenase E1-alpha (PDHA1) and glucokinase (GCK) resulting in altered dynamics of intracellular ATP/ADP ratio which affect ATP-requiring β-cell processes such as glucose-induced insulin secretion and insulin biosynthesis and processing [59]. Some of these miRNAs, e.g., miR-130a, miR-132, miR-212, and miR-335, appeared to be increased by hyperglycemia [57, 60], but regulatory mechanisms of others are still unknown. The contribution of reprogramming should also be taken into account because the maternal diabetic environment in the GK rat can influence DNA methylation [61]. Indeed, epigenetic modification in human T2D has been shown to reduce exocytosis [62].

### 1.1.5 Trend of Research

We were interested to know in which type of research GK rats are most utilized and performed a search on PubMed (https://www.ncbi.nlm.nih.gov/pubmed) at the end of 2017. A total of 840 original research publications was listed when we searched with the keywords "Goto," "Kakizaki," and "rat." We categorized them

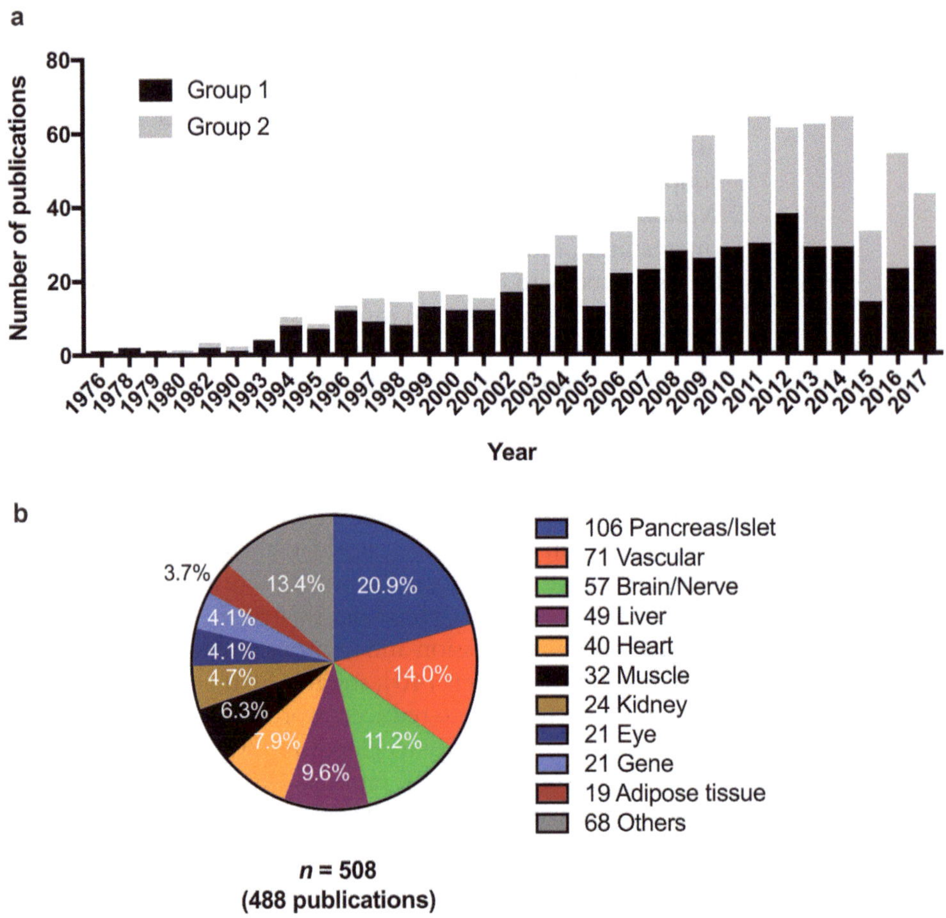


**Fig. 4** Trend of publications using Goto-Kakizaki (GK) rats. Number of publications using Goto-Kakizaki (GK) rats by year (**a**) and breakdown of research-targeted organs in the diabetes pathophysiological studies (**b**). The 840 original research publications about GK rats were listed in PubMed at the end of 2017. They were categorized into two groups by the titles and abstracts: *pathophysiological analyses for diabetes and its complications* (Group 1, black bars) and *interventional studies using hypoglycemic agents or other therapeutics* (Group 2, gray bars). The 488 papers in Group 1 were classified by the organ on which research was performed (total $n = 508$ organs in 488 publication)

into two groups: *pathophysiological analyses for diabetes and its complications* (Group 1) and *interventional studies using hypoglycemic agents or other therapeutics* (Group 2) (Fig. 4a). Then, we further classified the 488 reports in Group 1 by the main research target organ (Fig. 4b). As expected, the top researched organ was the pancreas including islets and the endocrine cells. The second was the vascular system, mainly functional analyses of the aorta or mesenteric arteries, indicating that the GK rat is a good model to address the pathophysiology of diabetic macrovascular complications. Indeed, GK rats have a mild cardiac functional disorder, characterized by diastolic dysfunction without ischemia [63–65]. Similar cardiac damage is well described in patients with diabetes, so-called diabetic cardiomyopathy [66]. It is therefore reasonable that many researchers have focused on the heart. The nerve, kidneys, and eyes are also ranked within the list. Indeed, Goto and colleagues demonstrated the usefulness of GK rats as an animal model of microvascular complications [67–70]. Of interest, Goto [71] described in his autobiographical book that another major aim of the selective breeding study was to produce an animal model which could contribute to the study of diabetes complications. Finally, insulin-targeting organs (the liver, muscles, and adipose tissues) are also listed, since insulin resistance plays a certain role for the development of hyperglycemia in GK rats. Overall, GK rats have served as a useful animal model for diabetes studies in various target organs, which corresponds to the clinical feature of diabetes as a systemic disease. Moreover, studies in the GK rat have led to important breakthroughs in overall knowledge and treatment strategies in human T2D, e.g., the discovery of genetic variants affecting the expression of *adra2a*/ADRA2A and insulin secretion [48, 72].

## 1.2 Nagoya-Shibata-Yasuda (NSY) Mouse

### 1.2.1 History

The NSY mice were founded by Masato Shibata and Bunji Yasuda around 1980 [73]. They initially treated outbred ICR mice with streptozotocin (80 mg/kg, i.v.) and kept them diabetic for one month, before mating. For the treated mice and their offspring, intraperitoneal glucose tolerance tests (IPGTT; glucose 2 g/kg) were performed, and the mice with higher blood glucose levels at time points of 60 and 120 min during the test were selected and bred continuously by sister-brother mating. The blood glucose levels during the IPGTT increased up to the seventh generation and have been stable thereafter. The colonies were previously kept at Nagoya University (Nagoya, Japan) and Osaka University (Osaka, Japan) [74]. Nowadays, NSY mice (NSY/Hos) can be purchased from a Japanese breeder (Japan SLC, Inc., Hamamatsu, Japan).

### 1.2.2 Phenotype

In the founders' early reports [73, 75], the ICR mouse was used as a control. When compared to ICR mice, male NSY mice show significantly higher non-fasting blood glucose levels (~17 mM vs. ~9 mM of ICR mice) with moderately greater body weight gain (~34.0 g vs. ~28.5 g) at 13–15 weeks of age [75]. After the colony was divided and placed at Osaka University at the 36th generation, researchers have used the inbred C3H/He mouse as a control because the outbred ICR mice have large variations in glucose tolerance and are not suitable for QTL studies [74]. The C3H/He mouse is considered as a glucose-tolerant strain as compared to the other inbred mouse strains [76]. Ueda et al. [74, 77] performed a longitudinal monitoring of glucose tolerance (using IPGTTs) in male NSY mice and demonstrated that the glucose tolerance is impaired in an age-dependent manner. While 8-week-old NSY mice show normal glucose tolerance as compared to

C3H/He mice, the post-challenge blood glucose levels increase up to 24 weeks of age [74]. In addition, the elevation of overnight fasting blood glucose levels in NSY mice becomes evident after 48 weeks of age [77]. The body weight of NSY mice increases more than that of control C3H/He mice from around 10 weeks of age, accompanied by increasing fat mass [77].

### 1.2.3 β-Cell Function

NSY mice show evident fasting hyperinsulinemia from 24 weeks of age. In contrast, insulin responses almost disappear during an IPGTT in these mice [74]. The data imply that aging NSY mice have both impaired insulin secretion from β-cells and insulin resistance. Islet morphology and islet insulin content of NSY mice are similar to those of control C3H/He mice even at 48 weeks of age [74, 77]. Interestingly, glucose-induced insulin secretion from isolated islets is impaired already at 12 weeks of age [74], suggesting that there are early defects in β-cell function in NSY mice. Studies on the β-cell defects in NSY mice, especially with regard to the stimulus-secretion coupling and insulin granular exocytosis, are still largely missing. In one study, Hamada et al. [78] treated isolated islets from 36-week-old NSY mice with the ATP-dependent $K^+$ channel blocker glibenclamide or with the voltage-activated $Ca^{2+}$ channel opener BayK8644 to investigate possible mechanisms affecting β-cell function and reported that insulin secretion in response to both stimuli is impaired in NSY mice as compared to control C3H/He mice [78], indicating that defect(s) would be located at the voltage-activated $Ca^{2+}$ channels or, more likely, the following process of insulin granular exocytosis.

### 1.2.4 Pathogenesis

Using QTL analyses of F2 mice crossed between male NSY mice and female C3H/He mice, Ueda et al. [79] identified three major susceptible loci (*Nidd1nsy*, *Nidd2nsy*, *Nidd3nsy*) on chromosomes 11, 14, and 6, respectively. In addition, *Nidd4nys* on chromosome 11 has been identified and suggested to affect insulin secretion only in young NSY mice [79]. As *Nidd4nys*, *Nidd1nys* on chromosome 11 is linked to impaired insulin secretion, but also in adult mice. *Nidd1nys* contains the genes *Tcf2* (encoding hepatocyte nuclear factor-1b) and *Nxn* (encoding nucleoredoxin) [80, 81]. Interestingly, the locus is overlapped with a susceptible locus for streptozotocin-induced diabetes [82]. Thus, the locus might be unintentionally selected at the initial step of the breeding study, since the origin of NSY mice is "streptozotocin-induced diabetic" ICR mice. In contrast to these two loci coupled to impaired insulin secretion, *Nidd2nsy* and *Nidd3nsy* are linked to insulin sensitivity. The loci *Fl1n* (chromosome 6) and *Bw1n* (chromosome 7), which are linked to fatty liver and body weight, respectively, have also been identified by the same group [83].

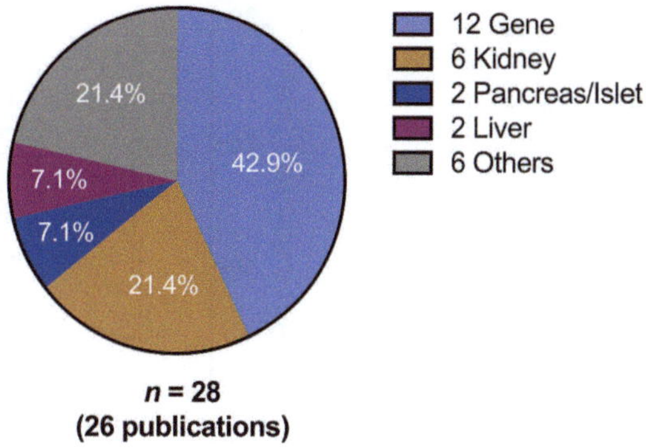


**Fig. 5** Breakdown of research-targeted organs in diabetes pathophysiological studies using Nagoya-Shibata-Yasuda (NSY) mice. By the survey of 36 research articles about Nagoya-Shibata-Yasuda (NSY) mice in PubMed, 26 papers were categorized as studies for the pathophysiology of diabetes and its complications. Then, these papers were further classified by the organ on which research was performed (total $n = 28$ organs in 26 publication)

#### 1.2.5 Trend of Research

Compared with the GK rat, fewer articles were listed on PubMed concerning NSY mice. We detected only 36 articles in total when searched by two sets of keywords: (1) "Nagoya," "Shibata," "Yasuda," and "mice" or (2) "NSY" and "mice." We categorized them into the same two groups as we did with the GK rat articles (Group 1 and Group 2). When the 26 papers in Group 1 were further classified, gene-related studies were the most numerous (Fig. 5). Phenotyping studies on consomic or congenic strains of NSY mouse chromosome 11 and/or 14 were also included in these studies [84, 85]. The second largest volume of the reports was performed on the kidneys. In addition to the glomerular basement membrane thickening and mesangial cell increase [86], NSY mice showed amyloid deposition in the renal glomerulus after 400 days of age [87]. Studies on the β-cell function and the cause of insulin resistance are limited, and there is room for additional investigations to better couple the defects in NSY mice to β-cell function. NSY mice have so far mainly been used as a polygenic animal model for investing the genetic background and the age-related etiology of T2D.

### 1.3 Oikawa-Nagao (ON) Mouse

#### 1.3.1 History

The selective breeding for two lines of ON mice, originally known as "Selectively bred Diet-induced Glucose intolerance-Prone/Resistant (SDG-P/R) mice" [88], was launched by Shinichi Oikawa in 2001. The original working hypothesis was to examine whether HFD-induced glucose intolerance is passed on to the next generations as a heritable trait. The selective breeding had been performed using hybrid mice derived from three inbred mouse lines, C57BL/6J, AKR/N, and C3H/HeJ, to extend genetic diversity which may play an important role in establishing polygenic

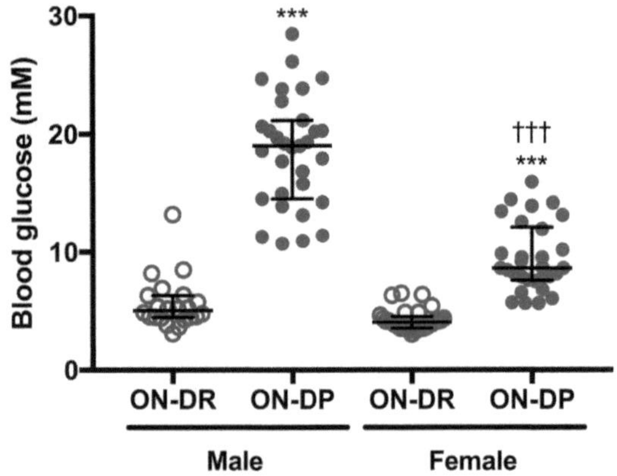


**Fig. 6** Distributions of blood glucose levels at 120 min during an oral glucose tolerance test in Oikawa-Nagao (ON) mice after a high-fat diet feeding. Blood glucose level at 120 min during an OGTT (glucose 2 g/kg) in Oikawa-Nagao Diabetes-Prone (ON-DP, 23th generation) and Diabetes-Resistant (ON-DR, 22th generation) mice after high-fat diet feeding, providing 32% energy as fat, for 5 weeks (from 5 to 10 weeks of age). The lines display the median with interquartile range of the plots. ON-DR mice, open circles (male $n = 22$, female $n = 29$). ON-DP mice, closed circles (male $n = 31$, female $n = 31$). $***P < 0.001$ in Student's *t*-test vs. ON-DR mice of the same gender. $^{\dagger\dagger\dagger}P < 0.001$ using Student's *t*-test, male vs. female mice in the same line

disease model animals by selective breeding. Initially, we fed the hybrid mice with HFD (providing 32% energy as fat) for 10 weeks and thereafter performed an OGTT (glucose 2 g/kg) on the mice. The mice with higher blood glucose levels at 120 min during the OGTT ($BG_{120\ min}$) were selected for breeding. In the early generations of the breeding, we found that some mice showed normal glucose tolerance even after HFD. Since 2005, therefore, HFD-fed mice showing both higher and lower $BG_{120\ min}$ were selected and bred repetitively. As a result, two mouse lines with different susceptibilities (prone and resistant) to HFD-induced glucose intolerance were established [88]. During the selective breeding, sister-brother mating was avoided to maintain the fertility even after the HFD feeding. The selection was terminated, when the distributions of $BG_{120\ min}$ reached levels that were distinctively different from each other (Fig. 6). This occurred at the 23rd generation of SDG-P and the 22nd generation of SDG-R mice. At the end of the selective breeding, we renamed the SDG-P and SDG-R lines to "Oikawa-Nagao Diabetes-Prone (ON-DP) and Diabetes-Resistant (ON-DR) mice," respectively. Both lines of ON mice are maintained in the Institute for Animal Reproduction (Kasumigaura, Japan). Also, new colonies are now developing at Lund University (Malmö, Sweden).

### 1.3.2 Phenotype

As compared to control ON-DR mice, even though male ON-DP mice have similar fasting and slightly higher post-challenge blood levels on a chow diet, they become diabetic after 5 weeks of HFD feeding with moderate fasting hyperglycemia and markedly higher post-challenge blood glucose levels during an OGTT [89]. While male ON-DP mice show evident post-challenge hyperglycemia after 5-week HFD feeding, female ON-DP mice show relatively lower $BG_{120\ min}$ (Fig. 6), indicating that female ON-DP mice are less prone to HFD-induced hyperglycemia relative to the male. Similar gender difference in the incidence of diabetes has been reported in NSY mice [74]. ON-DP mice gain more body weight during HFD feeding and develop more severe insulin resistance as compared to ON-DR mice [74, 89]. The excessive weight gain of ON-DP mice can be explained by hyperphagic behavior under HFD [88, 90]. As compared to ON-DR mice, ON-DP mice show higher inflammatory cytokine gene expression levels in the visceral fat and higher blood pressure after 5-week HFD feeding [91].

### 1.3.3 β-Cell Function

ON-DP mice have modestly higher post-challenge blood glucose levels and lower acute insulin response during an OGTT performed at 5 weeks of age (under normal chow feeding) as compared to ON-DR mice [89]. While islet morphology and β-cell mass remain the same in 5-week-old ON-DP and ON-DR mice on a chow diet, both glucose- and depolarization-induced insulin secretion are reduced in the ON-DP mouse islets [89]. The reduced insulin secretory response in ON-DP mouse islets is also observed after 5-week HFD feeding but here accompanied with a markedly increased β-cell mass as compared to ON-DR mice [89]. Altogether, these findings point toward that β-cell function is impaired downstream of ATP-dependent $K^+$ channel in ON-DP mice. Considering that insulin content is rather increased in the ON-DP mouse islets [89], the impairments most likely are at the level (s) of the voltage-activated $Ca^{2+}$ channels and/or the downstream exocytosis process. The latter hypothesis is supported by reduced gene expression levels of *Stx1a* and *Snap25* in ON-DP mouse islets compared to those in ON-DR [89].

### 1.3.4 Trend of Research

Studies in ON-DP/DR mice suggest that β-cell function and feeding behavior could be crucial determinants for the susceptibility to HFD-induced diabetes. Approaches to clarify the causes of β-cell dysfunction and hyperphagic behavior under HFD in ON-DP mice may therefore be meaningful for elucidating the etiology of T2D, especially related to the modern lifestyle of high-energy diet. In addition, female ON-DP mice have greater atherosclerotic lesion formation than ON-DR mice after 20-week atherogenic diet feeding [92], suggesting that ON mice may also be useful for elucidating the underlying mechanisms of diabetic macrovascular complications.

### 1.4 Comparison of the Selectively Bred Diabetes Models

Table 1 presents the phenotypic features of the selectively bred diabetes models. The GK rat is well-known as an inbred rat strain, but genetic diversity has been observed between the colonies [8]. Although ON mice are available as outbred strains so far, sister-brother mating has been introduced for both lines after the end of the selective breeding. The differences in the onset and pathogenesis of diabetes among the three strains are quite interesting. GK rats develop diabetes spontaneously and have a very short prediabetic window. On the other hand, NSY mice are within a prediabetic stage until 24 weeks of age [74]. ON-DP mice without HFD feeding may also serve as a prediabetic model because their

**Table 1**
**Phenotypic comparisons of the selectively bred diabetes models**

| | GK rats | NSY mice | ON-DP mice |
|---|---|---|---|
| Background strain | Wistar | ICR | C57BL6 × AKR × C3H |
| Breeding | Inbreeding* | Inbreeding | Closed colony breeding |
| *Control strain* | | | |
| Outbred | Wistar | ICR | ON-DR |
| Inbred | BN or Fisher | C3H | N/A |
| Pathogenesis | Polygene × glucotoxicity | Polygene × aging | Polygene? × diet |
| Diabetes onset | After weaning | Around 48 weeks old | After high-fat diet |
| Gender difference | N/C or male > female | Male > female | Male > female |
| Hyperglycemia | Severe | Mild | Mild to moderate |
| Obesity | Lean | Mild | Mild to moderate |
| Insulin resistance | + | + | + |
| β-Cell mass | N/C or ↓ | N/C? | ↑ |
| β-Cell function | ↓↓ | ↓ | ↓ |
| Glucose transport | ↓ | ? | ↓? |
| ATP production | N/C or ↓ | ? | ? |
| Ion channel activities | N/C | ↓? | ↓? |
| Exocytosis | ↓ | ↓? | ↓? |
| Complications | Vascular, heart, eye, kidney, nerve | Kidney | Atherosclerosis |

Phenotypic comparisons between Goto-Kakizaki (GK) rats, Nagoya-Shibata-Yasuda (NSY) mice, and Oikawa-Nagao Diabetes-Prone (ON-DP) and Diabetes-Resistant (ON-DR) mice. All comparisons were made at the onset of diabetes by the differences between each animal model and the control stain. *Genetic diversity has been observed between the colonies. *BN* Brown Norway; *N/A* not applied; *N/C* no change between the animal model and the control

fasting blood glucose are comparable to those in control ON-DR mice [88]. Such prediabetic rodent models could be useful for considering whether the phenotype is a cause or a consequence of hyperglycemia. NSY mice and ON mice may also provide valuable information on the etiologies of age-dependent diabetes and diet-induced diabetes, respectively.

The body weight phenotype of the three models varies in comparison with their corresponding control strains. However, insulin resistance is a common feature of the selectively bred diabetes models. Insulin resistance is considered as a phenotype accompanied by obesity in general. The presence of insulin resistance in lean GK rats is inconsistent with this general principle and suggests that insulin resistance might be controlled independently from obesity at the genetic level. This concept is supported by data from the NSY mice, presenting that their insulin resistance susceptible loci *Nidd2nsy* and *Nidd3nsy* were linked to neither body weight nor body mass index [79].

Another common feature of the three models is a defective stimulus-secretion coupling in the β-cells. It is of great interest that there is no causal relationship between the secretory dysfunction and the β-cell mass in these models. Lower β-cell mass has been suggested to be a part of the pathogenesis of T2D [93]. Therefore, the impaired insulin response during an OGTT in T2D has at least partly been considered as a consequence of decreased β-cell mass. The impaired insulin secretion in T2D has also been explained by defective β-cell function [93]. Interestingly, genetic association studies have reported that most of the variants associated with the risk of T2D are related to β-cell function [94]. Our recent study using human β-cells actually demonstrated that the presence and the multiplicity of genetic variants in major diabetes susceptible genes (in or near KCNQ1, ADRA2A, KCNJ11, HHEX/IDE, and SLC2A2) were associated with deteriorated granule docking and exocytosis [95]. These findings indicate that the genetic background principally regulates β-cell insulin secretion capacity and that the functional impairments may precede the β-cell loss. Further research using β-cells from these selectively bred animal models may provide useful information on the etiology of T2D in humans.

Based on the potential benefits of the selectively bred diabetes models for β-cell functional studies, we will introduce the methodologies for evaluating exocytosis using patch clamp technology in combination with capacitance measurements on single β-cells. Finally, we will explain how to measure islet expression of miRNAs and their putative targets and to analyze expression of exocytotic proteins involved in the exocytosis machinery using isolated islets.

## 2 Materials

### 2.1 Exocytosis Evaluation by Capacitance Measurement on Single *β*-Cells

#### 2.1.1 Dispersal and Plating of Islet Cells

1. $Ca^{2+}$-free solution: 138 mM NaCl, 5.6 mM KCl, 1.2 mM $MgCl_2$, 5 mM HEPES, 3 mM glucose, 0.1 mM EGTA (pH 7.4), 1 mg/mL BSA; sterilize by filtration, aliquot, and store in −20 °C.
2. Plastic 35 mm cell culture dishes.
3. 10 mL Tubes.

#### 2.1.2 Patch Clamp Solutions and Materials

1. Standard extracellular (EC) solution: 118 mM NaCl, 20 mM tetraethyl-ammonium chloride (TEA-Cl; to block voltage-gated $K^+$ currents), 5.6 mM KCl, 2.6 mM $CaCl_2$, 1.2 mM $MgCl_2$, 5 mM glucose, 5 mM HEPES (pH 7.4 using NaOH). RT during experiment; store at 4 °C.
2. Intracellular (IC) solution for trains of depolarizations: 125 mM Cs-Glut, 10 mM NaCl, 10 mM CsCl, 1 mM $MgCl_2$, 0.05 mM EGTA, 3 mM Mg-ATP, 5 mM HEPES, and 0.1 mM cAMP (pH 7.15 using CsOH). Keep on ice during the experiment; store in aliquots at −20 °C.
3. Intracellular (IC) solution for infusion of $Ca^{2+}$: 138 mM CsCl, 1 mM $MgCl_2$, 5 mM HEDTA, 3 mM Mg-ATP, 10 mM HEPES, 0.1 mM cAMP (pH 7.15 using CsOH), and 0.57 mM $CaCl_2$ (estimated free $Ca^{2+}$ concentration is 3 μM). Keep on ice during the experiment; store in aliquots at −20 °C.
4. Borosilicate glass.
5. Pipette puller.
6. Dental wax.
7. Pipette box with lid (use an insert with drilled holes or molding clay to store the pipettes).
8. Fire polisher.
9. Cell culture dish insert to minimize the volume of extracellular solution.
10. 1 mL Syringe with a nonmetallic syringe needle for filling micropipettes.
11. Patch clamp setup (Fig. 7a).
12. Heater.
13. Peristaltic pump.

#### 2.1.3 Patch Clamp Protocols

1. Trains of depolarization: Apply a train of ten 500 ms-long depolarizing pulses from −70 mV to 0 mV. $Ca^{2+}$ currents are measured during the depolarizations. Changes in cell membrane capacitance elicited by this protocol are measured by using the lock-in function of the software applying sine waves with a frequency of 500 Hz to the holding potential (−70 mV).

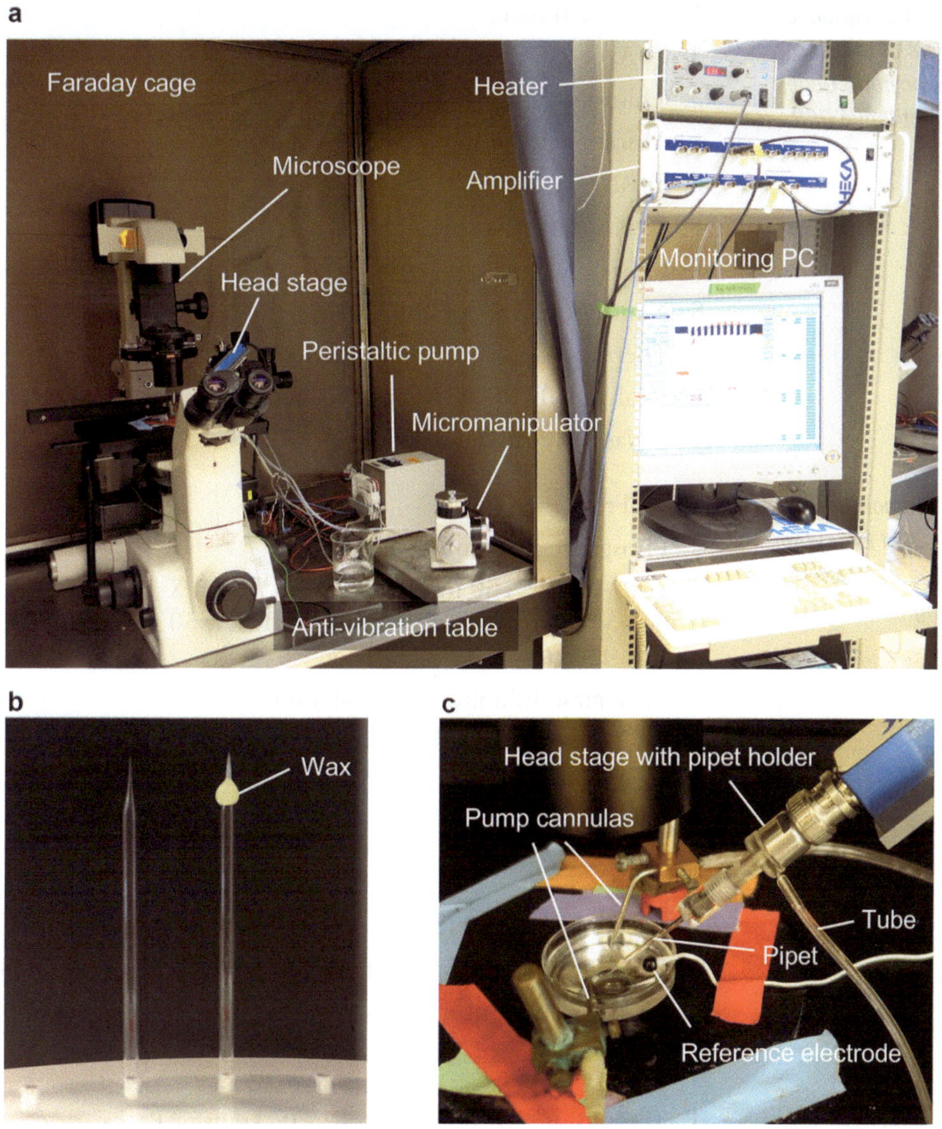


**Fig. 7** Representative images of patch clamp setup and pipettes. Representative patch clamp setup including heater and peristaltic pump (**a**). The microscope and head stage should be placed in the faraday cage for electronical shielding and on the anti-vibration table. Pipettes before (right) and after (left) coating the tip with dental wax (**b**). The pipette being filled with the intracellular solution is tightly attached to the pipette holder on the head stage before being moved into the extracellular solution in the dish and further on to touch the cell. This is followed by the formation of "giga-seal" through a small suction via the tube connected to the pipette holder (**c**)

2. $Ca^{2+}$ infusions: Perform a continuous recording at a holding potential of −70 mV. Changes in cell membrane capacitance are measured by using the lock-in function of the software applying continuous sine waves with a frequency of 500 Hz to the holding potential (−70 mV).

### 2.2 Transcriptomic Analysis of miRNAs and Exocytotic Genes by Quantitative PCR

1. 100 Fresh islets.
2. QIAzol Lysis Reagent.
3. Chloroform.
4. Absolute ethanol.
5. miRNeasy Mini Kit.
6. cDNA synthesis kits.
7. RNase/DNase-free water.
8. NanoDrop spectrophotometer.
9. Thermocyclers for reverse transcription and qPCR.
10. Real-time PCR instrument system.
11. RNase-free tubes.

### 2.3 Proteomic Analysis of Exocytotic Proteins by Western Blotting

#### 2.3.1 Sample Preparation

1. 50 Fresh islets.
2. Phosphate-buffered saline (PBS)
3. RIPA buffer: 150 mM NaCl, 1.0% Triton X-100, 0.5% sodium deoxycholate, 0.1% SDS, 2 mM EDTA, 50 mM NaF, 50 mM Tris–HCl (pH8.0).
4. Proteinase Inhibitor Cocktail Tablet.
5. BCA Protein Assay Kit.
6. 4× Loading buffer: 8% SDS, 40% glycerol, 0.4% bromophenol blue, 400 mM DTT, 200 mM Tris–HCl (pH 6.8).
7. Tubes.

#### 2.3.2 SDS-PAGE

1. 4–15% Mini-PROTEAN® TGX™ Precast Protein Gels.
2. Running buffer: 25 mM Tris, 200 mM glycine, 0.1% SDS.
3. Molecular weight markers.
4. Electrophoresis tank.

#### 2.3.3 Immunoblotting

1. Trans-Blot® Turbo™ Mini PVDF Transfer Packs.
2. Trans-Blot® Turbo™ System.
3. Tris-buffered saline with 0.1% Tween-20 (TBS-T): 150 mM NaCl, 20 mM Tris–HCl (pH 7.5), 0.1% Tween-20.
4. Blocking and antibody diluent solutions: 5% milk in TBS-T.
5. Primary antibodies (1:1000): STX1a, SNAP25, VAMP2, STXBP1, cyclophilin B.
6. Secondary antibodies (1:1000): Goat Anti-Rabbit IgG Horseradish Peroxidase Conjugate (HRP), Goat Anti-Mouse IgG HRP.

7. Clarity™ Western ECL Substrate.
8. CCD camera-based imager.
9. Plastic container.

## 3 Methods

### 3.1 Exocytosis Evaluation by Capacitance Measurement on Single β-Cells

#### 3.1.1 Dispersal and Plating of Islet Cells

1. Wash the islets once in $Ca^{2+}$-free solution.
2. Incubate the islets in 1 mL $Ca^{2+}$-free solution in a 10 mL tube for 12 min at 37 °C (in water bath).
3. Disperse the islet cells by pipetting up and down with a 1 mL pipette (blue tip; *see* **Note 1**).
4. Add ~8 mL preheated complete growth medium.
5. Centrifuge for 2 min at 250 rcf, RT.
6. Discard the supernatant, but leave ~0.5 mL (depending on the number of islets used; *see* **Note 2**).
7. Resuspend the islet cells in the leftover supernatant.
8. Add a drop of the cell suspension to the middle of a dish.
9. Put the dish in the cell incubator for ≥2 h to allow the cells to attach.
10. After ≥2 h, add complete growth medium, and keep the dish in the cell incubator until experiments (*see* **Note 3**).

#### 3.1.2 Preparation of Pipettes for Patch Clamp

1. Pull pipettes from borosilicate glass (pipette resistance ~3–6 MΩ when filled with intracellular solution).
2. Coat the pipette tip with dental wax (Fig. 7b).
3. Fire-polish (*see* **Note 4**).
4. Store in a box with lid.

#### 3.1.3 Patch Clamp Experiment

For most applications, the standard whole-cell configuration of the patch clamp technique can be used:

1. Take the cells out of the cell incubator.
2. Wash once with EC solution.
3. Place the insert in the dish, and add a drop of EC solution to avoid drying of the cells.
4. Mount the dish in the patch setup (Fig. 7c).
5. Set the temperature of the perfused EC solution to 32–34 °C (*see* **Notes 5** and **6**).
6. Under the microscope, localize a cell to patch (*see* **Note 7**).

7. Fill a pipette with IC solution using a nonmetallic syringe needle and attach it to the head stage (Fig. 7c).
8. Before touching the cell, remember to adjust the zero current potential.
9. Form a giga-seal by applying gentle suction.
10. Put the cells to a holding potential of −70 mV and compensate for fast capacitance transients.
11. Enter the whole-cell mode by applying a slight suction and compensate for slow capacitance transients (*see* **Note 8**).
12. Run the desired protocol to determine exocytosis (*see* **Note 9**).
13. Allow the ion channels to recover ~30 s before determining cell type (*see* below and **Note 10**).
14. Discard the pipette and move on to the next cell.

#### 3.1.4 Determination of Cell Type

Rodent β-cells can be identified based on the properties of their voltage-gated $Na^+$ channels using a standard two-pulse protocol:

1. Apply a conditioning pulse from −70 mV to voltages ranging from −130 mV to 40 mV.
2. Allow for a subsequent 1 ms resting period at −70 mV.
3. Apply a depolarizing pulse to 0 mV during which the $Na^+$ channel response is measured.
4. Analyze the relationship between the voltage of the conditioning pulse and the amplitude of the peak current (reflecting the $Na^+$ current). In mouse and rat β-cells, a half maximal inactivation of the $Na^+$ current occurs at ~−80 mV [96, 97].

### 3.2 Transcriptomic Analysis of miRNAs and Exocytotic Genes by Quantitative PCR

#### 3.2.1 Extraction of Total RNA Including Small RNAs

1. Transfer 100 islets into a RNase-free tube.
2. Centrifuge the tube briefly to collect islets and remove the supernatant.
3. Add 700 μL QIAzol into the tube.
4. Homogenize the islets by vortexing until all islets are dissolved indicating complete cell lysis.
5. Leave the homogenate for 5 min at RT.
6. Add 140 μL chloroform into the tube containing the homogenate.
7. Alternately vortex for 3–5 s, and shake vigorously to ensure complete mixing.
8. Leave the tube for 5 min at RT to allow initial phase separation (*see* **Note 11**).

9. Centrifuge the tube for 15 min at $12 \times 10^3$ rcf, 4 °C.
10. Carefully remove the tube without disturbing the phases (*see* **Note 12**).
11. Carefully transfer 300 µL of aqueous phase into a new RNase-free tube (*see* **Note 13**).
12. Add 450 µL 100% ethanol into the tube containing the aqueous phase.
13. Mix thoroughly by pipetting up and down.
14. Transfer all the mixture (750 µL) into an RNeasy mini spin column with a collection tube.
15. Centrifuge for 15 s at $8 \times 10^3$ rcf, RT.
16. Transfer the spin column into a new 2 mL collection tube.
17. Add 700 µL buffer RWT.
18. Centrifuge for 15 s at $8 \times 10^3$ rcf, RT.
19. Discard the flow through and reuse the collection tube.
20. Add 500 µL buffer RPE into the spin column.
21. Centrifuge for 15 s at $8 \times 10^3$ rcf, RT.
22. Discard the flow through and reuse the collection tube.
23. Add 500 µL buffer RPE into the spin column.
24. Centrifuge for 2 min at $8 \times 10^3$ rcf, RT.
25. Carefully transfer the spin column into a new collection tube.
26. Centrifuge for 1 min at $8 \times 10^3$ rcf, RT, to completely dry the membrane in the spin column (*see* **Note 14**).
27. Transfer the spin column into a new RNase-free tube.
28. Add 30 µL of nuclease-free water into the middle of the spin column. Make sure that the water gets into the membrane without touching the surroundings of the membrane.
29. Leave the tube for 1 min at RT to completely saturate the membrane.
30. Centrifuge for 1 min at $8 \times 10^3$ rcf, RT, to elute the RNA.
31. Quantify the concentration and purity of the RNA extract using NanoDrop spectrophotometer (a sample volume of 1.5 µL is necessary for one quantification).

#### 3.2.2 cDNA Preparation and qPCR

1. Synthesize cDNA using commercially available kits (*see* **Notes 15** and **16**).
2. Run qPCR in triplicates in 384 well plate at 10 µL reaction volume per well (*see* **Note 17**).

### 3.3 Proteomic Analysis of Exocytotic Proteins by Western Blotting

#### 3.3.1 Sample Preparation

1. Transfer 100 islets into a tube.
2. Centrifuge the tube for 30 s at 500 rpm, 4 °C.
3. Remove and discard the supernatant carefully.
4. Add ice-cold PBS 200 μL into the tube.
5. Centrifuge the tube for 30 s at 500 rpm, 4 °C.
6. Remove and discard the supernatant carefully.
7. Repeat **steps 4–6** once.
8. Add ice-cold RIPA buffer with proteinase inhibitor cocktail (one tablet for 25 mL RIPA buffer) 100 μL (*see* **Note 18**).
9. Rotate the tube for 30 min at 4 °C.
10. Centrifuge the tube for 10 min at $10 \times 10^3$ rcf, 4 °C.
11. Collect the supernatant and transfer it into a new tube.
12. Measure the protein concentration using BCA Protein Assay Kit.
13. Take an equal protein amount (10–20 μg) of sample into a new tube.
14. Add the ice-cold RIPA buffer to attain 30 μL of sample volume.
15. Add 10 μL 4× loading buffer and pipette up and down.
16. Heat the tube for 10 min at 70 °C.

#### 3.3.2 SDS-PAGE

1. Set the gel in an electrophoresis tank.
2. Add sufficient running buffer.
3. Load 40 μL samples onto the wells of the gel, along with molecular weight markers.
4. Run the gel for 35 min at 200 V.

#### 3.3.3 Immunoblotting

1. Remove the gel from the cover.
2. Assemble the transfer sandwich on a transfer cassette.
3. Make sure no air bubbles are trapped in the sandwich.
4. Place the cassette cover and set into Trans-Blot® Turbo™ System.
5. Transfer for 6 min at 2.5 A (constant), 25 V (*see* **Note 19**).
6. Rinse the blot with TBS-T for 10 min.
7. Block the blot for 40 min at RT.
8. Incubate overnight in the primary antibody solution against the target protein at 4 °C (*see* **Note 20**).
9. Rinse the blot with TBS-T for 10 min, two times.
10. Incubate in the HRP-conjugated secondary antibody solution for 1 h at RT.

11. Rinse the blot with TBS-T for 10 min, two times.
12. Incubate the blot in 10 mL conjugate of Clarity™ Western ECL Substrate.
13. Capture the chemiluminescent signals using a CCD camera-based imager.

## 4 Notes

1. The solution becomes milky in appearance when the islets are dispersed. Check under a microscope if you are unsure whether they have been dispersed or not.
2. We normally add the equivalent of 15–20 islets/dish.
3. We normally leave the cells overnight to allow them to firmly attach and recover before experiments.
4. To avoid dirt on the pipette tips, we fire-polish them the day of the experiment
5. At higher temperatures, stable seal formation is much more difficult.
6. It is important that the surface level of the perfused extracellular solution is kept stable during the experiment; otherwise, artifacts resembling capacitance changes can be evoked.
7. The dish contains a mixture of islet cells. However, rodent β-cells are normally larger ($\geq$5 pF) than the other islet cells.
8. Aim for an access (Rseries) of $\leq$20 MΩ and a leak current of $\leq$50 pA.
9. If trains of depolarizations are used, allow some time (~45 s) for the intracellular solution to wash in and the seal to stabilize before applying the protocol.
10. This program should be run last since the conditioning pulse triggers exocytosis.
11. The aqueous phase will be the top layer, while the organic phase will be the reddish-colored bottom layer.
12. The upper aqueous phase should be colorless and contains mostly RNA. The white interphase will contain large DNA fragments and some proteins, while most proteins and smaller DNA fragments will be in the bottom reddish-colored organic phase.
13. To ensure smooth fluid suction and to avoid phase contamination, use a pipette set at 150 μL volume and transfer aqueous phase two times. It is extremely important not to contaminate the aqueous phase with the white interphase or the red organic phase. If this happens, recentrifuge the tube again for 15 min at $12 \times 10^3$ rcf, 4 °C.

14. This step ensures residual ethanol is removed.
15. Depending on which RNAs are to be quantified, separate cDNA kits are used. For mRNAs, random primers are used, while for miRNAs, stem-loop primers (TaqMan miRNA assays) for specific miRNAs are used.
16. There are many commercially available cDNA kits, and we generally use 100 ng of total RNA for cDNA in a 20 μL reaction volume.
17. For qPCR, we mainly use TaqMan gene-specific assays for both miRNAs and mRNAs. Alternatively, SYBR green-based qPCR assays is also available for gene quantification.
18. The volume of RIPA buffer to lyse islets is recommended as "1 μL for one islet." In this case, the total protein concentration will be around 500 μg/mL in the BCA assay.
19. Using the transfer method, small amount of high-molecular-weight proteins (approx. >120 kD) was left in the transferred gel. However, proteins of 100 kD or less were transferred completely. We therefore consider the transfer method is available at least for these exocytosis proteins, because their molecular weights are from 13 (VAMP2) to 72 kD (STXBP1). If you want to check high-molecular-weight proteins, we recommend using a wet transfer method.
20. Beta-actin (ACTB) is not available for the endogenous control, because ACTB band intensity of GK rat islets tends to be lower than that of Wistar rat islets.

## Acknowledgment

All animal experiments were performed in accordance to ethical permits issued by the Malmö/Lund Ethical Committee of Animal Research (Malmö and Lund, Sweden) or the Nippon Medical School Animal Policy and Welfare Committee (Tokyo, Japan).

We thank Britt-Marie Nilsson, Anna-Maria Veljanovska Ramsay, and Neelanjan Vishnu (Lund University) for technical assistance of GK/LU rat studies, Holger Luthman (Lund University) for valuable discussion regarding GK/LU rats, and Momoyo Kawahara (Nippon Medical School), Miki Onodera, and Ryoji Hokao (Institute for Animal Reproduction) for technical assistance of ON mice studies.

The work is financially supported by the Swedish Foundation for Strategic Research (IRC-LUDC), Swedish Research Council (SFO-EXODIAB; LE, 2016-02124), Region Skåne-ALF (LE), Swedish Diabetes Foundation (LE; DIA2016-130), Albert Påhlsson Foundation (LE and JLSE), Japan Society for the Promotion of Science (MN, JLSE, and AA), European Foundation for the Study of Diabetes, Japan Diabetes Society (MN), Uehara Memorial Foundation (MN), Scandinavia-Japan Sasakawa Foundation (MN), Sumitomo Life Welfare Foundation (MN), Diabetes Wellness Sverige (MN, 720-2964 JDWG), and Lotte Shigemitsu Prize (AA).

## References


1. Goto Y, Kakizaki M, Masaki N (1975) Spontaneous diabetes produced by selective breeding of normal Wistar rats. Proc Jpn Acad 51 (1):80–85
2. Goto Y, Suzuki K, Sasaki M, Ono T, Abe S (1988) GK rats as a model of nonobese, noninsulin-dependent diabetes. Selective breeding over 35 generations. In: Shafrir E, Renold A (eds) Lessons from animal diabetes II. Libbey J, London, pp 490–492
3. Ostenson CG, Khan A, Abdel-Halim SM, Guenifi A, Suzuki K, Goto Y, Efendic S (1993) Abnormal insulin secretion and glucose metabolism in pancreatic islets from the spontaneously diabetic GK rat. Diabetologia 36 (1):3–8
4. Portha B, Serradas P, Bailbe D, Suzuki K, Goto Y, Giroix MH (1991) Beta-cell insensitivity to glucose in the GK rat, a spontaneous nonobese model for type II diabetes. Diabetes 40(4):486–491
5. Lewis BM, Ismail IS, Issa B, Peters JR, Scanlon MF (1996) Desensitisation of somatostatin, TRH and GHRH responses to glucose in the diabetic (Goto-Kakizaki) rat hypothalamus. J Endocrinol 151(1):13–17
6. Hughes SJ, Suzuki K, Goto Y (1994) The role of islet secretory function in the development of diabetes in the GK Wistar rat. Diabetologia 37(9):863–870
7. Lagerholm S, Park HB, Luthman H, Nilsson M, McGuigan F, Swanberg M, Akesson K (2010) Genetic loci for bone architecture determined by three-dimensional CT in crosses with the diabetic GK rat. Bone 47 (6):1039–1047. https://doi.org/10.1016/j.bone.2010.08.003
8. Bihoreau MT, Dumas ME, Lathrop M, Gauguier D (2017) Genomic regulation of type 2 diabetes endophenotypes: contribution from genetic studies in the Goto-Kakizaki rat. Biochimie 143:56–65. https://doi.org/10.1016/j.biochi.2017.08.012
9. Östenson C (2007) The Goto-Kakizaki rat. In: Shafrir E (ed) Animal models of diabetes, Frontiers in research, 2nd edn. CRC Press, New York, pp 119–138
10. Goto Y (1991) Foundation of the GK rat. J Jpn Diabetes Soc 34(11):939–941
11. Kimura K, Toyota T, Kakizaki M, Kudo M, Takebe K, Goto Y (1982) Impaired insulin secretion in the spontaneous diabetes rats. Tohoku J Exp Med 137(4):453–459
12. Suzuki K, Goto Y, Toyoda T (1993) Spontaneously diabetic GK (Goto-Kakizaki) rats. In: Shafrir E (ed) Lessons from animal diabetes IV. Smith-Gordon, London, pp 107–116
13. Portha B, Giroix MH, Tourrel-Cuzin C, Le-Stunff H, Movassat J (2012) The GK rat: a prototype for the study of non-overweight type 2 diabetes. Methods Mol Biol 933:125–159. https://doi.org/10.1007/978-1-62703-068-7_9
14. Ostenson CG, Efendic S (2007) Islet gene expression and function in type 2 diabetes; studies in the Goto-Kakizaki rat and humans. Diabetes Obes Metab 9(Suppl 2):180–186. https://doi.org/10.1111/j.1463-1326.2007.00787.x
15. Bisbis S, Bailbe D, Tormo MA, Picarel-Blanchot F, Derouet M, Simon J, Portha B (1993) Insulin resistance in the GK rat: decreased receptor number but normal kinase activity in liver. Am J Physiol 265(5 Pt 1):E807–E813. https://doi.org/10.1152/ajpendo.1993.265.5.E807
16. Tourrel C, Bailbe D, Lacorne M, Meile MJ, Kergoat M, Portha B (2002) Persistent improvement of type 2 diabetes in the Goto-Kakizaki rat model by expansion of the beta-cell mass during the prediabetic period with glucagon-like peptide-1 or exendin-4. Diabetes 51(5):1443–1452
17. Movassat J, Calderari S, Fernandez E, Martin MA, Escriva F, Plachot C, Gangnerau MN, Serradas P, Alvarez C, Portha B (2007) Type 2 diabetes - a matter of failing beta-cell neogenesis? Clues from the GK rat model. Diabetes Obes Metab 9(Suppl 2):187–195. https://doi.org/10.1111/j.1463-1326.2007.00786.x
18. Movassat J, Saulnier C, Serradas P, Portha B (1997) Impaired development of pancreatic beta-cell mass is a primary event during the

progression to diabetes in the GK rat. Diabetologia 40(8):916–925. https://doi.org/10.1007/s001250050768
19. Guenifi A, Abdel-Halim SM, Hoog A, Falkmer S, Ostenson CG (1995) Preserved beta-cell density in the endocrine pancreas of young, spontaneously diabetic Goto-Kakizaki (GK) rats. Pancreas 10(2):148–153
20. Momose K, Nunomiya S, Nakata M, Yada T, Kikuchi M, Yashiro T (2006) Immunohistochemical and electron-microscopic observation of beta-cells in pancreatic islets of spontaneously diabetic Goto-Kakizaki rats. Med Mol Morphol 39(3):146–153. https://doi.org/10.1007/s00795-006-0324-9
21. Hoog A, Sandberg-Nordqvist AC, Abdel-Halim SM, Carlsson-Skwirut C, Guenifi A, Tally M, Ostenson CG, Falkmer S, Sara VR, Efendic S, Schalling M, Grimelius L (1996) Increased amounts of a high molecular weight insulin-like growth factor II (IGF-II) peptide and IGF-II messenger ribonucleic acid in pancreatic islets of diabetic Goto-Kakizaki rats. Endocrinology 137(6):2415–2423. https://doi.org/10.1210/endo.137.6.8641194
22. Homo-Delarche F, Calderari S, Irminger JC, Gangnerau MN, Coulaud J, Rickenbach K, Dolz M, Halban P, Portha B, Serradas P (2006) Islet inflammation and fibrosis in a spontaneous model of type 2 diabetes, the GK rat. Diabetes 55(6):1625–1633. https://doi.org/10.2337/db05-1526
23. Salehi A, Henningsson R, Mosen H, Ostenson CG, Efendic S, Lundquist I (1999) Dysfunction of the islet lysosomal system conveys impairment of glucose-induced insulin release in the diabetic GK rat. Endocrinology 140(7):3045–3053. https://doi.org/10.1210/endo.140.7.6862
24. Frese T, Bazwinsky I, Muhlbauer E, Peschke E (2007) Circadian and age-dependent expression patterns of GLUT2 and glucokinase in the pancreatic beta-cell of diabetic and nondiabetic rats. Horm Metab Res 39(8):567–574. https://doi.org/10.1055/s-2007-984471
25. Ohneda M, Johnson JH, Inman LR, Chen L, Suzuki K, Goto Y, Alam T, Ravazzola M, Orci L, Unger RH (1993) GLUT2 expression and function in beta-cells of GK rats with NIDDM. Dissociation between reductions in glucose transport and glucose-stimulated insulin secretion. Diabetes 42(7):1065–1072
26. Ling ZC, Efendic S, Wibom R, Abdel-Halim SM, Ostenson CG, Landau BR, Khan A (1998) Glucose metabolism in Goto-Kakizaki rat islets. Endocrinology 139(6):2670–2675. https://doi.org/10.1210/endo.139.6.6053
27. Hughes SJ, Faehling M, Thorneley CW, Proks P, Ashcroft FM, Smith PA (1998) Electrophysiological and metabolic characterization of single beta-cells and islets from diabetic GK rats. Diabetes 47(1):73–81
28. Fabregat ME, Novials A, Giroix MH, Sener A, Gomis R, Malaisse WJ (1996) Pancreatic islet mitochondrial glycerophosphate dehydrogenase deficiency in two animal models of non-insulin-dependent diabetes mellitus. Biochem Biophys Res Commun 220(3):1020–1023. https://doi.org/10.1006/bbrc.1996.0525
29. Matsuoka T, Kajimoto Y, Watada H, Umayahara Y, Kubota M, Kawamori R, Yamasaki Y, Kamada T (1995) Expression of CD38 gene, but not of mitochondrial glycerol-3-phosphate dehydrogenase gene, is impaired in pancreatic islets of GK rats. Biochem Biophys Res Commun 214(1):239–246
30. Ostenson CG, Abdel-Halim SM, Rasschaert J, Malaisse-Lagae F, Meuris S, Sener A, Efendic S, Malaisse WJ (1993) Deficient activity of FAD-linked glycerophosphate dehydrogenase in islets of GK rats. Diabetologia 36(8):722–726
31. Tsuura Y, Ishida H, Okamoto Y, Kato S, Horie M, Ikeda H, Seino Y (1994) Reduced sensitivity of dihydroxyacetone on ATP-sensitive K+ channels of pancreatic beta cells in GK rats. Diabetologia 37(11):1082–1087
32. Rasschaert J, Giroix MH, Conget I, Mercan D, Leclercq-Meyer V, Sener A, Portha B, Malaisse WJ (1994) Pancreatic islet response to dicarboxylic acid esters in rats with type 2 diabetes: enzymatic, metabolic and secretory aspects. J Mol Endocrinol 13(2):209–217
33. Metz SA, Meredith M, Vadakekalam J, Rabaglia ME, Kowluru A (1999) A defect late in stimulus-secretion coupling impairs insulin secretion in Goto-Kakizaki diabetic rats. Diabetes 48(9):1754–1762
34. Tsuura Y, Ishida H, Okamoto Y, Kato S, Sakamoto K, Horie M, Ikeda H, Okada Y, Seino Y (1993) Glucose sensitivity of ATP-sensitive K+ channels is impaired in beta-cells of the GK rat. A new genetic model of NIDDM. Diabetes 42(10):1446–1453
35. Marie JC, Bailbe D, Gylfe E, Portha B (2001) Defective glucose-dependent cytosolic Ca2+ handling in islets of GK and nSTZ rat models of type 2 diabetes. J Endocrinol 169(1):169–176
36. Zaitsev S, Efanova I, Ostenson CG, Efendic S, Berggren PO (1997) Delayed Ca2+ response to glucose in diabetic GK rat. Biochem Biophys Res Commun 239(1):129–133. https://doi.org/10.1006/bbrc.1997.7441

37. Rose T, Efendic S, Rupnik M (2007) Ca2+-secretion coupling is impaired in diabetic Goto Kakizaki rats. J Gen Physiol 129 (6):493–508. https://doi.org/10.1085/jgp.200609604
38. Ohara-Imaizumi M, Nishiwaki C, Kikuta T, Nagai S, Nakamichi Y, Nagamatsu S (2004) TIRF imaging of docking and fusion of single insulin granule motion in primary rat pancreatic beta-cells: different behaviour of granule motion between normal and Goto-Kakizaki diabetic rat beta-cells. Biochem J 381 (Pt 1):13–18. https://doi.org/10.1042/BJ20040434
39. Eliasson L, Abdulkader F, Braun M, Galvanovskis J, Hoppa MB, Rorsman P (2008) Novel aspects of the molecular mechanisms controlling insulin secretion. J Physiol 586(14):3313–3324. https://doi.org/10.1113/jphysiol.2008.155317
40. Gaisano HY, Ostenson CG, Sheu L, Wheeler MB, Efendic S (2002) Abnormal expression of pancreatic islet exocytotic soluble N-ethylmaleimide-sensitive factor attachment protein receptors in Goto-Kakizaki rats is partially restored by phlorizin treatment and accentuated by high glucose treatment. Endocrinology 143(11):4218–4226. https://doi.org/10.1210/en.2002-220237
41. Zhang W, Khan A, Ostenson CG, Berggren PO, Efendic S, Meister B (2002) Down-regulated expression of exocytotic proteins in pancreatic islets of diabetic GK rats. Biochem Biophys Res Commun 291(4):1038–1044. https://doi.org/10.1006/bbrc.2002.6555
42. Ohara-Imaizumi M, Nishiwaki C, Nakamichi Y, Kikuta T, Nagai S, Nagamatsu S (2004) Correlation of syntaxin-1 and SNAP-25 clusters with docking and fusion of insulin granules analysed by total internal reflection fluorescence microscopy. Diabetologia 47 (12):2200–2207. https://doi.org/10.1007/s00125-004-1579-0
43. Qin T, Liang T, Zhu D, Kang Y, Xie L, Dolai S, Sugita S, Takahashi N, Ostenson CG, Banks K, Gaisano HY (2017) Munc18b increases insulin granule fusion, restoring deficient insulin secretion in Type-2 diabetes human and Goto-Kakizaki rat islets with improvement in glucose homeostasis. EBioMedicine 16:262–274. https://doi.org/10.1016/j.ebiom.2017.01.030
44. Seino S, Shibasaki T (2005) PKA-dependent and PKA-independent pathways for cAMP-regulated exocytosis. Physiol Rev 85 (4):1303–1342. https://doi.org/10.1152/physrev.00001.2005
45. Abdel-Halim SM, Guenifi A, He B, Yang B, Mustafa M, Hojeberg B, Hillert J, Bakhiet M, Efendic S (1998) Mutations in the promoter of adenylyl cyclase (AC)-III gene, overexpression of AC-III mRNA, and enhanced cAMP generation in islets from the spontaneously diabetic GK rat model of type 2 diabetes. Diabetes 47 (3):498–504
46. Portela-Gomes GM, Abdel-Halim SM (2002) Overexpression of Gs proteins and adenylyl cyclase in normal and diabetic islets. Pancreas 25(2):176–181
47. Wierup N, Bjorkqvist M, Kuhar MJ, Mulder H, Sundler F (2006) CART regulates islet hormone secretion and is expressed in the beta-cells of type 2 diabetic rats. Diabetes 55 (2):305–311
48. Rosengren AH, Jokubka R, Tojjar D, Granhall C, Hansson O, Li DQ, Nagaraj V, Reinbothe TM, Tuncel J, Eliasson L, Groop L, Rorsman P, Salehi A, Lyssenko V, Luthman H, Renstrom E (2010) Overexpression of alpha2A-adrenergic receptors contributes to type 2 diabetes. Science 327 (5962):217–220. https://doi.org/10.1126/science.1176827
49. Abdel-Halim SM, Guenifi A, Khan A, Larsson O, Berggren PO, Ostenson CG, Efendic S (1996) Impaired coupling of glucose signal to the exocytotic machinery in diabetic GK rats: a defect ameliorated by cAMP. Diabetes 45(7):934–940
50. Dolz M, Bailbe D, Giroix MH, Calderari S, Gangnerau MN, Serradas P, Rickenbach K, Irminger JC, Portha B (2005) Restitution of defective glucose-stimulated insulin secretion in diabetic GK rat by acetylcholine uncovers paradoxical stimulatory effect of beta-cell muscarinic receptor activation on cAMP production. Diabetes 54(11):3229–3237
51. Gauguier D, Froguel P, Parent V, Bernard C, Bihoreau MT, Portha B, James MR, Penicaud L, Lathrop M, Ktorza A (1996) Chromosomal mapping of genetic loci associated with non-insulin dependent diabetes in the GK rat. Nat Genet 12(1):38–43. https://doi.org/10.1038/ng0196-38
52. Nobrega MA, Solberg Woods LC, Fleming S, Jacob HJ (2009) Distinct genetic regulation of progression of diabetes and renal disease in the Goto-Kakizaki rat. Physiol Genomics 39 (1):38–46. https://doi.org/10.1152/physiolgenomics.90389.2008
53. Galli J, Li LS, Glaser A, Ostenson CG, Jiao H, Fakhrai-Rad H, Jacob HJ, Lander ES, Luthman H (1996) Genetic analysis of non-insulin dependent diabetes mellitus in the GK rat. Nat

Genet 12(1):31–37. https://doi.org/10.1038/ng0196-31
54. Wallace KJ, Wallis RH, Collins SC, Argoud K, Kaisaki PJ, Ktorza A, Woon PY, Bihoreau MT, Gauguier D (2004) Quantitative trait locus dissection in congenic strains of the Goto-Kakizaki rat identifies a region conserved with diabetes loci in human chromosome 1q. Physiol Genomics 19(1):1–10. https://doi.org/10.1152/physiolgenomics.00114.2004
55. Atanur SS, Diaz AG, Maratou K, Sarkis A, Rotival M, Game L, Tschannen MR, Kaisaki PJ, Otto GW, Ma MC, Keane TM, Hummel O, Saar K, Chen W, Guryev V, Gopalakrishnan K, Garrett MR, Joe B, Citterio L, Bianchi G, McBride M, Dominiczak A, Adams DJ, Serikawa T, Flicek P, Cuppen E, Hubner N, Petretto E, Gauguier D, Kwitek A, Jacob H, Aitman TJ (2013) Genome sequencing reveals loci under artificial selection that underlie disease phenotypes in the laboratory rat. Cell 154(3):691–703. https://doi.org/10.1016/j.cell.2013.06.040
56. Granhall C, Rosengren AH, Renstrom E, Luthman H (2006) Separately inherited defects in insulin exocytosis and beta-cell glucose metabolism contribute to type 2 diabetes. Diabetes 55(12):3494–3500. https://doi.org/10.2337/db06-0796
57. Esguerra JL, Bolmeson C, Cilio CM, Eliasson L (2011) Differential glucose-regulation of microRNAs in pancreatic islets of non-obese type 2 diabetes model Goto-Kakizaki rat. PLoS One 6(4):e18613. https://doi.org/10.1371/journal.pone.0018613
58. Salunkhe VA, Ofori JK, Gandasi NR, Salo SA, Hansson S, Andersson ME, Wendt A, Barg S, Esguerra JLS, Eliasson L (2017) MiR-335 overexpression impairs insulin secretion through defective priming of insulin vesicles. Physiol Rep 5(21). https://doi.org/10.14814/phy2.13493
59. Ofori JK, Salunkhe VA, Bagge A, Vishnu N, Nagao M, Mulder H, Wollheim CB, Eliasson L, Esguerra JL (2017) Elevated miR-130a/miR130b/miR-152 expression reduces intracellular ATP levels in the pancreatic beta cell. Sci Rep 7:44986. https://doi.org/10.1038/srep44986
60. Malm HA, Mollet IG, Berggreen C, Orho-Melander M, Esguerra JL, Goransson O, Eliasson L (2016) Transcriptional regulation of the miR-212/miR-132 cluster in insulin-secreting beta-cells by cAMP-regulated transcriptional co-activator 1 and salt-inducible kinases. Mol Cell Endocrinol 424:23–33. https://doi.org/10.1016/j.mce.2016.01.010
61. Chavey A, Ah Kioon MD, Bailbe D, Movassat J, Portha B (2014) Maternal diabetes, programming of beta-cell disorders and intergenerational risk of type 2 diabetes. Diabetes Metab 40(5):323–330. https://doi.org/10.1016/j.diabet.2014.02.003
62. Dayeh T, Volkov P, Salo S, Hall E, Nilsson E, Olsson AH, Kirkpatrick CL, Wollheim CB, Eliasson L, Ronn T, Bacos K, Ling C (2014) Genome-wide DNA methylation analysis of human pancreatic islets from type 2 diabetic and non-diabetic donors identifies candidate genes that influence insulin secretion. PLoS Genet 10(3):e1004160. https://doi.org/10.1371/journal.pgen.1004160
63. El-Omar MM, Yang ZK, Phillips AO, Shah AM (2004) Cardiac dysfunction in the Goto-Kakizaki rat. A model of type II diabetes mellitus. Basic Res Cardiol 99(2):133–141. https://doi.org/10.1007/s00395-004-0440-4
64. Devanathan S, Nemanich ST, Kovacs A, Fettig N, Gropler RJ, Shoghi KI (2013) Genomic and metabolic disposition of non-obese type 2 diabetic rats to increased myocardial fatty acid metabolism. PLoS One 8(10): e78477. https://doi.org/10.1371/journal.pone.0078477
65. Korkmaz-Icoz S, Lehner A, Li S, Vater A, Radovits T, Brune M, Ruppert M, Sun X, Brlecic P, Zorn M, Karck M, Szabo G (2016) Left ventricular pressure-volume measurements and myocardial gene expression profile in type 2 diabetic Goto-Kakizaki rats. Am J Physiol Heart Circ Physiol 311(4): H958–H971. https://doi.org/10.1152/ajpheart.00956.2015
66. Nagao M, Asai A, Oikawa S (2013) FoxO1 breaks diabetic heart. J Diabetes Investig 4(1):37–38. https://doi.org/10.1111/jdi.12022
67. Yagihashi S, Goto Y, Kakizaki M, Kaseda N (1978) Thickening of glomerular basement membrane in spontaneously diabetic rats. Diabetologia 15(4):309–312
68. Yagihashi S, Kaseda N, Kakizaki M, Goto Y (1979) Evolution of glomerular lesions in rats with spontaneous diabetes. Tohoku J Exp Med 127(4):359–367
69. Oikawa S, Kakizaki M, Goto Y (1982) Inhibitory effect of pancreatic elastase on thickening of the renal glomerular basement membrane in the spontaneously diabetic rat. Tohoku J Exp Med 138(1):103–109
70. Yagihashi S, Tonosaki A, Yamada K, Kakizaki M, Goto Y (1982) Peripheral neuropathy in selectively-inbred spontaneously diabetic rats: electrophysiological,

morphometrical and freeze-replica studies. Tohoku J Exp Med 138(1):39–48

71. Goto Y (2009) Our diabetes studies over 50 years. Soshinsya, Tokyo
72. Tang Y, Axelsson AS, Spegel P, Andersson LE, Mulder H, Groop LC, Renstrom E, Rosengren AH (2014) Genotype-based treatment of type 2 diabetes with an alpha2A-adrenergic receptor antagonist. Sci Transl Med 6(257):257ra139. https://doi.org/10.1126/scitranslmed.3009934
73. Shibata M, Yasuda B (1980) New experimental congenital diabetic mice (N.S.Y. mice). Tohoku J Exp Med 130(2):139–142
74. Ueda H, Ikegami H, Yamato E, Fu J, Fukuda M, Shen G, Kawaguchi Y, Takekawa K, Fujioka Y, Fujisawa T et al (1995) The NSY mouse: a new animal model of spontaneous NIDDM with moderate obesity. Diabetologia 38(5):503–508
75. Fushimi H, Shibata M, Tarui S (1980) Glycosidase activities in the liver and kidney of hereditary diabetic mice. J Biochem 87(3):941–949
76. Kaku K, Fiedorek FT Jr, Province M, Permutt MA (1988) Genetic analysis of glucose tolerance in inbred mouse strains. Evidence for polygenic control. Diabetes 37(6):707–713
77. Ueda H, Ikegami H, Kawaguchi Y, Fujisawa T, Nojima K, Babaya N, Yamada K, Shibata M, Yamato E, Ogihara T (2000) Age-dependent changes in phenotypes and candidate gene analysis in a polygenic animal model of Type II diabetes mellitus; NSY mouse. Diabetologia 43(7):932–938. https://doi.org/10.1007/s001250051472
78. Hamada Y, Ikegami H, Ueda H, Kawaguchi Y, Yamato E, Nojima K, Yamada K, Babaya N, Shibata M, Ogihara T (2001) Insulin secretion to glucose as well as nonglucose stimuli is impaired in spontaneously diabetic Nagoya-Shibata-Yasuda mice. Metabolism 50 (11):1282–1285. https://doi.org/10.1053/meta.2001.27198
79. Ueda H, Ikegami H, Kawaguchi Y, Fujisawa T, Yamato E, Shibata M, Ogihara T (1999) Genetic analysis of late-onset type 2 diabetes in a mouse model of human complex trait. Diabetes 48(5):1168–1174
80. Babaya N, Ikegami H, Fujisawa T, Nojima K, Itoi-Babaya M, Inoue K, Ohno T, Shibata M, Ogihara T (2005) Susceptibility to streptozotocin-induced diabetes is mapped to mouse chromosome 11. Biochem Biophys Res Commun 328(1):158–164. https://doi.org/10.1016/j.bbrc.2004.12.149
81. Ueda H, Ikegami H, Kawaguchi Y, Fujisawa T, Nojima K, Babaya N, Yamada K, Shibata M, Yamato E, Ogihara T (2001) Mapping and promoter sequencing of HNF-1beta gene in diabetes-prone and -resistant mice. Diabetes Res Clin Pract 53(2):67–71
82. Gonzalez C, Cuvellier S, Hue-Beauvais C, Levi-Strauss M (2003) Genetic control of non obese diabetic mice susceptibility to high-dose streptozotocin-induced diabetes. Diabetologia 46(9):1291–1295. https://doi.org/10.1007/s00125-003-1168-7
83. Itoi-Babaya M, Ikegami H, Fujisawa T, Ueda H, Nojima K, Babaya N, Kobayashi M, Noso S, Kawaguchi Y, Yamaji K, Shibata M, Ogihara T (2007) Fatty liver and obesity: phenotypically correlated but genetically distinct traits in a mouse model of type 2 diabetes. Diabetologia 50(8):1641–1648. https://doi.org/10.1007/s00125-007-0700-6
84. Babaya N, Fujisawa T, Nojima K, Itoi-Babaya-M, Yamaji K, Yamada K, Kobayashi M, Ueda H, Hiromine Y, Noso S, Ikegami H (2010) Direct evidence for susceptibility genes for type 2 diabetes on mouse chromosomes 11 and 14. Diabetologia 53 (7):1362–1371. https://doi.org/10.1007/s00125-010-1737-5
85. Babaya N, Ueda H, Noso S, Hiromine Y, Itoi-Babaya M, Kobayashi M, Fujisawa T, Ikegami H (2014) Genetic dissection of susceptibility genes for diabetes and related phenotypes on mouse chromosome 14 by means of congenic strains. BMC Genet 15:93. https://doi.org/10.1186/s12863-014-0093-8
86. Shibata M, Kishi T, Yasuda B, Kuno T (1986) The inhibitory effect of lysozyme on the glomerular basement membrane thickening in spontaneous diabetic mice (NSY mice). Tohoku J Exp Med 149(1):39–46
87. Shimizu K, Morita H, Niwa T, Maeda K, Shibata M, Higuchi K, Takeda T (1993) Spontaneous amyloidosis in senile NSY mice. Acta Pathol Jpn 43(5):215–221
88. Nagao M, Asai A, Kawahara M, Nakajima Y, Sato Y, Tanimura K, Okajima F, Takaya M, Sudo M, Takemitsu S, Harada T, Sugihara H, Oikawa S (2012) Selective breeding of mice for different susceptibilities to high fat diet-induced glucose intolerance: development of two novel mouse lines, Selectively bred Diet-induced Glucose intolerance-Prone and -Resistant. J Diabetes Investig 3(3):245–251. https://doi.org/10.1111/j.2040-1124.2011.00175.x
89. Nagao M, Asai A, Inaba W, Kawahara M, Shuto Y, Kobayashi S, Sanoyama D, Sugihara H, Yagihashi S, Oikawa S (2014) Characterization of pancreatic islets in two selectively bred mouse lines with different

susceptibilities to high-fat diet-induced glucose intolerance. PLoS One 9(1):e84725. https://doi.org/10.1371/journal.pone.0084725
90. Nagao M, Asai A, Sugihara H, Oikawa S (2015) Transgenerational changes of metabolic phenotypes in two selectively bred mouse colonies for different susceptibilities to diet-induced glucose intolerance. Endocr J 62 (4):371–378. https://doi.org/10.1507/endocrj.EJ14-0241
91. Nagao M, Asai A, Sugihara H, Oikawa S (2015) Fat intake and the development of type 2 diabetes. Endocr J 62(7):561–572. https://doi.org/10.1507/endocrj.EJ15-0055
92. Asai A, Nagao M, Kawahara M, Shuto Y, Sugihara H, Oikawa S (2013) Effect of impaired glucose tolerance on atherosclerotic lesion formation: an evaluation in selectively bred mice with different susceptibilities to glucose intolerance. Atherosclerosis 231 (2):421–426. https://doi.org/10.1016/j.atherosclerosis.2013.10.009
93. Halban PA, Polonsky KS, Bowden DW, Hawkins MA, Ling C, Mather KJ, Powers AC, Rhodes CJ, Sussel L, Weir GC (2014) Beta-cell failure in type 2 diabetes: postulated mechanisms and prospects for prevention and treatment. Diabetes Care 37(6):1751–1758. https://doi.org/10.2337/dc14-0396
94. Groop L, Pociot F (2014) Genetics of diabetes – are we missing the genes or the disease? Mol Cell Endocrinol 382(1):726–739. https://doi.org/10.1016/j.mce.2013.04.002
95. Rosengren AH, Braun M, Mahdi T, Andersson SA, Travers ME, Shigeto M, Zhang E, Almgren P, Ladenvall C, Axelsson AS, Edlund A, Pedersen MG, Jonsson A, Ramracheya R, Tang Y, Walker JN, Barrett A, Johnson PR, Lyssenko V, McCarthy MI, Groop L, Salehi A, Gloyn AL, Renstrom E, Rorsman P, Eliasson L (2012) Reduced insulin exocytosis in human pancreatic beta-cells with gene variants linked to type 2 diabetes. Diabetes 61(7):1726–1733. https://doi.org/10.2337/db11-1516
96. Hiriart M, Matteson DR (1988) Na channels and two types of Ca channels in rat pancreatic B cells identified with the reverse hemolytic plaque assay. J Gen Physiol 91(5):617–639
97. Gopel S, Kanno T, Barg S, Galvanovskis J, Rorsman P (1999) Voltage-gated and resting membrane currents recorded from B-cells in intact mouse pancreatic islets. J Physiol 521 (Pt 3):717–728

# Chapter 4

# A Review of Mouse Models of Monogenic Diabetes and ER Stress Signaling

**Paraskevi Salpea, Cristina Cosentino, and Mariana Igoillo-Esteve**

## Abstract

Diabetes is a major public health problem: it is estimated that 420 million people are affected globally. Monogenic forms of diabetes are less common, but variants in monogenic diabetes genes have been shown to contribute to type 2 diabetes risk. In vitro and in vivo models of monogenic forms of diabetes related to the endoplasmic reticulum (ER) stress response provided compelling evidence on the role of ER stress and dysregulated ER stress signaling on β cell demise in type 1 and type 2 diabetes. In this chapter, we describe the genetics, background, and phenotype of ER stress-related monogenic diabetes mouse models, and we comment on their advantages and disadvantages. We conclude that these mouse models are very useful tools for monogenic diabetes molecular pathogenesis studies, although there is a variability on the methodology that is used. Regarding the use of these models for therapeutic testing of ER stress modulators, a specific consideration should be given to the fact that they recapitulate some, but not all, the phenotypic characteristics of the human disease.

**Key words** Monogenic diabetes, Endoplasmic reticulum stress, β cell, Akita, Mouse model

## 1 Introduction

Diabetes is a major public health problem: it is estimated that 420 million people are affected globally [1]. Most types of diabetes are polygenic: 10–15% of patients have type 1 diabetes, where the immune system selectively targets insulin-producing pancreatic β cells, and 80% of them have type 2 diabetes that is characterized by β cell dysfunction in a context of insulin resistance. Monogenic forms of diabetes, caused by single-gene mutations, are less common and represent 2–3% of all cases. Variants in monogenic diabetes genes have been shown to contribute to type 2 diabetes risk [2]. Monogenic forms of diabetes can therefore serve as experiments of nature that provide us with "human knockout models," to obtain insight into the pathogenesis of monogenic and polygenic diabetes. These diseases can further serve as models to develop therapies tailored to specific genetic defects or pathways [3].

Aileen J. F. King (ed.), *Animal Models of Diabetes: Methods and Protocols*, Methods in Molecular Biology, vol. 2128, https://doi.org/10.1007/978-1-0716-0385-7_4, 

ER stress is defined as an imbalance between the protein-folding capacity of the organelle and the functional demand that is placed on it. This imbalance leads to accumulation of unfolded or misfolded proteins in the ER lumen. In order to restore ER homeostasis, cells trigger the ER stress response, also known as the unfolded protein response (UPR) [4]. There are three main ER stress signaling branches as part of the UPR: the protein kinase RNA (PKR)-like ER kinase (PERK), the inositol-requiring protein-1 (IRE1), and a third branch that is activated by activating transcription factor 6 (ATF6). In addition to these classical ER stress transducers, the CREB3 and CREB3L1–4 transcription factors elicit UPR signaling in a cell type and context-specific manner [5]. Studies on monogenic forms of diabetes related to the ER stress response produced compelling evidence on the role of ER stress and dysregulated ER stress signaling in β cell demise in type 1 and type 2 diabetes [6, 7].

In this chapter, we describe mouse models currently used for the study of ER stress-related monogenic forms of diabetes. These models are powerful tools to elucidate disease mechanisms and to assess the role of ER stress in diabetes development. They are also valuable for in vivo testing of new therapeutic approaches. In this chapter, we will describe the genetic modifications, phenotypic characteristics, and advantages and disadvantages of each of these in vivo models.

## 2 Akita Mouse Model

The Akita insulin mutation was described for the first time in C57BL/6NJcl mice [8] and became soon one of the most representative model of ER stress-mediated diabetes. Interestingly, *INS* mutations were subsequently described to be responsible of neonatal diabetes in man [9, 10]. Neonatal diabetes is a rare disorder characterized by primary β cell dysfunction, typically developed and diagnosed within the first 6 months from birth, even though some patients present a delayed onset [11].

Akita mice are characterized by early-age-onset diabetes, inherited in autosomal dominant mode, not accompanied by obesity or insulitis [8]. The genetic locus identified by linkage analysis was mapped on chromosome 7 distal to D7Mit189 and was named *Mody4* due to phenotypic and genetic similarities of this model with the human syndrome of maturity-onset diabetes of the young (MODY). The authors reported hyperglycemia in mutated mice just after weaning (7 weeks old). The Akita diabetic mice have significantly lower levels of circulating insulin when compared to nondiabetic animals. Further investigations showed a severe impairment of in vitro insulin secretion already at 4 weeks of age. Histological and morphometric analysis revealed a decrease of active β cells and lower insulin content in islets from diabetic mice. Significant gender differences in the severity of the disease were reported. When compared to females, males showed higher glycemia at 7 weeks of age and a continuous progression of the disease and had significantly shorter survival time than unaffected males. Female mice exhibited milder phenotype, and their life span was unaffected by the mutation.

Further investigation on the *Mody4* locus showed that this area includes the important candidate gene *Ins2* [12]. Direct amplification and sequencing revealed the presence of the missense mutation G1907A in the exon 3 of one of the *Ins2* alleles that led to the protein variant Cys96Tyr (Ins2$^{C96Y}$) in Akita mice. The amino acid substitution occurs in the seventh residue of the A chain of mature insulin (A7). In silico analysis predicted deleterious consequences of the Cys96Tyr substitution due to the disruption of disulfide bonds normally formed between the Cys96 and the Cys31 located on the seventh residue of the insulin B chain (B7). Although deleterious effect of the heterozygous mutation is possible, mice have another functional insulin gene *Ins1*. This means that Akita mice still have three functional insulin alleles that can encode for the correct protein. In line with that, it was shown that the overall mRNA expression levels of *Ins1* and *Ins2* genes are unchanged in Akita diabetic mice compared to wild-type animals. However, immunofluorescent analysis of pancreatic sections of Akita mice showed smaller islets and a prominent reduction of insulin-positive cells when compared to wild-type animals [8, 12–14]. Additionally, electron microscopy analysis of Akita β cells showed mature insulin granules reduced in size and number and expanded ER. Further observations confirmed an impaired proinsulin processing and the accumulation of high-molecular-weight complexes. Finally, Wang and colleagues have shown that the molecular chaperones PDI and BIP, important for correct protein folding, were overexpressed in Akita mouse islets in the attempt to restore ER homeostasis [12]. These evidences suggest that, in Akita mice, the activation of unfolded protein response (UPR) and the incapacity of β cells to resolve ER stress lead to activation of pro-apoptotic molecules and cell death. The link between UPR and cell death can be found in gene expression changes downstream the pathway. While the overall protein translation is inhibited during ER stress, the translation of certain mRNAs is actually induced. The C/EBP homologous protein (CHOP) is a transcription factor expressed during ER stress that promotes cell death. To investigate whether CHOP was responsible for β cell loss in Akita diabetes, a double-mutant mouse model with mutation in *Ins2* and disruption of *Chop* gene was established by crossing Chop$^{-/-}$ mice with Ins2$^{WT/C96Y}$ mice [15]. Islets derived from Ins2$^{WT/C96Y}$ mice at 4 weeks showed overexpression of CHOP in β cells. The disruption of CHOP expression protected β cells from apoptosis and consequently preserved β cell mass. Furthermore, the development of diabetes was delayed in Ins2$^{WT/C96Y}$ Chop$^{-/-}$ mice that showed normoglycemia until 8 weeks of age.

Homozygous Akita mice were also characterized [13]. These mice have a more severe phenotype and develop diabetes at 2 weeks of age. Homozygous mice present hypoplastic islets and very low β cell density (only 20% of wild type).

While in Akita mice the insulin secretion is profoundly impaired, the human MODY present a wider variability of the phenotype. At the time of diagnosis, most of *INS* patients present severe hyperglycemia, often accompanied by ketoacidosis that indicates insulin deficiency [10]. However, some *INS* mutated patients are characterized by initial normal or in some cases elevated immunoreactive C-peptide that decreases in time until reaching undetectable levels. Initial hyperinsulinemia and hyperproinsulinemia were also reported in some cases of neonatal diabetes [10, 16]. These observations, together with the absence of gender differences, differentiate the human MODY syndrome from the Akita mouse diabetes. Nevertheless, in *INS* patients, the progressive decrease of circulating C-peptide suggests a proteotoxic mechanism with consequent ER stress and β cell apoptosis. These deleterious events have been studied and can be further characterized in Akita mice. Indeed, thanks to the extensive investigation performed since the first description of the Akita mice, this model can be considered a prototype of ER stress-induced diabetes.

## 3 Munich Mouse Model

The so-called Munich mouse model derives from the genomic missense mutation T1903A in the exon 3 of *Ins2*, identified by a mutagenesis screening performed in N-ethyl-N-nitrosourea-treated mice [17]. This mutation results in the protein variant C95S present in a neighboring site of the Akita mutation. The substitution of a Cys with a Ser disrupts the intra-chain A6-A11 disulfide bond. After the identification of the predicted deleterious mutation, the authors used different parameters to fully characterize the heterozygous *Ins2*$^{C95S}$ mutant mice. They observed that body weight was not changed in these animals at 4 weeks of age, but it was significantly lower in 3–6-month-old mice compared to wild type. Fasted and postprandial blood glucose were higher in 1-, 3-, and 6-month-old mutant mice with respect to aged-matched wild-type animals. However, their fasted and postprandial insulin levels were indistinguishable. Male mice showed a more pronounced progression of the disease and a more severe phenotype at 6 months of age. Insulin tolerance tests performed in Munich and wild-type mice showed a smaller decrease in glycemia after 10 min of insulin injection in the mutant animals, suggestive of insulin resistance.

Oral glucose tolerance tests (OGTT) confirmed the hyperglycemic phenotype in heterozygous Munich mice and a more severe progression of the disease in males. Mutant mice showed reduced insulinemia during the OGTT compared to controls. This observation was supported by immunohistochemical analysis that evidenced a disruption of pancreatic islet organization with very few insulin-positive cells and a bigger proportion of glucagon-expressing cells within the islet of Munich mice. Electron microscopy analysis of Munich β cells evidenced ultrastructural rearrangement caused by enlarged ER and almost missing insulin granules. Homozygous mutant mice were also characterized. They showed a more severe phenotype and died within 3 months of age. The full metabolic and phenotypical characterization of heterozygous Munich mice describes an ideal model of early-onset monogenic diabetes not accompanied from obesity or insulitis.

## 4 Eif2s1 (S51A) Mouse Model

The Akita and the Munich diabetes mouse models highlighted the crucial role of ER homeostasis in the maintenance of β cell function and survival. The accumulation of misfolded proteins in the ER lumen, due to genetic mutations or a higher functional demand, can disrupt the ER homeostasis leading to UPR activation that, when prolonged, leads to ER stress. One of the key mechanisms to counteract ER stress is to decrease the total protein synthesis. The ER stress transducer PERK plays an important role in this process. Indeed, it phosphorylates the eukaryotic translation initiation factor 2 (eIF2) in its α subunit in Ser51 (S51) inhibiting its translation initiation function. To further investigate the role of eIF2α (also called Eif2s1) in diabetes development, Scheuner and colleagues [18] used gene targeting to establish a mouse model bearing a S51A replacement in eIF2α to prevent eIF2 phosphorylation [18]. For this purpose, the mutated sequence of eIF2α exon 2 was inserted in a loxP-flanked NEO cassette to allow in vivo ZP3-Cre recombinase-mediated deletion. The construct was electroporated into embryonic stem (ES) cells. The ES cells were then injected into C57BL/6 blastocysts, and the colony was established by breeding chimeric males with females.

Homozygous Eif2s1 (S51A) mice had an important reduction in β cell mass already at birth. These animals died within 18 h after birth due to severe hypoglycemia associated with decreased gluconeogenesis. Mouse embryonic fibroblast (MEF) derived from homozygous Eif2s1 (S51A) mice was used for in vitro studies that demonstrated the importance of eIF2α phosphorylation and translational control for cell survival in response to ER stress. Heterozygous Eif2s1 (S51A) mice were viable and fertile and presented normal phenotype at birth [19]. However, when challenged with a high-fat diet (HFD), these mice developed insulin resistance, obesity, and diabetes. HFD is an environmental stress that leads to a higher β cell functional demand. β cell function was also compromised in heterozygous mice, with a loss of glucose-stimulated insulin secretion. These animals also exhibited an abnormal distension of the ER lumen. Defective folding and trafficking of proinsulin could account for the lower number of insulin granules and impairment in glucose-stimulated insulin secretion in heterozygous Eif2s1 (S51A) mice compared to wild type.

Homozygous and heterozygous Eif2s1 (S51A) mice constitute a powerful tool for the study of transitional control in ER stress-mediated diabetes. The inhibition of eIF2 function is essential to ensure a correct response to ER stress and cell survival; however, it was shown that prolonged activation of this response leads to β cell apoptosis. Agents that prevent eIF2α dephosphorylation, such as salubrinal [20] and guanabenz [21], have been described to be protective against ER stress-induced apoptosis in rat pheochromocytoma cells, HeLa cells, and primary Akita mutated β cells. However, both salubrinal and guanabenz have been shown to sensitize human and rat β cells to fatty acid-induced ER stress and apoptosis [22–24]. For this reason, eIF2 has been described as the Achilles heel of pancreatic β cells [7]. Mutations in eIF2α have not been described in man, but mutations in *EIF2S3* (eIF2γ) have been linked to MEHMO, a multisystemic syndrome characterized by intellectual disability, epileptic seizures, hypogonadism and hypogenitalism, microcephaly, and obesity, in some cases accompanied by impaired glucose homeostasis [25–27]. Through in vitro functional studies, Skopkova and colleagues showed that eIF2γ frameshift mutations lead to reduced translation fidelity, increased ATF4 translation, and CHOP expression [27]. There are no studies on mouse eIF2γ mutant models. C57BL/6 N mouse embryonic stem cell lines with a gene trap insertion at the locus of the *EIF2S3* gene are available from a random mutagenesis study, but no functional data has been published [28].

## 5 Wolcott-Rallison Syndrome Mouse Model

Recessive mutations in EIF2AK3, encoding PERK, give rise to Wolcott-Rallison syndrome [29]. The disease is characterized by neonatal or early-infancy-onset non-autoimmune, insulin-dependent diabetes, often with a ketoacidotic presentation. Associated features are growth retardation, epiphyseal dysplasia, hepatic steatosis and dysfunction, exocrine pancreas insufficiency, intellectual disability, and microcephaly [30].

Harding et al. [31] first constructed and described a mouse model of Wolcott-Rallison syndrome, the PERK knockout (KO) mouse (PERK$^{-/-}$). For the construction of this model, the whole genomic region encoding the PERK transmembrane domain was replaced by a neomycin-resistance cassette in ES cells. These mice showed significantly reduced eIF2α phosphorylation and ER stress. This dysregulated ER stress response led to β cell dysfunction and loss. Since the main symptom of human Wolcott-Rallison cases is diabetes, the glucose metabolism of these mice was studied. By 4 weeks of age, both male and female $PERK^{-/-}$ mice showed hyperglycemia, low body weight, and reduced insulin levels compared to wild-type animals indicative of endocrine pancreas failure. Pancreas immunostaining showed pancreatic acinar cell loss, along with exocrine pancreas insufficiency. The $PERK^{-/-}$ mice also presented growth retardation and high mortality. Heterozygous PERK mutant mice ($PERK^{+/-}$) had a milder phenotype characterized by impaired glucose tolerance that was evidenced by intraperitoneal glucose tolerance tests.

An osteoblast-specific PERK KO ($ObPERK^{-/-}$) was generated. LoxP sites were inserted into the intronic sequences flanking the three exons that encode part of the lumenal domain, the transmembrane domain, and part of the catalytic domain [32]. This floxed PERK allele strain was crossed with a transgenic strain carrying the Cre recombinase under the control of the Col2.3 promoter to drive Cre expression at the mature osteoblast stage [33]. This mouse model revealed impaired osteoblast growth and function due to PERK deficiency, another resemblance with the human Wolcott-Rallison syndrome [34].

Additional tissue- and cell-specific PERK KO mice were produced by crossing floxed PERK allele strains with the following Cre mice: RIP-Cre to generate "bPKO" (β cell specific), Ngn3-Cre to generate "enPKO" (endocrine pancreas specific), and Pdx-1-Cre to generate "pcPKO" (pancreas specific) [35]. These mouse models presented similar phenotypic characteristics than the whole-body PERK KO, notably hyperglycemia, low insulin levels, and β cell death. However, there was no evidence in these models of uncontrolled protein synthesis, activation of the UPR, and apoptosis. Instead, these mice showed impaired ER-to-Golgi trafficking and endoplasmic reticulum-associated protein degradation (ERAD). These findings were confirmed in additional in vitro cell studies [36].

Mouse models of Wolcott-Rallison syndrome importantly contributed, and still contribute, to understanding the importance of the PERK-eIF2α branch on glucose metabolism and pancreatic β cell homeostasis. The Wolcott-Rallison mouse models present similar phenotypic characteristics with Wolcott-Rallison syndrome patients, suggesting that they can be good tools for testing possible therapeutic approaches. An additional advantage of these models is that they do not present phenotypic differences between genders or mouse strain backgrounds [31, 35]. However, some divergences between them must be highlighted, such as the mild phenotype of the heterozygous perk$^{+/-}$ mice in contrast with human heterozygous carriers of *PERK* mutations that do not present diabetes.

## 6 P58$^{IPK\ -/-}$ Mouse Model

Synofzik et al. [37] described recessive DNAJC3 loss-of-function mutations in two families with diabetes and neurodegenerative features, including ataxia, upper motor neuron damage, peripheral neuropathy, hearing loss, and cerebral atrophy. DNAJC3 encodes p58IPK, a BiP co-chaperone that inhibits PERK activity and downstream signaling and thus exerts a negative feedback on PERK-eIF2α signaling.

Ladiges et al. [38] created a p58$^{IPK-/-}$ on a C57BL/6 background. In order to construct this mouse, the P58$^{IPK}$ exon 1 was replaced by an actin/enhanced GFP/galactosidase/neomycin cassette in ES cells.

The P58$^{IPK-/-}$ mice have decreased adiposity and are smaller than WT and heterozygous P58$^{IPK+/-}$ mice. They can be easily distinguished from heterozygous and wild-type mice as early as 2 weeks of age. Four-month-old p58$^{IPK-/-}$ mice show polydipsia, polyuria, and evidence of glycosuria. They also have increased glycemia and hypoinsulinemia, when compared to WT littermates; however, they are not outwardly sick. Intraperitoneal glucose tolerance tests (IPGTT) performed in 8-week-old mice showed that p58$^{IPK-/-}$ had normal glucose tolerance, since they cleared glucose just as efficiently as wild-type controls when challenged with a standard glucose load. Immunohistological studies of pancreatic sections from 6-month-old mice, though, showed pathological morphology with significantly decreased insulin-producing cells and increased β cell apoptosis. There are no phenotypic differences between genders in p58$^{IPK-/-}$ mice. The p58$^{IPK-/-}$ phenotype was less severe than the phenotype of the eIF2α S51A-mutated mouse [18] or the PERK$^{-/-}$ mouse [31], probably because p58IPK plays a regulatory role rather than a direct involvement in translational attenuation. The heterozygous p58IPK$^{+/-}$, like human DNAJC3 heterozygous carriers, have no obvious phenotype. DNAJC3 loss of function diabetes is a newly identified monogenic form of diabetes. Therefore, the p58$^{IPK-/-}$ mouse can serve as a tool to further examine the role of p58$^{IPK}$ in ER stress signaling regulation and to elucidate the molecular pathogenesis of this disease. However, it is important to highlight that the p58$^{IPK-/-}$ mouse has a β cell-specific phenotype, while DNAJC3 patients and PERK-eIF2α mutants/patients also have defects in other tissues. This drawback must be considered in molecular pathogenesis studies and in vivo therapeutic tests associated with this model.

## 7 Wolfram Syndrome Mouse Model

Wolfram syndrome is a rare autosomal recessive orphan disease. The clinical manifestations are young-onset diabetes, optic nerve atrophy, and deafness. Most Wolfram patients carry mutations in *WFS1*. WFS1 deficiency results in ER stress, leading to neurodegeneration and pancreatic β cell dysfunction and death [39, 40].

The first Wolfram mouse model was created by Ishihara et al. [41]. In this model, the WFS1 exon 2 was replaced with a neomycin-resistance cassette in 129Sv ES cells. The exon 2 ablation resulted in a WFS1 N-terminus-truncated nonfunctional protein. Phenotypic studies performed in WFS1$^{-/-}$ mice in a mixed (129Sv/C57BL/6) background revealed increased non-fasting blood glucose levels starting at the age of 16 weeks in the KO animals with respect to wild types. More than 60% of these mice developed overt diabetes by the age of 36 weeks. WFS$^{-/-}$ mice in pure C57BL/6 J background showed a less severe phenotype with no apparent hyperglycemia by the age of 36 weeks. However, these animals presented impaired glucose tolerance when subjected to OGTTs and significantly reduced insulin plasma levels when compared to wild-type littermates. Intraperitoneal insulin sensitivity tests (IPIST) did not show insulin resistance in WFS1$^{-/-}$ mice. These data indicate that impaired glucose homeostasis in WFS1$^{-/-}$ animals is caused by insulin secretory defects. Immunostaining studies in pancreatic sections revealed a progressive β cell loss. In good correlation, DNA fragmentation experiments showed WFS1$^{-/-}$ islets to be more susceptible to apoptosis especially induced by synthetic ER stress.

In 2005, Riggs et al. [42] created a conditional mouse with a β cell-specific WFS1 exon 8 deletion. For this purpose, Wfs1 exon-8-floxed mice were crossed with RIP2-Cre animals. This gave rise to β cell-specific conditional WFS1 KO mice in a mixed 129Sv/C57BL/6 background. Phenotypic characterization of these animals revealed impaired glucose tolerance and decreased insulin secretion after IPGTT. Immunostaining of pancreatic sections revealed disrupted islet morphology and increased β cell apoptosis. Electron microscopy analysis of isolated islets from WFS1 KO mice and wild-type littermates revealed distended ER in β cells from the WFS1-deficient mice. In good correlation with these findings, islets from WFS1 KO mice showed increased BiP and CHOP mRNA expression and caspase 3 activation. This model clearly suggests that diabetes in Wolfram syndrome results from ER stress-induced β cell apoptosis.

In 2009, Kõks and colleagues [43] generated a whole-body WFS1 exon 8 null mouse. These animals, originally on a mixed 129Sv/C57BL/6 background, were glucose intolerant and showed growth retardation.

Additional studies with WFS1$^{-/-}$ exon-8-deleted mice showed that male animals have worse glycemic phenotype than females [44]. In Wolfram syndrome patients, however, there are no clear gender differences. Other studies also revealed impaired behavioral adaptation in stress environment for the WFS1 KO animals [45].

WFS2 (also known as CISD2) mutations lead to Wolfram syndrome 2 which has similar phenotypic characteristics than Wolfram syndrome 1 but is even less frequent [46]. CISD2 knockout mice, in C57BL6J background, have been generated to study Wolfram syndrome 2 [47]. To create these animals, exons 2 and 3 of *Cisd2* were replaced with a puromycin cassette. The *Cisd2* deficiency triggers premature aging, nerve and muscle degeneration as a result of mitochondrial dysfunction, and autophagic cell death. The *Cisd2* knockout mice develop optic nerve degeneration at 23 weeks of age; however, their glucose tolerance and insulin secretion are only mildly impaired, with no clear changes in β cell mass in contrast with the Wfs1-deficient mouse models [47].

The Wolfram syndrome mouse models have been very important to understand the implication of ER stress in β cell loss and the pathogenesis of diabetes in Wolfram syndrome. These models recapitulate some, but not all, of the phenotypic characteristics of Wolfram syndrome. In the case of Wolfram syndrome mouse models, some shortcomings have to be taken into consideration when working with these animals: (1) the fact that Wolfram syndrome mice present gender differences in the severity of the disease that have not been reported in Wolfram syndrome patients and (2) the severity of the disease depends on the mouse strain background [41], with a more severe phenotype reported in C57BL/6 animals and a milder one in 129Sv. This can cause variations between studies and especially if a mixed mouse strain background is used.

## 8 Conclusions

Several mouse models of monogenic forms of diabetes associated with ER stress have been developed and recognized as valuable tools for the elucidation and understanding of disease mechanisms. Over the years, these models have provided compelling evidence that unresolvable ER stress is ill tolerated by β cells and that this leads to diabetes. In addition, these models can serve as valuable tools for in vivo testing of possible therapeutic agents.

There are some concerns regarding the use of monogenic diabetes mouse models that are mostly related with differences between the human and mouse phenotype at homozygous or heterozygous state, strain-associated phenotype severity, and gender differences as occurs in the Akita and Wolfram syndrome mouse models.

Due to the lack of universal guidelines on the methodology to be used for the characterization of diabetes in mouse models, several tests are currently applied for the metabolic phenotyping of diabetic mice. In line with that, we found that the in vivo metabolic studies on the different monogenic diabetes mouse models analyzed in this chapter were performed using diverse approaches. Indeed, even if the glucose tolerance tests (GTT) and insulin tolerance tests (ITT) appeared as the gold standard methods used for the evaluation of the impaired glucose metabolism, some divergences in the tests were detected, such as route of glucose administration, time of the test, and period of fasting, that made difficult the interpretation of the data along studies. In a recent review, Alquier and Poitout [48] report the most common experimental tests to assess the diabetic phenotype and propose some guidelines for the design, analysis, and interpretation of the available tests. In the future, a more organized effort is needed, such as the formation of a consortium of leading scientist in the field of diabetes and mouse models in order to create guidelines for diabetic mouse model phenotyping.

## References


1. IDF (2017) IDF diabetes atlas, 8th edn International Diabetes Federation, Brussels, Belgium
2. Flannick J, Johansson S, Njølstad PR (2016) Common and rare forms of diabetes mellitus: towards a continuum of diabetes subtypes. Nat Rev Endocrinol 12:394–406. https://doi.org/10.1038/nrendo.2016.50
3. Murphy R, Ellard S, Hattersley AT (2008) Clinical implications of a molecular genetic classification of monogenic β-cell diabetes. Nat Clin Pract Endocrinol Metab 4:200–213. https://doi.org/10.1038/ncpendmet0778
4. Ron D, Walter P (2007) Signal integration in the endoplasmic reticulum unfolded protein response. Nat Rev Mol Cell Biol 8:519–529
5. Asada R, Kanemoto S, Kondo S et al (2011) The signalling from endoplasmic reticulum-resident bZIP transcription factors involved in diverse cellular physiology. J Biochem 149:507–518. https://doi.org/10.1093/jb/mvr041
6. Cnop M, Foufelle F, Velloso LA (2012) Endoplasmic reticulum stress, obesity and diabetes. Trends Mol Med 18:59–68
7. Cnop M, Toivonen S, Igoillo-Esteve M, Salpea P (2017) Endoplasmic reticulum stress and eIF2α phosphorylation: the Achilles heel of pancreatic β cells. Mol Metab 6:1024–1039
8. Yoshioka M, Kayo T, Ikeda T, Koizumi A (1997) A novel locus, Mody4, distal to D7Mit189 on chromosome 7 determines early-onset NIDDM in nonobese C57BL/6 (Akita) mutant mice. Diabetes 46:887–894
9. Stoy J, Edghill EL, Flanagan SE et al (2007) Insulin gene mutations as a cause of permanent neonatal diabetes. Proc Natl Acad Sci U S A 104:15040–15044. https://doi.org/10.1073/pnas.0707291104
10. Colombo C, Porzio O, Liu M et al (2008) Seven mutations in the human insulin gene linked to permanent neonatal/infancy-onset diabetes mellitus. J Clin Invest 118:2148–2156. https://doi.org/10.1172/JCI33777
11. Polak M, Shield J (2004) Neonatal diabetes mellitus – genetic aspects 2004. Pediatr Endocrinol Rev 2:193–198
12. Wang J, Takeuchi T, Tanaka S et al (1999) A mutation in the insulin 2 gene induces diabetes with severe pancreatic beta-cell dysfunction in the Mody mouse. J Clin Invest 103:27–37. https://doi.org/10.1172/JCI4431
13. Kayo T, Koizumi A (1998) Mapping of murine diabetogenic gene mody on chromosome 7 at D7Mit258 and its involvement in pancreatic islet and beta cell development during the perinatal period. J Clin Invest 101:2112–2118. https://doi.org/10.1172/JCI1842

14. Barbetti F, Colombo C, Haataja L et al (2016) Hyperglucagonemia in an animal model of insulin- deficient diabetes: what therapy can improve it? Clin Diabetes Endocrinol 2:11. https://doi.org/10.1186/s40842-016-0029-5
15. Oyadomari S, Koizumi A, Takeda K et al (2002) Targeted disruption of the Chop gene delays endoplasmic reticulum stress-mediated diabetes. J Clin Invest 109:525–532. https://doi.org/10.1172/JCI14550
16. Stoy J, Steiner DF, Park S-Y et al (2010) Clinical and molecular genetics of neonatal diabetes due to mutations in the insulin gene. Rev Endocr Metab Disord 11:205–215. https://doi.org/10.1007/s11154-010-9151-3
17. Herbach N, Rathkolb B, Kemter E et al (2007) Dominant-negative effects of a novel mutated Ins2 allele causes early-onset diabetes and severe beta-cell loss in Munich Ins2C95S mutant mice. Diabetes 56:1268–1276. https://doi.org/10.2337/db06-0658
18. Scheuner D, Song B, McEwen E et al (2001) Translational control is required for the unfolded protein response and in vivo glucose homeostasis. Mol Cell 7:1165–1176. https://doi.org/10.1016/S1097-2765(01)00265-9
19. Scheuner D, Vander Mierde D, Song B et al (2005) Control of mRNA translation preserves endoplasmic reticulum function in beta cells and maintains glucose homeostasis. Nat Med 11:757–764. https://doi.org/10.1038/nm1259
20. Boyce M, Bryant KF, Jousse C et al (2005) A selective inhibitor of eIF2alpha dephosphorylation protects cells from ER stress. Science 307:935–939. https://doi.org/10.1126/science.1101902
21. Tsaytler P, Harding HP, Ron D, Bertolotti A (2011) Selective inhibition of a regulatory subunit of protein phosphatase 1 restores proteostasis. Science 332:91–94. https://doi.org/10.1126/science.1201396
22. Cnop M, Ladriere L, Hekerman P et al (2007) Selective inhibition of eukaryotic translation initiation factor 2 alpha dephosphorylation potentiates fatty acid-induced endoplasmic reticulum stress and causes pancreatic beta-cell dysfunction and apoptosis. J Biol Chem 282:3989–3997. https://doi.org/10.1074/jbc.M607627200
23. Abdulkarim B, Hernangomez M, Igoillo-Esteve M et al (2017) Guanabenz sensitizes pancreatic beta cells to lipotoxic endoplasmic reticulum stress and apoptosis. Endocrinology 158:1659–1670. https://doi.org/10.1210/en.2016-1773
24. Ladriere L, Igoillo-Esteve M, Cunha DA et al (2010) Enhanced signaling downstream of ribonucleic acid-activated protein kinase-like endoplasmic reticulum kinase potentiates lipotoxic endoplasmic reticulum stress in human islets. J Clin Endocrinol Metab 95:1442–1449. https://doi.org/10.1210/jc.2009-2322
25. Borck G, Shin B-S, Stiller B et al (2012) eIF2-gamma mutation that disrupts eIF2 complex integrity links intellectual disability to impaired translation initiation. Mol Cell 48:641–646. https://doi.org/10.1016/j.molcel.2012.09.005
26. Moortgat S, Desir J, Benoit V et al (2016) Two novel EIF2S3 mutations associated with syndromic intellectual disability with severe microcephaly, growth retardation, and epilepsy. Am J Med Genet A 170:2927–2933. https://doi.org/10.1002/ajmg.a.37792
27. Skopkova M, Hennig F, Shin B-S et al (2017) EIF2S3 mutations associated with severe X-linked intellectual disability syndrome MEHMO. Hum Mutat 38:409–425. https://doi.org/10.1002/humu.23170
28. Hansen GM, Markesich DC, Burnett MB et al (2008) Large-scale gene trapping in C57BL/6N mouse embryonic stem cells. Genome Res 18:1670–1679. https://doi.org/10.1101/gr.078352.108
29. Delépine M, Nicolino M, Barrett T et al (2000) EIF2AK3, encoding translation initiation factor 2-alpha kinase 3, is mutated in patients with Wolcott-Rallison syndrome. Nat Genet 25:406–409. https://doi.org/10.1038/78085
30. Julier C, Nicolino M (2010) Wolcott-Rallison syndrome. Orphanet J Rare Dis 5:29
31. Harding HP, Zeng H, Zhang Y et al (2001) Diabetes mellitus and exocrine pancreatic dysfunction in Perk−/− mice reveals a role for translational control in secretory cell survival. Mol Cell 7:1153–1163. https://doi.org/10.1016/S1097-2765(01)00264-7
32. Zhang P, McGrath B, Li S et al (2002) The PERK eukaryotic initiation factor 2 alpha kinase is required for the development of the skeletal system, postnatal growth, and the function and viability of the pancreas. Mol Cell Biol 22:3864–3874. https://doi.org/10.1128/MCB.22.11.3864
33. Wei J, Sheng X, Feng D et al (2008) PERK is essential for neonatal skeletal development to regulate osteoblast proliferation and differentiation. J Cell Physiol 217:693–707. https://doi.org/10.1002/jcp.21543

34. Harding HP, Zhang Y, Bertolotti A et al (2000) Perk is essential for translational regulation and cell survival during the unfolded protein response. Mol Cell 5:897–904. S1097-2765(00)80330-5 [pii]
35. Zhang W, Feng D, Li Y et al (2006) PERK EIF2AK3 control of pancreatic β cell differentiation and proliferation is required for postnatal glucose homeostasis. Cell Metab 4:491–497. https://doi.org/10.1016/j.cmet.2006.11.002
36. Gupta S, McGrath B, Cavener DR (2010) PERK (EIF2AK3) regulates proinsulin trafficking and quality control in the secretory pathway. Diabetes 59:1937–1947. https://doi.org/10.2337/db09-1064
37. Synofzik M, Haack TB, Kopajtich R et al (2014) Absence of BiP Co-chaperone DNAJC3 causes diabetes mellitus and multisystemic neurodegeneration. Am J Hum Genet 95:689–697. https://doi.org/10.1016/j.ajhg.2014.10.013
38. Ladiges WC, Knoblaugh SE, Morton JF et al (2005) Pancreatic ??-cell failure and diabetes in mice with a deletion mutation of the endoplasmic reticulum molecular chaperone gene P58IPK. Diabetes 54:1074–1081. https://doi.org/10.2337/diabetes.54.4.1074
39. Barrett TG, Bundey SE, Macleod AF (1995) Neurodegeneration and diabetes: UK nationwide study of Wolfram (DIDMOAD) syndrome. Lancet 346:1458–1463. https://doi.org/10.1016/S0140-6736(95)92473-6
40. Kinsley BT, Swift M, Dumont RH, Swift RG (1995) Morbidity and mortality in the Wolfram syndrome. Diabetes Care 18:1566–1570. 5p
41. Ishihara H, Takeda S, Tamura A et al (2004) Disruption of the WFS1 gene in mice causes progressive β-cell loss and impaired stimulus - Secretion coupling in insulin secretion. Hum Mol Genet 13:1159–1170. https://doi.org/10.1093/hmg/ddh125
42. Riggs AC, Bernal-Mizrachi E, Ohsugi M et al (2005) Mice conditionally lacking the Wolfram gene in pancreatic islet beta cells exhibit diabetes as a result of enhanced endoplasmic reticulum stress and apoptosis. Diabetologia 48:2313–2321. https://doi.org/10.1007/s00125-005-1947-4
43. Kõks S, Soomets U, Paya-Cano JL et al (2009) Wfs1 gene deletion causes growth retardation in mice and interferes with the growth hormone pathway. Physiol Genomics 37:249–259. https://doi.org/10.1152/physiolgenomics.90407.2008
44. Noormets K, Kõks S, Muldmaa M et al (2011) Sex differences in the development of diabetes in mice with deleted wolframin (Wfs1) gene. Exp Clin Endocrinol Diabetes 119:271–275. https://doi.org/10.1055/s-0030-1265163
45. Luuk H, Plaas M, Raud S et al (2009) Wfs1-deficient mice display impaired behavioural adaptation in stressful environment. Behav Brain Res 198:334–345. https://doi.org/10.1016/j.bbr.2008.11.007
46. Rigoli L, Di Bella C (2012) Wolfram syndrome 1 and Wolfram syndrome 2. Curr Opin Pediatr 24:512–517
47. Chen YF, Kao CH, Chen YT et al (2009) Cisd2 deficiency drives premature aging and causes mitochondria-mediated defects in mice. Genes Dev 23:1183–1194. https://doi.org/10.1101/gad.1779509
48. Alquier T, Poitout V (2018) Considerations and guidelines for mouse metabolic phenotyping in diabetes research. Diabetologia 61:526–538

# Chapter 5

## Rat Models of Human Type 1 Diabetes

**Sigurd Lenzen, Tanja Arndt, Matthias Elsner, Dirk Wedekind, and Anne Jörns**

### Abstract

Rat models of human type 1 diabetes have been shown to be of great importance for the elucidation of the mechanisms underlying the development of autoimmune diabetes. The three major well-established spontaneous rat models are the BioBreeding (BB) diabetes-prone rat, the Komeda diabetes-prone (KDP) rat, and the IDDM (LEW.1AR1-*iddm*) rat. Their distinctive features are described with special reference to their pathology, immunology, and genetics and compared with the situation in patients with type 1 diabetes mellitus. For all three established rat models, a distinctive genetic mutation has been identified that is responsible for the manifestation of the diabetic syndrome in these rat strains.

**Key words** Type 1 diabetes mellitus (T1DM), Rat models of human type 1 diabetes, BioBreeding (BB) diabetes-prone rat, IDDM (LEW.1AR1-*iddm*) rat, Komeda diabetes-prone (KDP) rat

## 1 Introduction

Three well-established rat models of human type 1 diabetes mellitus (T1DM) exist, which were all discovered serendipitously by attentive observers. All three models are the result of a spontaneous mutation each, the BioBreeding (BB) rat in a Wistar colony, the Komeda diabetes-prone (KDP) rat in a Long-Evans colony, and the IDDM (LEW.1AR1-*iddm*) rat in a colony (LEW.1AR1) with a Lewis background. They all develop a severe diabetic syndrome with a high incidence (>50%) characterized clinically by weight loss, polydipsia, polyuria, glycosuria, ketonuria, hyperglycemia, and hypoinsulinemia. The syndrome is so severe that insulin supplementation (i.e., by implantation of a subcutaneous slow-release insulin pellet) is required to prevent the development of life-threatening ketoacidosis and to allow breeding with diabetic animals. Disease frequency is roughly comparable in males and females in the three strains. So, there is no significant sex bias, and this is analogous to humans with autoimmune diabetes.

Aileen J. F. King (ed.), *Animal Models of Diabetes: Methods and Protocols*, Methods in Molecular Biology, vol. 2128, https://doi.org/10.1007/978-1-0716-0385-7_5, 

BB and KDP rats often develop, in addition to insulitis, a lymphocytic thyroiditis [1–3]. Prevalence is variable among BB sublines [4]. Other associated autoimmune diseases are uncommon [2]. In the IDDM rat, affections of other organs have not been observed [5].

In all strains, the diabetes incidence was initially around 20%. Upon systematic breeding with matings between sibs, the incidence increased to more than 50% in subsequent generations in all models [6–8]. Besides the genotype, quite a number of viruses can induce autoimmune diabetes in rats [9, 10]. However, since the three rat strains are kept under specific pathogen-free conditions [8, 9], an effect of these viruses on diabetes manifestation and incidence can be excluded.

Islet cell autoantibodies (ICA), specifically against glutamic acid decarboxylase (GAD), insulin (IAA), and islet antigen-2 (IA-2), which are characteristic for autoimmune diabetes in patients with T1DM, have not been observed in the BB rat [2, 11] and in the IDDM rat [8].

## 2 BioBreeding (BB) Rat Model

The BioBreeding (BB) rat was discovered in 1974 in a commercial colony in outbred Wistar rats at the BioBreeding Laboratories in Ottawa, Canada [12]. The original colony was transferred from the BioBreeding Laboratories, hence the name BB, to the Health Protection Branch Laboratories of Health Canada in Ottawa/Ontario in 1977. This diabetes-prone BB stock has been maintained since as a partially inbred colony through matings between sibs [6]. Affected animals of the Ottawa colony were also founders of a colony in Worcester (USA), which has been inbred and is now called BBDP/Wor [2].

Detailed epidemiological data are available for the BB rat [6, 13–15]. Disease manifestation is typically between 60 and 120 days of life [6, 9, 13, 14]. The mean age of manifestation has been reported to be around 90 days [6, 13, 14]. Less than 0.5% of the animals become diabetic before 60 days of age [13], and less than 10% of the cases are diagnosed in rats older than 120 days [13]. Cumulative incidence is 50–70% [6, 13]. There exist quite a number of sublines, also called tertiary colonies, of the diabetes-prone BB rats in North America and Europe [2, 13], and for some of them higher diabetes incidence rates of up to 80–90% have been reported.

### 2.1 Lymphopenia

Lymphopenia (lymphocyte reduction of >90%) is a characteristic of the BB rat [16–19]. It is already present at birth [20]. The extent of the T lymphocyte reduction has been reported to be somewhat variable [19]. Typically, $CD4^+$ T cell numbers are reduced by around 90% and $CD8^+$ T cells are virtually absent [19]. The

ART2.1 (formerly RT6.1)-positive T cells are also particularly reduced. Depletion of ART2.1-positive T cells has been shown to induce diabetes indicating that the loss of regulatory T cells is apparently associated with the development of insulitis and diabetes in the BB rat [21, 22]. Lymphopenia in the BB rat model is due to a mutation in the *Gimap5* gene [18].

While diabetes-prone BB (BBDP) rats are lymphopenic, diabetes-resistant BB (BBDR) rats are not [23]. But this distinction is not absolute. The diabetes incidence in the different BBDP lines is variable, and there have also been reports of a very low incidence in some lymphopenic lines [23]. On the other hand, non-lymphopenic BBDR lines have been described, which develop diabetes, albeit at a very low incidence (<3%) [24]. Incidences of insulitis and thyroiditis are also low (around 10%) in these colonies. Nevertheless, it can be concluded that a high incidence of insulitis as well as thyroiditis occurs as a result of T cell lymphopenia and is always associated with a high diabetes incidence in the BB rat [25]. This conclusion is also supported by the BBDR$^{lyp/lyp}$ (lymphopenia and diabetes) rat. This model has been generated through transplantation of bone marrow from congenic BBDR rats, which gain both lymphopenia and diabetes through a manoeuver, which consists of an introgression on RNO4 of about 0.33 Mb DNA from BBDP rats [26]. This allows the conclusion that it is this critical region in the congenic BBDR$^{lyp/lyp}$ model that is responsible for the lymphopenia phenotype as well as the diabetes phenotype [27]. This is in agreement with the conclusion that the same mutation is responsible for both phenotypes (see below). Nevertheless, it is not clear why the penetrance of lymphopenia is 100% in contrast to that of diabetes. But the main conclusion is that a high incidence of diabetes is always associated with lymphopenia. Thus, the BB rat is immunodeficient, which is not a characteristic of human T1DM [28, 29]. It has not been observed in other animal models of human T1DM such as the IDDM rat [28, 30] and the KDP rat [31].

## 3 Komeda Diabetes-Prone (KDP) Rat Model

The Komeda diabetes-prone (KDP) rat, a non-lymphopenic model, formerly called Long-Evans Tokushima Lean (LETL) rat, was discovered in 1983 in the Tokushima Research Institute of Otsuka Pharmaceutical Company (Tokushima, Japan) [31]. The strain had been obtained a year before as an outbred Long-Evans strain from Charles River in Canada. In 1990, inbreeding was started, which resulted in two substrains, the Komeda diabetes-prone (KDP) and the Komeda nondiabetic (KND) substrains [7, 32]. The mean age of disease manifestation is between 90 and 100 days of life [32]. Cumulative incidence of 80% has been reported [32].

## 4 LEW.1AR1-*iddm* (IDDM) Rat Model

The LEW.1AR1-*iddm* rat (short name, IDDM rat) arose in 1997 in the MHC congenic LEW.1AR1 strain with a defined genetic background (LEW.1AR1/Ztm) ($RT1.A^{a}B/D^{u}C^{u}$) (also designed $RT1^{r2}$) (with an intra-MHC recombination of a *u* haplotype) in the Institute for Laboratory Animal Science of Hannover Medical School (Ztm) (Hannover, Germany) [8]. Therefore, this rat model is also known as the LEW.1AR1/Ztm-*iddm* rat.

Detailed data on the epidemiology are available. Disease manifestation in the IDDM rat is between 40 and 90 days with a maximum around 60 days of life (Fig. 1) [8]. Less than 1% of cases were diagnosed in rats older than 90 days (Fig. 1). No rats become diabetic before day 40 of life (Fig. 1). The mean manifestation was originally reported to be at day 58 of life [8]. The daily diabetes manifestation rate is documented in Fig. 1, based on data from a cohort of 4926 diabetic rats during the 20 year period since the model arose in 1997. For the animals that manifested diabetes during this 20-year period between 1997 and 2017 ($n = 2386$), mean age of diabetes manifestation was 62 days, both in male and female rats. Of these rats, 44% were male and 56% female. In the age group between 40 and 55 days of life, there is a slight preference of diabetes manifestation in female rats, and the same was the case in the age group between 80 and 90 days (Fig. 1). In the age range between 56 and 79 days, there is a slight preference for male rats (Fig. 1).

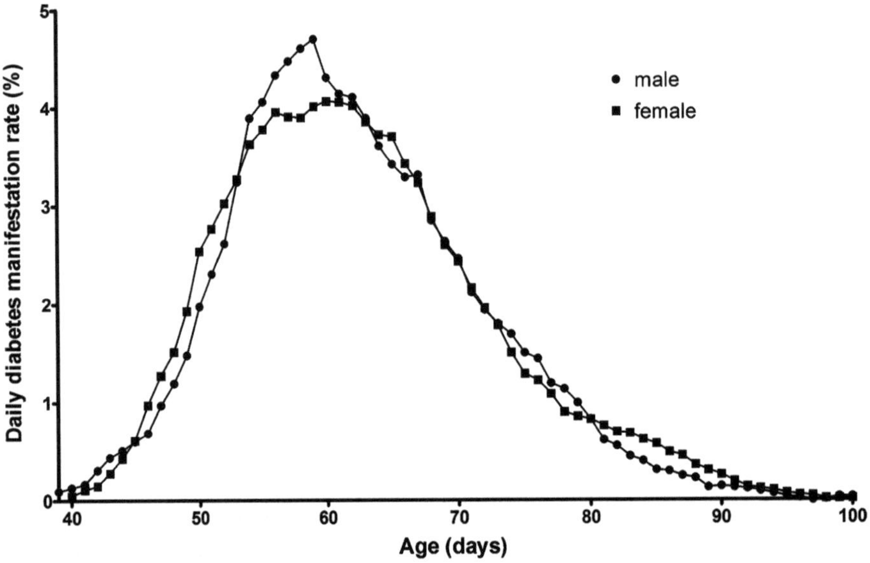


**Fig. 1** Age-dependent diabetes manifestion in the cohort of male and female IDDM rats (seven day walking mean)

#### 4.1 Variable T Lymphocyte Frequency

The T lymphocyte frequency in peripheral blood is somewhat more variable in the IDDM rat with a reduction in $CD3^+$ T cells of around 10% in nondiabetic and 20% in diabetic rats as compared to the background strain [28]. This marginal reduction is responsible for a decreased $CD4^+/CD8^+$ ratio, which results from a slight reduction of the $CD4^+$ T cell number rather than the $CD8^+$ T cell number [28]. The T cells express the ART2.1 (formerly RT6.1) differentiation antigen [8]. Nonetheless, variability was greater for both $CD4^+$ and $CD8^+$ T cells [28]. The variable T cell content in this model is due to the *Dock8* mutation, which is also responsible for diabetes manifestation [28] (see below).

## 5 LEW.1WR1 Rat Model

The LEW.1WR1 rat is another MHC congenic LEW rat strain ($RT1.A^uB/D^uC^a$) (also designed $RT1^{r4}$) that manifests spontaneous autoimmune diabetes in around 2% of animals [15]. Much higher frequencies can also be induced in this strain [15]. Disease manifestation is at 59 days of life on average [15]. The MHC haplotype also includes a class II *u* allele. This strain is interesting for purposes of comparison, since it also contains the diabetes predisposing haplotype, but not a mutation, such as it is the case in the three well-established T1DM rat models with high diabetes incidence.

## 6 The Role of the Different Immune Cell Types in T1DM Development

The initiating event responsible for beta cell autoimmunity that leads to pancreatic islet immune cell infiltration is unknown. A presentation of an autoantigen by $RT1^u$ molecules is a possibility, but no primary autoantigen has been identified so far in any of the rat models of human T1DM [2].

T cells from diabetic BB rats [2, 33, 34] and diabetic IDDM rats [35, 36] can adoptively transfer diabetes to athymic recipients. Transfer of autoreactive potential to the LEW.1AR1 background strain was not effective [35]. But it was possible to confer protection against diabetes through adoptive transfer of immune cells from the LEW.1AR1 background strain into young prediabetic IDDM rats [35]. Thus, regulatory elements of the cellular immune system apparently protect LEW.1AR1 rats against beta cell autoimmunity [35]. Only transfer of $CD4^+$ T cells induces diabetes [36, 37]. Transfer of $CD8^+$ T cells into athymic rats does not cause diabetes [36]. Rather, $CD8^+$ T cells provide protection against beta cell destruction [36]. This protection went along with an accumulation of regulatory T cells in the pancreas-draining lymph nodes [36].

Treatment with poly I:C, a synthetic double-stranded RNA molecule that mimics a viral infection and activates the autoaggressive potential of T lymphocytes, stimulates the innate immune system in a dose-dependent manner both in BB and in IDDM rats as well as in LEW.1WR1 rats, as documented by an increased incidence of autoimmune diabetes [2, 8, 21, 35]. Transfer of Con A-activated lymphocytes to poly I:C-treated IDDM rats before diabetes manifestation further increased the incidence of diabetes [35].

All these results prove that autoreactive T cells cause beta cell destruction and induce a diabetic metabolic state in the rat models of human T1DM, while regulatory T cells confer protection. The balance between autoreactive T cells and regulatory T cells ultimately decides whether diabetes is induced or whether its manifestation is prevented in an animal.

### 6.1 Pancreatic Islet Immune Cell Infiltration

Islet immune cell infiltration is the key characteristic of the autoimmune process in all rat models of human T1DM. The infiltration starts from the islet periphery during the prediabetic phase before diabetes manifestation. This prediabetic phase lasts up to 1 month in the BB rat [2, 13, 38, 39] and 1 week only in the IDDM rat [40]. The macrophages are the first immune cell subpopulation invading the islet periphery [40–44]. They apparently act as initiators of the immune process in the pancreas.

This initial phase is followed by a gradual shift to a predominance of $CD8^+$ T cells (>50%) in the infiltrate along with a continuously decreasing beta cell mass due to apoptosis, while the infiltrate gradually migrates from the periphery into the center of the islet ultimately covering the whole islet [40]. A peri-islet infiltration is not a characteristic of the immune cell infiltration process in the rat models of T1DM [2, 5]. The $CD4^+$ T cell proportion remains stable (at ~10%) during this progression of infiltration. Nevertheless, $CD4^+$ T cells apparently play an important role in the recruitment of $CD8^+$ T cells [40]. The portion of B cells remains small (<5%). The portion of NK cells is highest (~10%) in the initial phase of infiltration into the islet core [40], parallel to a high rate of apoptosis, indicating a role in the beta cell destruction process [40, 45–47].

In all rat models of human T1DM, the final outcome of the autoimmune-mediated beta cell-destructive process are so-called residual end-stage islets with very few or no immune cells. These end-stage islets are also virtually free of beta cells and composed only of non-beta cells, mainly alpha cells, but also of all other known islet cell types [2, 5].

After diabetes manifestation, at that moment when beta cell loss approaches 70% [40], the immune cell composition of the islet infiltrate is comparable in all three rat models of spontaneous autoimmune diabetes and also to that of the human diabetic

**Table 1**
**Immune cell subpopulations in the infiltrate of pancreatic islets after diabetes manifestation of the BB rat, the IDDM rat, and the Komeda rat**

| Immune cell types | Diabetic IDDM rat | Diabetic BB rat | Diabetic KDP rat |
|---|---|---|---|
| MØ (CD68) | ++ | ++ | ++ |
| $CD4^+$ T cells | + | + | + |
| $CD8^+$ T cells | +++ | + | +++ |
| NK cells (Pen5) | + | + | + |
| B cells (CD20) | (+) | (+) | (+) |

+++ = very strong expression, ++ = strong expression, + = expression, (+) = mild expression
Data compiled in this table were derived from information obtained from [48]

pancreas [48]. Characteristic is always a dominance of T lymphocytes, in particular $CD8^+$ T cells and to a lesser extent of $CD4^+$ T cells over macrophages along with fewer percentages of various other immune cell types (Table 1). In the BB rat, this dominance is less prominent, due to the lymphopenia in this model (Table 1).

### *6.2 Pro-inflammatory Cytokine Expression Profile in the Islet-Infiltrating Immune Cells*

All immune cell types in the islet infiltrate are activated and produce cytotoxic pro-inflammatory cytokines, in particular TNFα and IL-1β and to a lesser extent IFNγ [48] (Table 2), which mediate beta cell death during the T1DM disease process, as documented by caspase 3 activation. The exception is the KDP model. This rat model exhibits an expression profile of pro-inflammatory cytokines in the immune cell infiltrate [48], which differs from that in the BB and the IDDM rat models as well as from the situation in the human T1DM pancreas [48] (Table 2). In addition to TNFα, IFNγ is the other pro-inflammatory cytokine predominating in the KDP pancreas [48]. The lack of IL-1β expression in the KDP rat model is in agreement with the lack of iNOS expression in the infiltrating immune cells [48], at variance from the BB rat and the IDDM rat, which express both IL-1β and iNOS in the infiltrating immune cells [48, 49]. Thus, iNOS expression can be classified as a marker for IL-1β expression. This allows the conclusion that, as proven also in in vitro studies [50], TNFα is the pro-inflammatory cytokine in the islet infiltrate crucial for beta cell death in T1DM. IL-1β expression alone is not sufficient to induce significant beta cell death.

The infiltration with a lesser number of activated immune cells in the islets in the BB rat, due to the severe lymphopenia, as compared to the other rat models, reduces also beta cell toxic pro-inflammatory cytokine production and can provide an explanation for the longer prodromal phase of islet immune cell infiltration before outbreak of the disease.

**Table 2**
**Cytokine protein and gene expression pattern in the immune cells of the infiltrate after diabetes manifestation of the BB rat, the IDDM rat, and the Komeda rat**

| Cytokines | Diabetic BB rat | Diabetic IDDM rat | Diabetic KDP rat |
|---|---|---|---|
| IL-1β | + | ++ | (+) |
| TNFα | ++ | ++ | ++ |
| IFNγ | (+) | (+) | ++ |
| Caspase 3 | + | + | ++ |

++ = strong expression, + = expression, (+) = mild expression
Data compiled in this table were derived from information obtained from [48]

BB rats have recently been shown to develop early deficits before diabetes manifestation, in particular reduced insulin secretion and beta cell mass [27]. Thus, the possibility exists that these early changes may be an element in a predisposition of the BB rats to the later development of diabetes. On the other hand, it cannot be excluded that the lymphopenia, present in the BB rats since birth, may be a contributory factor, for example, to a reduction in the number of medium- and small-sized islets in the pancreas, as recently reported [27]. In the pancreas of the IDDM rat, we have not observed such changes morphologically prior to diabetes manifestation (unpublished observation). In contrast to the IDDM rat, these disparities limit the suitability of the BB rat and KDP rat models for prevention studies.

Pancreas-draining lymph nodes are important for the islet-specific antigen presentation and immune cell activation in the prediabetic phase during diabetes development and the maintenance of the inflammation in the pancreas after diabetes manifestation. The activated immune cell types in the lymph nodes are comparable to those in the pancreatic islet infiltrate with the same cytokine expression profile [51–53]. After a specific priming process, the activated immune cells of the pancreas-draining lymph nodes migrate into the pancreas either via the circulation or the lymphatic vessels. Ongoing islet infiltration allows a circulation of immune cells from the infiltrated islets to the pancreas-draining lymph nodes [51–53]. This process maintains the infiltration through a continuous recirculation of the immune cells.

### 6.3 Gut Dysfunction in T1DM Rat Models

The gut system is a bridge between environment and metabolic state. Environmental variables can act as promoters or even triggers of autoimmune diabetes manifestation. This has been documented experimentally also in two of the T1DM rat models, the BB rat [54–57] and the IDDM rat [58]. Diet composition has major effects on the gut immune system and consequently affects diabetes development and incidence. Compared with a traditional diet, modified diets can reduce the extent of insulitis and diabetes incidence. The challenge is to identify the diet components that promote diabetes development and possible diet modifications with protective potential [54]. Particularly prominent has been the protective effect of low antigen hydrolyzed casein-based diets as also documented in studies in rat models of T1DM [55, 58]. These studies provide evidence for a reduced immunoregulatory capacity due to gut immune cell deficits [56] and are likely to be related to the mutations in these rats, which are responsible for not only diabetes manifestation but also for functional changes in the immune system. These studies document the usefulness of these rat models for elucidating the bridging function of the gut between environment and diabetes development. It is important, however, to emphasize that environmental factors have a modulatory function but do not completely prevent diabetes manifestation in these rat models. Environment along with lifestyle and diet changes is likely to be a significant contributory factor in the steadily increasing incidence of T1DM in humans, as documented by positive effects of sensible modulation of the intestinal microbiome [59]. Since, in contrast to humans, laboratory rats live in a well-defined environment and receive a uniform diet, T1DM rat models are well suited for analyses of distinct effects of diet components on diabetes development.

## 7 Genetics of Autoimmune Diabetes in T1DM Rat Models

The major histocompatibility complex (MHC) (also called the *RT1* complex) plays a crucial role for the manifestation of autoimmune diabetes. The primary genetic prerequisite for disease manifestation is the diabetes permitting MHC class II haplotype $RT1\ B/D^u$ (designed *IDDM1*) on rat chromosome 20 (RNO20). This is the MHC II haplotype that all three T1DM rat models have in common, and it is the principle prerequisite for the development of autoimmune diabetes [3, 8, 60–64] (Table 3). Even the LEW.1WR1 strain with its very low 2% diabetes incidence has this MHC II haplotype [15]. This haplotype determines also the tissue specificity of the disease. The MHC-dependent genetic predisposition is responsible for an imbalance between beta cell toxic T cells and regulatory T cells [65]. This disequilibrating genetic element results in an activation of autoreactive T cells. However, this genetic factor is sufficient only for a very low diabetes incidence (no more than 2%) [15].

The reason for the high disease incidence in the three rat models of spontaneous autoimmune diabetes is in each case the result of an additional single spontaneous mutation [3, 18, 64, 66, 67] (Table 3). This mutation causes an aggravation of this disequilibrium [65]. It is a further perturbation of this balance that results

**Table 3**
**Mutations in the BB rat, the IDDM rat, and the Komeda rat**

| Model | BB rat (BioBreeding) | IDDM rat (LEW.1AR1-*iddm)* | KDP rat (Komeda) |
|---|---|---|---|
| Mean day of diabetes manifestation | 90 | *62* | 90–100 |
| MHC II (*Iddm1*) | $RT1^{u/u/u}$ | $RT1^{a/u/u}$ (also designed $RT1^{r2}$) | $RT1^{u/u/u}$ |
| *Vwa2* variant (*Iddm8* ) | T (Trp) | T (Trp) | T (Trp) |
| Main mutation | *Gimap5* (*Iddm2*) (RNO4) | *Dock8* (*Iddm8*) (RNO1) | *Cblb* (*Iddm/kdp1*) (RNO11) |

All three rat models have (a) the MHC *II B/D*$^{u}$ ($RT1^{u}$) haplotype, (b) carry the *Vwa2* variant T, and (c) have a unique mutation each on different chromosomes responsible for the high incidence of the diabetic phenotype

in a diabetes manifestation in at least half of the animals. The LEW.1WR1 rat with a low spontaneous diabetes incidence of only 2% lacks such a specific mutation.

The mutation in the diabetes-prone BB rat has been identified in the *Gimap5* (GTPase of the immune-associated protein 5) (formerly also called *IAN4* or *IAN5*) gene on chromosome 4 (RNO4) (Table 3), being due to a single-nucleotide deletion [66]. The resulting frameshift causes a large truncation of the protein [18]. It is the mutation in this single locus, which is responsible for autoimmune diabetes as well as the severe T cell lymphopenia. The consequence of the mutation is a virtually complete loss of this encoded GTPase function [18].

The mutation in the IDDM rat has been identified as a single base exchange [C/G] in the *Dock8* gene on chromosome 1 (RNO1) at the proximal end of *Iddm8* [68, 69], resulting in an amino acid exchange in the protein from glutamine to glutamate [67] (Table 3). At variance from the mutations in the BB rat and the KDP rat, it is not a nonsense mutation and therefore does not cause a truncation of the protein. The mutation in the IDDM rat is functionally critical for the *Dock8* gene and has an influence on the binding properties of the Dock8 protein with the Rho GTPase Cdc42 affecting the normal function of this protein [67]. This influences T cell development negatively, resulting in an enhanced apoptotic T cell death and a reduced proliferation [70]. The mutation in the IDDM rat causes in addition to autoimmune diabetes a second phenotype due to the effect on T cell development. This is a somewhat more variable $CD3^{+}$ T cell frequency in the peripheral blood of the animals, but in contrast to the BB rat without lymphopenia [28]. Both phenotypes are ultimately due to the loss of function of the regulatory guanosine nucleotide exchange factor (GEF) of the Rho GTPase Cdc42 [67].

The disease mediating mutation in the *Cblb* (Casitas B-lineage lymphoma b, a member of the Cbl/Sli family of ubiquitin-protein ligases) gene on chromosome 11 (RNO11) of the KDP rat is due to a single base pair exchange [C/T] [64]. It is not present in the nondiabetic KND strain [64]. This nonsense mutation results in a removal of 484 amino acids from the protein [64, 71]. The mutation is entirely different from the mutations in the two other rat models (Table 3), though the gene may also have T cell regulatory potential [71] and apparently facilitates the activation of autoreactive T cells [72]. But data on possible changes of T lymphocyte numbers in peripheral blood are not available. The mutation in the *Cblb* gene may have an influence on the binding capacity to GTPases. This mutation may also explain the very different pro-inflammatory cytokine expression profile in the immune cell infiltrate in the KDP rat model [48] (Table 2).

## 8 Comparisons of the Mutations in the Different Rat Models and in Humans with T1DM

When comparing the mutations in the BB rat and in the IDDM rat, there are remarkable similarities: in one case, the loss of function of the GTPase [18] and, in the other case, the compromised function of the regulatory guanine nucleotide exchange factor (GEF) [28]. In both rat models of human T1DM, the respective crucial mutation confers susceptibility to a spontaneous outbreak of autoimmune diabetes. However, the magnitude of the effects on the T lymphocytes in the peripheral blood differs very considerably in that the mutation in the BB rat causes a severe lymphopenia with a loss of >80–90% of all T lymphocyte subsets [18], while the mutation in the IDDM rat causes a greater variability of the T lymphocyte frequency in the peripheral blood. This greater variability goes along with a small decrease in the T lymphocyte number (~ 20% in diabetic and ~10% in nondiabetic rats), mainly in the subset of the $CD4^+$ T cells but not of the $CD8^+$ T cells [28]. This observation documents that a second phenotype with minor changes in the T lymphocyte profile is sufficient to accompany the development of a diabetic metabolic syndrome in the majority of the animals. In this sense, the diabetic syndrome in the IDDM rat model is reminiscent of the human T1DM situation, where changes of limited extent in the T lymphocyte distribution have also been reported and considered to contribute to T1DM manifestation [28]. This milder effect of the mutation on the T lymphocyte status when compared with the BB rat may also explain the somewhat lower diabetes incidence in the IDDM rat.

It has been known for many years that a number of genes can be involved in the manifestation of autoimmune diabetes in the different rat models [2, 32, 68]. The detailed knowledge that has been obtained about the mutations in these models over the last years clearly documents that mutations in various genes can be involved in the manifestation of insulitis and diabetes. So, it is possible nowadays to specify in the rat models of human T1DM the traditional generalizing statement that calls T1DM simply a "polygenic" disease [73].

The mutation in the *Dock8* gene of the IDDM rat, a single nucleotide exchange, causes a milder phenotype. It only reduces the function of the GEF through a reduction of its activation capacity of the G protein Rho-GTPase Cdc42. This results in a greater variability of the T cell content along with a slight T cell reduction in the range of 10–20% [36]. The likelihood of diabetes manifestation tends to increase along with the extent of T cell reduction in this 10–20% range in the IDDM rat with its diabetes predisposing MHC haplotype [36].

So, the genetic traits in the IDDM rat model allow something that has not been achieved with the BB rat. It is possible to associate a mutation causing a primary immunodeficiency disease with an immune cell dysfunction due to a mutation in the *Dock8* gene with a negative influence on the ability of the G protein Rho-GTPase Cdc42 to be activated by the regulatory GEF [74]. But for the manifestation of autoimmune diabetes, a predisposing MHC haplotype is required in addition.

The main mutations in the BB rat and in the KDP rat could not be associated with a susceptibility to human T1DM [71]. Mutations in the *DOCK8* gene have also been reported, both in mice and humans, causing a DOCK8 deficiency that results in a slight lymphocyte dysfunction; however, a diabetic metabolic state has not been reported [74]. This is not surprising since the additional presence of a diabetes predisposing MHC haplotype is an indispensable prerequisite for diabetes manifestation. A possible autoimmune diabetes manifestation in humans with a *DOCK8* gene mutation along with a predisposing MHC haplotype has not been reported so far.

Since the incidence of T1DM in humans is around 100 times lower than in the established rat models of T1DM and even lower than in rats with a predisposing MHC haplotype only, it is not surprising that a simple transfer of the knowledge obtained through the elucidations of the mutations responsible for autoimmune diabetes manifestation in the rat models is not possible in a one to one manner to patients with T1DM. The only possible explanation for this higher diabetes incidence in the rat models of autoimmune diabetes is that in humans with a predisposing MHC haplotype additional mutations fostering diabetes manifestation may be of milder penetrance than in the rat models of human T1DM.

#### 8.1 Other Mutations in the Rat Models of Human T1DM

In addition, a base exchange [C/T] in the von Willebrand factor A domain-containing protein 2 (*Vwa2*) gene on chromosome 1 (RNO1) at the distal end of *Iddm8* is present in all three diabetes-prone rat strains (Table 3) resulting in an arginine/tryptophan polymorphism [28]. The fact that it is present also in many other rat strains excludes the possibility that this genotype is of major importance for diabetes manifestation in the three diabetes-prone rat strains [28]. But the fact that the LEW.1WR1 rat strain with its very low diabetes incidence is also homozygous for this genotype can be interpreted as an indication that this variation in the *Vwa2* gene may contribute, even though to a very minor extent, to the pathogenesis of autoimmune diabetes in the rat models of human T1DM [28].

## 9 Rat Models of Human T1DM as Tools for the Development of New Therapies for Patients with T1DM

Reliable animal models of human T1DM are of crucial importance for the preclinical testing of new antidiabetic therapies before clinical studies can be performed. The NOD mouse has been the most popular T1DM animal model used for this purpose [75]. However, though many hundred therapy schemes have shown success in the NOD mouse model [76], none proved successful in reversing the disease in clinical studies in patients with T1DM, even in the early disease phase [75].

The BB rat has been used in the past to assess preventive intervention strategies (for review, *see* ref. 2). The most prominent agent found to provide protection against diabetes manifestation was the immunosuppressant cyclosporine [2]. But the BB rat has lost attractiveness for this purpose during the last decade, due to the problem with the lymphopenia in this model. The KDP rat has been used rarely for this purpose. The IDDM rat, however, is used increasingly often both in primary and secondary prevention studies and has provided reliable results [52, 77].

When administered before diabetes manifestation, FTY720 (fingolimod) has been shown to prevent an outbreak of the disease by promoting retention of activated immune cells in the lymph nodes [51]. This protective effect of FTY720 has also been described in the BB rat [78]. FTY720 may also be an effective adjuvant when combined in prevention therapies with therapeutic antibodies, i.e., with anti-TCR [77].

Monotherapy treatments with a variety of therapeutic antibodies, even with anti-TCR [77], failed to reverse the diabetic metabolic state, when administered after outbreak of the disease in the IDDM rat. Therapy success was achieved only in combination therapies (for review, *see* ref. 75). A therapy with recombinant

TNFα protein, when initiated before islet immune cell infiltration and diabetes outbreak, has also been shown to provide protection in the BB rat [79].

A combination of anti-TCR with anti-TNFα proved to be the most effective one so far in reversing the diabetic metabolic state in the IDDM rat model of human T1DM [52]. Further effective combinations of antibodies that show therapeutic effects in other autoimmune diseases in humans are awaiting publication showing curative potential not only in the IDDM rat model but also show promise of successful translation to patients with T1DM in the early disease phase (for discussion, *see* ref. 75). Thus, there is hope that the IDDM rat, the spontaneous rat model of T1DM with the closest similarity to the human disease, will be valuable in the translation of new combination therapies with curative potential to patients with T1DM (for discussion, *see* ref. 75). For this purpose, models of chemically induced (i.e., alloxan or streptozotocin) insulin-dependent diabetes that are not of autoimmune nature are not suited [80].

## Acknowledgment

This book chapter is dedicated to the memory of my old friend, the distinguished biochemist and diabetologist Professor Eleazar Shafrir (1924–2016), Jerusalem.

### References

1. Yanagisawa M, Hara Y, Satoh K et al (1986) Spontaneous autoimmune thyroiditis in Bio Breeding/Worcester (BB/W) rat. Endocrinol Jpn 33:851–861
2. Mordes JP, Poussier P, Rossini AA, Blankenhorn EP, Greiner DL (2007) Rat models of type 1 diabetes: genetics, environment, and autoimmunity. In: Shafrir E (ed) Animal models of diabetes. CRC Press, Boca Raton, FL
3. Yokoi N, Hayashi C, Fujiwara Y, Wang HY, Seino S (2007) Genetic reconstitution of autoimmune type 1 diabetes with two major susceptibility genes in the rat. Diabetes 56:506–512
4. Rajatanavin R, Appel MC, Reinhardt W, Alex S, Yang YN, Braverman LE (1991) Variable prevalence of lymphocytic thyroiditis among diabetes-prone sublines of BB/W or rats. Endocrinology 128:153–157
5. Jörns A, Kubat B, Tiedge M et al (2004) Pathology of the pancreas and other organs in the diabetic LEW.1AR1/Ztm- *iddm* rat, a new model of spontaneous insulin-dependent diabetes mellitus. Virchows Arch 444:183–189
6. Scott FW (1996) Food-induced type 1 diabetes in the BB rat. Diabetes Metab Rev 12:341–359
7. Komeda K, Noda M, Terao K, Kuzuya N, Kanazawa M, Kanazawa Y (1998) Establishment of two substrains, diabetes-prone and non-diabetic, from Long-Evans Tokushima Lean (LETL) rats. Endocr J 45:737–744
8. Lenzen S, Tiedge M, Elsner M et al (2001) The LEW.1AR1/Ztm-*iddm* rat: a new model of spontaneous insulin-dependent diabetes mellitus. Diabetologia 44:1189–1196
9. Guberski DL, Thomas VA, Shek WR et al (1991) Induction of type I diabetes by Kilham's rat virus in diabetes-resistant BB/Wor rats. Science 254:1010–1013
10. Yoon JW, Jun HS (2006) Viruses cause type 1 diabetes in animals. Ann N Y Acad Sci 1079:138–146
11. Markholst H, Klaff LJ, Klöppel G, Lernmark A, Mordes JP, Palmer J (1990) Lack of

systematically found insulin autoantibodies in spontaneously diabetic BB rats. Diabetes 39:720–727

12. Nakhooda AF, Like AA, Chappel CI, Murray FT, Marliss EB (1977) The spontaneously diabetic Wistar rat. Metabolic and morphologic studies. Diabetes 26:100–112
13. Mordes JP, Desemone J, Rossini AA (1987) The BB rat. Diabetes Metab Rev 3:725–750
14. Scott J (1990) The spontaneously diabetic BB rat: sites of the defects leading to autoimmunity and diabetes mellitus. A review. Curr Top Microbiol Immunol 156:1–14
15. Mordes JP, Guberski DL, Leif JH et al (2005) LEW.1WR1 rats develop autoimmune diabetes spontaneously and in response to environmental perturbation. Diabetes 54:2727–2733
16. Jackson R, Rassi N, Crump T, Haynes B, Eisenbarth GS (1981) The BB diabetic rat. Profound T-cell lymphocytopenia. Diabetes 30:887–889
17. Elder ME, Maclaren NK (1983) Identification of profound peripheral T lymphocyte immunodeficiencies in the spontaneously diabetic BB rat. J Immunol 130:1723–1731
18. MacMurray AJ, Moralejo DH, Kwitek AE et al (2002) Lymphopenia in the BB rat model of type 1 diabetes is due to a mutation in a novel immune-associated nucleotide (Ian)-related gene. Genome Res 12:1029–1039
19. Ramanathan S, Poussier P (2001) BB rat lyp mutation and type 1 diabetes. Immunol Rev 184:161–171
20. Yale JF, Grose M, Marliss EB (1985) Time course of the lymphopenia in BB rats. Relation to the onset of diabetes. Diabetes 34:955–959
21. Nakamura N, Tsutsumi Y, Kimata S et al (1991) Induction of diabetes by poly I:C and anti-RT6.1 antibody treatment in DR-BB rats. Endocrinol Jpn 38:523–526
22. Bortel R, Waite DJ, Whalen BJ et al (2001) Levels of Art$^{2+}$ cells but not soluble Art2 protein correlate with expression of autoimmune diabetes in the BB rat. Autoimmunity 33:199–211
23. Joseph S, Diamond AG, Smith W, Baird JD, Butcher GW (1993) BB-DR/Edinburgh: a lymphopenic, non-diabetic subline of BB rats. Immunology 78:318–328
24. Like AA, Guberski DL, Butler L (1986) Diabetic BioBreeding/Worcester (BB/Wor) rats need not be lymphopenic. J Immunol 136:3254–3258
25. Awata T, Guberski DL, Like AA (1995) Genetics of the BB rat: association of autoimmune disorders (diabetes, insulitis, and thyroiditis) with lymphopenia and major histocompatibility complex class II. Endocrinology 136:5731–5735
26. Hawkins T, Fuller J, Olson K, Speros S, Lernmark A (2005) DR.lyp/lyp bone marrow maintains lymphopenia and promotes diabetes in lyp/lyp but not in +/+ recipient DR.lyp BB rats. J Autoimmun 25:251–257
27. Medina A, Parween S, Ullsten S et al (2017) Early deficits in insulin secretion, beta cell mass and islet blood perfusion precede onset of autoimmune type 1 diabetes in BioBreeding rats. Diabetologia 61:896–905
28. Arndt T, Jörns A, Weiss H et al (2013) A variable CD3(+) T-cell frequency in peripheral blood lymphocytes associated with type 1 diabetes mellitus development in the LEW.1AR1-*iddm* rat. PLoS One 8:e64305
29. Wilcox NS, Rui J, Hebrok M, Herold KC (2016) Life and death of beta cells in type 1 diabetes: a comprehensive review. J Autoimmun 71:51–58
30. Arndt T, Jörns A, Hedrich HJ, Lenzen S, Wedekind D (2014) Variable immune cell frequencies in peripheral blood of LEW.1AR1-*iddm* rats over time compared to other congenic LEW strains. Clin Exp Immunol 177:168–178
31. Kawano K, Hirashima T, Mori S, Saitoh Y, Kurosumi M, Natori T (1991) New inbred strain of Long-Evans Tokushima lean rats with IDDM without lymphopenia. Diabetes 40:1375–1381
32. Yokoi N, Namae M, Fuse M et al (2003) Establishment and characterization of the Komeda diabetes-prone rat as a segregating inbred strain. Exp Anim 52:295–301
33. McKeever U, Mordes JP, Greiner DL et al (1990) Adoptive transfer of autoimmune diabetes and thyroiditis to athymic rats. Proc Natl Acad Sci U S A 87:7618–7622
34. Whalen BJ, Greiner DL, Mordes JP, Rossini AA (1994) Adoptive transfer of autoimmune diabetes mellitus to athymic rats: synergy of CD4$^{+}$ and CD8$^{+}$ T cells and prevention by RT6$^{+}$ T cells. J Autoimmun 7:819–831
35. Wedekind D, Weiss H, Jörns A, Lenzen S, Tiedge M, Hedrich HJ (2005) Effects of polyinosinic-polycytidylic acid and adoptive transfer of immune cells in the Lew.1AR1-*iddm* rat and in its coisogenic LEW.1AR1 background strain. Autoimmunity 38:265–275
36. Arndt T, Wedekind D, Weiss H et al (2009) Prevention of spontaneous immune-mediated diabetes development in the LEW.1AR1-*iddm* rat by selective CD8$^{+}$ T cell transfer is associated with a cytokine shift in the pancreas-

draining lymph nodes. Diabetologia 52:1381–1390
37. Metroz-Dayer MD, Mouland A, Brideau C, Duhamel D, Poussier P (1990) Adoptive transfer of diabetes in BB rats induced by CD4 T lymphocytes. Diabetes 39:928–932
38. Logothetopoulos J, Valiquette N, Madura E, Cvet D (1984) The onset and progression of pancreatic insulitis in the overt, spontaneously diabetic, young adult BB rat studied by pancreatic biopsy. Diabetes 33:33–36
39. Lally FJ, Ratcliff H, Bone AJ (2001) Apoptosis and disease progression in the spontaneously diabetic BB/S rat. Diabetologia 44:320–324
40. Jörns A, Günther A, Hedrich HJ, Wedekind D, Tiedge M, Lenzen S (2005) Immune cell infiltration, cytokine expression, and beta-cell apoptosis during the development of type 1 diabetes in the spontaneously diabetic LEW.1AR1/Ztm-*iddm* rat. Diabetes 54:2041–2052
41. Oschilewski U, Kiesel U, Kolb H (1985) Administration of silica prevents diabetes in BB-rats. Diabetes 34:197–199
42. Walker R, Bone AJ, Cooke A, Baird JD (1988) Distinct macrophage subpopulations in pancreas of prediabetic BB/E rats. Possible role for macrophages in pathogenesis of IDDM. Diabetes 37:1301–1304
43. Voorbij HA, Jeucken PH, Kabel PJ, De Haan M, Drexhage HA (1989) Dendritic cells and scavenger macrophages in pancreatic islets of prediabetic BB rats. Diabetes 38:1623–1629
44. Hanenberg H, Kolb-Bachofen V, Kantwerk-Funke G, Kolb H (1989) Macrophage infiltration precedes and is a prerequisite for lymphocytic insulitis in pancreatic islets of pre-diabetic BB rats. Diabetologia 32:126–134
45. Sobel DO, Azumi N, Creswell K et al (1995) The role of NK cell activity in the pathogenesis of poly I:C accelerated and spontaneous diabetes in the diabetes prone BB rat. J Autoimmun 8:843–857
46. Iwakoshi NN, Greiner DL, Rossini AA, Mordes JP (1999) Diabetes prone BB rats are severely deficient in natural killer T cells. Autoimmunity 31:1–14
47. Todd DJ, Forsberg EM, Greiner DL, Mordes JP, Rossini AA, Bortell R (2004) Deficiencies in gut NK cell number and function precede diabetes onset in BB rats. J Immunol 172:5356–5362
48. Jörns A, Arndt T, Meyer zu Vilsendorf A et al (2014) Islet infiltration, cytokine expression and beta cell death in the NOD mouse, BB rat, Komeda rat, LEW.1AR1-*iddm* rat and humans with type 1 diabetes. Diabetologia 57:512–521
49. Kolb H, Worz-Pagenstert U, Kleemann R, Rothe H, Rowsell P, Scott FW (1996) Cytokine gene expression in the BB rat pancreas: natural course and impact of bacterial vaccines. Diabetologia 39:1448–1454
50. Kacheva S, Lenzen S, Gurgul-Convey E (2011) Differential effects of proinflammatory cytokines on cell death and ER stress in insulin-secreting INS1E cells and the involvement of nitric oxide. Cytokine 55:195–201
51. Jörns A, Rath KJ, Terbish T et al (2010) Diabetes prevention by immunomodulatory FTY720 treatment in the LEW.1AR1-*iddm* rat despite immune cell activation. Endocrinology 151:3555–3565
52. Jörns A, Ertekin UG, Arndt T, Terbish T, Wedekind D, Lenzen S (2015) TNF-alpha antibody therapy in combination with the T-cell-cpecific antibody anti-TCR reverses the diabetic metabolic state in the LEW.1AR1-*iddm* rat. Diabetes 64:2880–2891
53. Ferraro A, Socci C, Stabilini A et al (2011) Expansion of Th17 cells and functional defects in T regulatory cells are key features of the pancreatic lymph nodes in patients with type 1 diabetes. Diabetes 60:2903–2913
54. Scott FW, Mongeau R, Kardish M, Hatina G, Trick KD, Wojcinski Z (1985) Diet can prevent diabetes in the BB rat. Diabetes 34:1059–1062
55. Graham S, Courtois P, Malaisse WJ, Rozing J, Scott FW, Mowat AM (2004) Enteropathy precedes type 1 diabetes in the BB rat. Gut 53:1437–1444
56. Scott FW, Pound LD, Patrick C, Eberhard CE, Crookshank JA (2017) Where genes meet environment-integrating the role of gut luminal contents, immunity and pancreas in type 1 diabetes. Transl Res 179:183–198
57. Hara N, Alkanani AK, Ir D et al (2013) The role of the intestinal microbiota in type 1 diabetes. Clin Immunol 146:112–119
58. Crookshank JA, Patrick C, Wang GS, Noel JA, Scott FW (2015) Gut immune deficits in LEW.1AR1-*iddm* rats partially overcome by feeding a diabetes-protective diet. Immunology 145:417–428
59. Knip M, Honkanen J (2017) Modulation of type 1 diabetes risk by the intestinal microbiome. Curr Diab Rep 17:105
60. Chao NJ, Timmerman L, McDevitt HO, Jacob CO (1989) Molecular characterization of MHC class II antigens (beta 1 domain) in the BB diabetes-prone and -resistant rat. Immunogenetics 29:231–234

61. Colle E (1990) Genetic susceptibility to the development of spontaneous insulin-dependent diabetes mellitus in the rat. Clin Immunol Immunopathol 57:1–9
62. Ellerman KE, Like AA (2000) Susceptibility to diabetes is widely distributed in normal class IIu haplotype rats. Diabetologia 43:890–898
63. Mordes JP, Bortell R, Blankenhorn EP, Rossini AA, Greiner DL (2004) Rat models of type 1 diabetes: genetics, environment, and autoimmunity. ILAR J 45:278–291
64. Yokoi N, Komeda K, Wang HY et al (2002) Cblb is a major susceptibility gene for rat type 1 diabetes mellitus. Nat Genet 31:391–394
65. Mordes JP, Bortell R, Doukas J et al (1996) The BB/Wor rat and the balance hypothesis of autoimmunity. Diabetes Metab Rev 12:103–109
66. Hornum L, Rømer J, Markholst H (2002) The diabetes-prone BB rat carries a frameshift mutation in Ian4, a positional candidate of Iddm1. Diabetes 51:1972–1979
67. Arndt T, Wedekind D, Jörns A et al (2015) A novel Dock8 gene mutation confers diabetogenic susceptibility in the LEW.1AR1/Ztm-*iddm* rat, an animal model of human type 1 diabetes. Diabetologia 58:2800–2809
68. Weiss H, Bleich A, Hedrich HJ et al (2005) Genetic analysis of the LEW.1AR1-*iddm* rat: an animal model for spontaneous diabetes mellitus. Mamm Genome 16:432–441
69. Weiss H, Arndt T, Jörns A et al (2008) The mutation of the LEW.1AR1-*iddm* rat maps to the telomeric end of rat chromosome 1. Mamm Genome 19:292–297
70. Smits K, Iannucci V, Stove V et al (2010) Rho GTPase Cdc42 is essential for human T-cell development. Haematologica 95:367–375
71. Payne F, Smyth DJ, Pask R et al (2004) Haplotype tag single nucleotide polymorphism analysis of the human orthologues of the rat type 1 diabetes genes Ian4 (Lyp/Iddm1) and Cblb. Diabetes 53:505–509
72. Yokoi N (2005) Identification of a major gene responsible for type 1 diabetes in the Komeda diabetes-prone rat. Exp Anim 54:111–115
73. Wallis RH, Wang K, Marandi L et al (2009) Type 1 diabetes in the BB rat: a polygenic disease. Diabetes 58:1007–1017
74. Su HC, Jing H, Zhang Q (2011) DOCK8 deficiency. Ann N Y Acad Sci 1246:26–33
75. Lenzen S (2017) Animal models of human type 1 diabetes for evaluating combination therapies and successful translation to the patient with type 1 diabetes. Diabetes Metab Res Rev 33. https://doi.org/10.1002/dmrr.2915
76. Ben Nasr M, D'Addio F, Usuelli V, Tezza S, Abdi R, Fiorina P (2015) The rise, fall, and resurgence of immunotherapy in type 1 diabetes. Pharmacol Res 98:31–38
77. Jörns A, Akin M, Arndt T et al (2014) Anti-TCR therapy combined with fingolimod for reversal of diabetic hyperglycemia by beta cell regeneration in the LEW.1AR1-*iddm* rat model of type 1 diabetes. J Mol Med (Berl) 92:743–755
78. Popovic J, Kover KL, Moore WV (2004) The effect of immunomodulators on prevention of autoimmune diabetes is stage dependent: FTY720 prevents diabetes at three different stages in the diabetes-resistant biobreeding rat. Pediatr Diabetes 5:3–9
79. Satoh J, Seino H, Shintani S et al (1990) Inhibition of type 1 diabetes in BB rats with recombinant human tumor necrosis factor-alpha. J Immunol 145:1395–1399
80. Lenzen S (2008) The mechanisms of alloxan- and streptozotocin-induced diabetes. Diabetologia 51:216–226

# Chapter 6

# Mouse Models of Autoimmune Diabetes: The Nonobese Diabetic (NOD) Mouse

**Dawei Chen, Terri C. Thayer, Li Wen, and F. Susan Wong**


## Abstract

There are now a number of different mouse models for type 1 diabetes. The best known is the nonobese diabetic (NOD) mouse which has a genetic susceptibility to autoimmune diabetes with some features that are similar to human type 1 diabetes. The mice also have a propensity to other autoimmune diatheses, including autoimmune thyroid disease and sialadenitis. In addition, it is well known that environmental factors affect the incidence of disease in these mice. While there are other rodent models, including numerous transgenic and knockout models, as well as those that express human proteins, none of these develop spontaneous diabetes over a period of time, when the natural history can be studied. We focus here on the unmanipulated NOD mouse and discuss features of the husbandry and investigation of the mice that allow for use of these long-studied mice in the pathogenesis of an autoimmune type of diabetes.


**Key words** NOD mice, Type 1 diabetes, Animal models, Autoimmunity, Genetic susceptibility, Environment

## 1 Introduction

In humans and mice, type 1 diabetes (T1D) is an autoimmune disease characterized by the infiltration of immune cells, which results in the destruction of insulin-producing pancreatic β cells. The nonobese diabetic (NOD) mouse strain was originally developed by Makino and colleagues during the selection of a cataract-prone strain (cataract Shionogi mice) derived from the outbred Jcl: ICR line of mice [1, 2]. In the pancreatic islets, immune cell infiltration starts at about 5–6 weeks of age in females, with a delay in the male mice. Mononuclear cell infiltrates surround the islet at this point (peri-insulitis), followed by subsequent invasion of the islets and the onset of overt diabetes [3]. Typically, female NOD mice start to develop diabetes between 10 and 14 weeks of age, whereas diabetes in male mice occurs later [3], and they have a lower incidence of disease. For this reason, many investigators choose to study female mice. However, it should be noted that

Aileen J. F. King (ed.), *Animal Models of Diabetes: Methods and Protocols*, Methods in Molecular Biology, vol. 2128,
https://doi.org/10.1007/978-1-0716-0385-7_6, 

the age at which diabetes first develops varies between colonies in different research institutions. Although B cells, NK cells, dendritic cells, neutrophils, and macrophages can also be identified in the lesions, the destruction of insulin-producing β cells in islets is primarily dependent on autoreactive CD8+ T cells helped by CD4+ T cells.

The most important genetic factor contributing to T1D in NOD mice is the MHC. NOD mice have the haplotype $H2^{g7}$ comprising $K^d$, $Aa^d$, $Ab^{g7}$, $E^{null}$, and $D^b$; the MHC-I molecules $H\text{-}2K^d$ *and* $H\text{-}2D^b$ are found in other strains of mice, but it is the MHC-II molecule $I\text{-}A^{g7}$ (ortholog of HLA-DQ) which is unique, and the mice do not express I-E (ortholog of HLA-DR) because of a defective Eα locus [3]. $I\text{-}A^{g7}$ contributes significant susceptibility to developing diabetes [4]. In addition to the importance of the MHC to the progression of T1D, non-MHC genes are also found to contribute to the development of T1D. To date, at least fifty *insulin-dependent diabetes* (*Idd*) loci on at least 11 mouse chromosomes have been identified (*Idd2-Idd12*, *Idd13a*, *Idd13b*, and *Idd14-Idd19*) through crosses of NOD mice with various diabetes-resistant strains [4, 5]. These *Idd* loci contribute to the development of T1D at either initiation of insulitis or progression to overt diabetes or both. These loci act in a collective, recessive manner to determine development of insulitis as well as disease frequency and severity [6].

The environment is also an important factor, as it is very clear that NOD mice in different colonies have different incidences of diabetes [7], and the length of time over which this develops also varies, as discussed below.

A number of antigen-specific T cells have been identified and cloned from the NOD mouse, and T cell receptor (TCR) transgenic mice on the NOD genetic background have been generated (reviewed in ref. 4).

In addition to these various strains, there are also a number of immunodeficient derivative strains of NOD mice, which may be useful for studying individual components of the immune response. These include those which are globally immunodeficient that include NOD.scid mice and NOD.RAG2−/− mice as well as alpha-beta TCR-deficient, B cell-deficient, and more specialized models where individual immune subsets can be depleted. It should be noted that this chapter will provide an introductory overview of care of NOD mice and will not be a comprehensive account of NOD and derivative mice, which now number in the hundreds, many of which are available commercially.

## 2 Materials

### 2.1 Housing of Animals

1. Specific pathogen-free facilities.
2. Isolator or scantainer with appropriate cages.
3. Bedding.
4. Diet.
5. Personal protective equipment, including gloves, scrubs or overalls, hair net, face mask (if needed), and shoe covers.

### 2.2 Monitoring Onset of Diabetes

1. Diastix urine test strips.
2. Glucometer and test strips.
3. Scalpel or surgical scissors.
4. Anesthetic spray, if needed.

## 3 Methods

### 3.1 Housing of Animals

The incidence of diabetes in NOD mice varies between different institutions and laboratories. It has been suggested that environmental factors are as important as genetic susceptibilities in shaping the risk of T1D development [4]. Many environmental factors in NOD mice have been implicated in altering diabetes susceptibility, including exposure to dietary factors such as wheat [8, 9] and gluten [9], exposure to infectious organisms [10–12], and changes to the gut microbiota [4]. The incidence of diabetes is highest when NOD mice are maintained in a relatively germ-free environment; however, this dramatically decreases when mice are exposed to dirty environments and/or infections [3]. Exposure to pinworms [13], *Salmonella typhimurium* [11], mouse hepatitis virus (MHV) [14], and *Schistosoma mansoni* [10] has all been associated with abrogation of diabetes development. Accordingly, consideration should be given to routine monitoring of colonies for health monitoring of breeding, and experimental animals should be undertaken according to Federation for Laboratory Animal Science Associations (FELASA) or similar recommendations [15].

Appropriate housing and dietary requirements need to be carefully considered within the host institution and animal housing facilities:

1. NOD mice are generally bred and maintained in specific pathogen-free (SPF) facilities that are carefully monitored for the presence of mouse pathogens. Thus, breeding pairs and small colonies can be kept in micro-isolators, although, more commonly, mice of this type are housed in individually ventilated cages or scantainers.

2. It is good practice to rotate breeders, and because of the propensity of the mice to develop diabetes, it is advisable to restrict the breeding to 2–3 litters, as the females will not be able to care for the litters if they develop diabetes (which can occur any time after the age of 10–14 weeks as above).
3. A new colony can be established with 2–4 breeding pairs. NOD mice are generally very good breeders, regularly producing litter sizes of eight mice or more. The colony may be maintained by brother-sister mating.
4. Limiting the number of staff members to access to the animal facilities may be useful, and all individuals involved in the care and use of the mice should be educated as to the needs for housing and monitoring experimental and potentially diabetic animals. Dependent on local institutional rules, researchers and caretakers should wear protective equipment, including hair coverings, gloves, scrubs, shoe covers or overalls, and masks, to prevent the spread of pathogens to the mice and exposure to allergens to humans.
5. Food, water, bedding, cages, and other materials that mice will contact should be sterilized or disinfected. Water should be acidified to pH 2.5–3.0 to control *Pseudomonas* species contamination.
6. Mice should be assessed for diabetes regularly, depending on the experimental conditions and strain (*see* **Note 1**). NOD female mice may develop diabetes at 10–14 weeks old. Wet bedding should be changed immediately.
7. Sentinel cages can be set up within each isolator and scantainer to monitor for infection. Fecal pellets may be collected and screened by PCR for infection by pinworm, for example.

### 3.2 Monitoring for Diabetes

Regular monitoring of mice for diabetes is important to maintain colony incidence and highlight any potential contamination that may interfere with studies. Monitoring the welfare of the mice is also critical to prevent undue stress to diabetic animals:

1. Animals are tested daily to weekly, depending upon experimental conditions (*see* **Note 1**) by urinalysis using Diastix test strips, in the first instance. A small drop of urine is placed on the test strip.
2. Following manufacturer's directions, the test strip is assessed for any color change and compared to the scale provided.
3. If there is no color change, the mouse is likely to be nondiabetic.
4. If glycosuria is noted by a color change on the urine test, indicating positive urinalysis, the mouse should be retested 24 h later. If a positive urinalysis is confirmed, then blood glucose is measured.

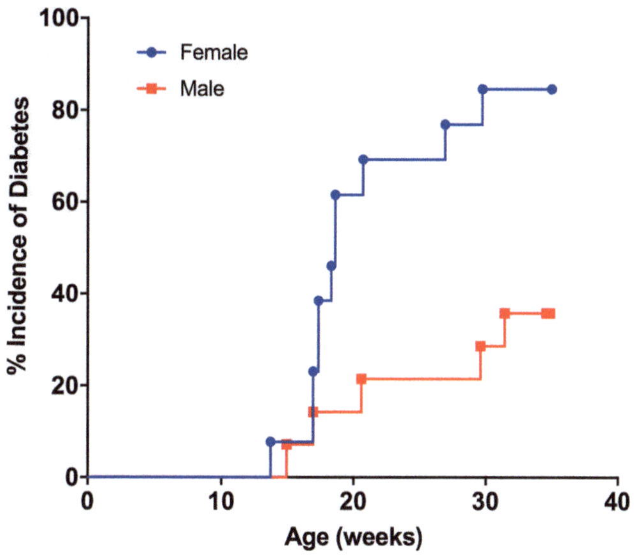


**Fig. 1** Incidence of spontaneous diabetes in NOD mice. NOD mice were observed over a period of 35 weeks and tested weekly for glycosuria with diabetes confirmed by a blood glucose of >13.9 mmol/l (250 mg/dl)

5. Positive diabetes diagnosis is determined by blood glucose measurement:
   (a) Spray tail with anesthetic spray if needed.
   (b) Using a clean scalpel, the tip of the mouse tail is cut.
   (c) A small drop of blood is placed on the test strip mounted in a glucometer; set up according to manufacturer's directions (*see* **Note 2**).
   (d) Diabetes is generally diagnosed with a blood glucose over 13.9 mmol/L (250 mg/dL).
6. Mouse gender and age are recorded to establish an incidence curve (Fig. 1) and should be monitored regularly to assess maintenance of a high incidence of diabetes. A decrease in the incidence of diabetes may be an early warning of infection.
7. Diabetic mice can be used for further experiments or culled by approved methods (*see* **Note 3**).

## 4 Notes

1. NOD mice monitored for development of spontaneous diabetes are checked weekly. Mice receiving immunization or adoptive transfer of diabetogenic cells should be monitored daily, as onset may be rapid (within days).

2. Blood is drawn up by capillary action. Ensure there is enough blood to fill the test strip; otherwise, an error will occur, and test will need to be repeated. Carefully place the blood in the area where the test strip indicated, and avoid blood spill over the area, which will also lead to error readings.

3. Approximately 70–90% of females and 30–40% of males may develop diabetes by 30 weeks, dependent on the individual colony; in the example shown, median female incidence was 19 weeks (Fig. 1).

## References


1. Kikutani H, Makino S (1992) The murine autoimmune diabetes model: NOD and related strains. Adv Immunol 51:285–322
2. Makino S, Kunimoto K, Muraoka Y, Mizushima Y, Katagiri K, Tochino Y (1980) Breeding of a non-obese, diabetic strain of mice. Jikken Dobutsu 29(1):1–13
3. Anderson MS, Bluestone JA (2005) The NOD mouse: a model of immune dysregulation. Annu Rev Immunol 23:447–485. https://doi.org/10.1146/annurev.immunol.23.021704.115643
4. Pearson JA, Wong FS, Wen L (2016) The importance of the non obese diabetic (NOD) mouse model in autoimmune diabetes. J Autoimmun 66:76–88. https://doi.org/10.1016/j.jaut.2015.08.019
5. Steward CA, Gonzalez JM, Trevanion S, Sheppard D, Kerry G, Gilbert JG, Wicker LS, Rogers J, Harrow JL (2013) The non-obese diabetic mouse sequence, annotation and variation resource: an aid for investigating type 1 diabetes. Database (Oxford) 2013:bat032. https://doi.org/10.1093/database/bat032
6. Grattan M, Mi QS, Meagher C, Delovitch TL (2002) Congenic mapping of the diabetogenic locus Idd4 to a 5.2-cM region of chromosome 11 in NOD mice: identification of two potential candidate subloci. Diabetes 51 (1):215–223
7. Pozzilli P, Signore A, Williams AJ, Beales PE (1993) NOD mouse colonies around the world – recent facts and figures. Immunol Today 14(5):193–196. https://doi.org/10.1016/0167-5699(93)90160-M
8. Schmid S, Koczwara K, Schwinghammer S, Lampasona V, Ziegler AG, Bonifacio E (2004) Delayed exposure to wheat and barley proteins reduces diabetes incidence in non-obese diabetic mice. Clin Immunol 111 (1):108–118. https://doi.org/10.1016/j.clim.2003.09.012
9. Maurano F, Mazzarella G, Luongo D, Stefanile R, D'Arienzo R, Rossi M, Auricchio S, Troncone R (2005) Small intestinal enteropathy in non-obese diabetic mice fed a diet containing wheat. Diabetologia 48 (5):931–937. https://doi.org/10.1007/s00125-005-1718-2
10. Cooke A, Tonks P, Jones FM, O'Shea H, Hutchings P, Fulford AJ, Dunne DW (1999) Infection with Schistosoma mansoni prevents insulin dependent diabetes mellitus in non-obese diabetic mice. Parasite Immunol 21(4):169–176
11. Zaccone P, Raine T, Sidobre S, Kronenberg M, Mastroeni P, Cooke A (2004) Salmonella typhimurium infection halts development of type 1 diabetes in NOD mice. Eur J Immunol 34(11):3246–3256. https://doi.org/10.1002/eji.200425285
12. Drescher KM, Kono K, Bopegamage S, Carson SD, Tracy S (2004) Coxsackievirus B3 infection and type 1 diabetes development in NOD mice: insulitis determines susceptibility of pancreatic islets to virus infection. Virology 329 (2):381–394. https://doi.org/10.1016/j.virol.2004.06.049
13. Gale EA (2002) A missing link in the hygiene hypothesis? Diabetologia 45(4):588–594. https://doi.org/10.1007/s00125-002-0801-1
14. Wilberz S, Partke HJ, Dagnaes-Hansen F, Herberg L (1991) Persistent MHV (mouse hepatitis virus) infection reduces the incidence of diabetes mellitus in non-obese diabetic mice. Diabetologia 34(1):2–5
15. FELASA, Mahler Convenor M, Berard M, Feinstein R, Gallagher A, Illgen-Wilcke B, Pritchett-Corning K, Raspa M (2014) FELASA recommendations for the health monitoring of mouse, rat, hamster, guinea pig and rabbit colonies in breeding and experimental units. Lab Anim 48(3):178–192. https://doi.org/10.1177/0023677213516312

# Chapter 7

# Mouse Models of Virus-Induced Type 1 Diabetes

**Gustaf Christoffersson and Malin Flodström-Tullberg**


## Abstract

Virus infections have been linked to the induction of autoimmunity and disease development in human type 1 diabetes. Experimental models have been instrumental in deciphering processes leading to break of immunological tolerance and type 1 diabetes development. Animal models have also been useful for proof-of-concept studies and for preclinical testing of new therapeutic interventions. This chapter describes two robust and clinically relevant mouse models for virus-induced type 1 diabetes; acceleration of disease onset in prediabetic nonobese diabetic (NOD) mice following Coxsackievirus infection and diabetes induction by lymphocytic choriomeningitis virus (LCMV) infection of transgenic mice expressing viral neo-antigens under control of the rat insulin promoter (RIP).


**Key words** Coxsackievirus, Enterovirus, Diabetes, Immunization, Infection, Lymphocytic choriomeningitis virus, Peptides, T lymphocytes, Type 1 diabetes, Virus-induced type 1 diabetes

## 1 Introduction

Type 1 diabetes results from a selective destruction of the insulin-producing pancreatic beta cells and requires daily insulin injections for survival. The disease is one of the most common chronic diseases in children but can also develop in adulthood. Although the disease processes in human type 1 diabetes remain incompletely known, many observations point to a critical role played by autoreactive CD4+ and CD8+ T lymphocytes in causing beta cell destruction. The risk of developing type 1 diabetes is modulated by both genetic and environmental factors. Among the environmental factors proposed to contribute to disease development, infections with common viruses (e.g., enteroviruses, EV) have been hypothesized to play an important role either in inducing or accelerating the autoimmune reaction to the pancreatic beta cell [1–4].

Experimental animal models have proven useful in studies on the etiology and pathogenesis of type 1 diabetes [5–7]. This is particularly true for virus-induced type 1 diabetes, for which

Aileen J. F. King (ed.), *Animal Models of Diabetes: Methods and Protocols*, Methods in Molecular Biology, vol. 2128,
https://doi.org/10.1007/978-1-0716-0385-7_7, 

models have assisted in the identification of several potential scenarios for how a viral infection can either contribute to or prevent disease development (e.g., [8–13]). These type of models have also been employed for proof-of-concept studies and preclinical assessment of novel therapeutics (e.g., [10, 14–17]). Two rat models for virus-induced diabetes are discussed in a separate chapter in this book. In this chapter, we describe murine models of type 1 diabetes induced by virus infection: one model of spontaneous disease development in which an EV infection can alter the kinetics of disease development and another model which is based on genetically modified mice that are susceptible to diabetes induced by lymphocytic choriomeningitis virus (LCMV).

The nonobese diabetes (NOD) mouse model is a commonly used animal model for the study of type 1 diabetes (reviewed in refs. 5–7). Starting around 10–14 weeks of age, female NOD mice develop severe and chronic hyperglycemia with a diabetes incidence reaching 65–95% by 30 weeks of age. There are numerous similarities between type 1 diabetes in humans and the disease in NOD mice including the appearance of islet autoantibodies and recruitment of immune cells (primarily lymphocytes) to the pancreas in the prediabetic stage and hyperglycemia and glucosuria at the time of disease onset. A difference is that male NOD mice have a lower susceptibility to disease development than female NOD mice and seldom reach an incidence above 40% by the age of 30 weeks. In humans, there is no gender difference in the risk for disease development besides a slightly higher risk for adolescent males to develop the disease as compared to age-matched females (e.g., [18]).

Diabetes development in the NOD mouse is preceded by a symptom-free, prediabetic phase that starts around 3–4 weeks of age. At this age, immune cells (e.g., dendritic cells and T cells) begin to infiltrate the pancreas and surround the islets of Langerhans (commonly denoted "peri-insulitis"). Despite steadily increasing in numbers, T cells do not attack the beta cells until weeks later. Once the T cell attack commences, beta cells are rapidly destroyed, and when beta cell loss reaches above 70–75%, the animals develop severe hyperglycemia [5–7].

Coxsackie B viruses (CVBs) belong to the EV genus and consist of six serotypes (CVB1–6). Infections with CVBs have been linked to the induction of islet autoimmunity and type 1 diabetes in humans [2–4]. Early studies by Serreze et al. demonstrated that a CVB4 infection in prediabetic NOD mice led to faster disease development [13]. The majority of the infected animals developed diabetes within two weeks of viral challenge. Later studies have shown a similar acceleration of disease onset if female NOD mice are infected with other serotypes of CVBs, including CVB1 [16] or CVB3 [19]. A requirement for the faster disease onset appears to be the presence of a critical mass of autoreactive T cells within the islets of Langerhans [13, 20], which may become activated by so-called bystander activation [9]. A faster disease progression has also been observed among autoantibody-positive humans who encountered an enterovirus infection prior to clinical disease onset, as compared to autoantibody-positive individuals who did not (the DAISY study) [21]. Although the study was small and should be repeated in a larger cohort of autoantibody-positive cases, the results increased the validity of the CVB-induced accelerated disease model in NOD mice as a model of the human disease.

Transgenic mice expressing a viral neo-antigen under control of the rat insulin promoter (RIP) have for the past three decades also provided a valuable model for understanding the pathogenesis of type 1 diabetes [11, 12, 22–24]. These mice express the glycoprotein (GP) or nucleoprotein (NP) of the LCMV on their β-cells. This enables the induction of autoimmune diabetes by activation of the immune system by either infection with LCMV or activation by immunization with the corresponding antigen.

There are two strains of RIP-GP mice: the strain commonly referred to as RIP-GP Berlin expresses the glycoprotein from the LCMV strain WE [11], and the strain commonly referred to as RIP-GP Armstrong expresses the glycoprotein from the LCMV strain Armstrong [12]. These strains develop diabetes with different kinetics and incidence following LCMV infection due to differences in amount of antigen expressed on the β-cells. The Berlin strain reaches 100% incidence in 7–8 days, whereas the Armstrong strain reaches 75–80% incidence in 10–12 days [25]. The RIP-GP Armstrong strain is also less susceptible to adoptive transfer/peptide immunization-induced diabetes where only about 10% of the mice have sustained hyperglycemia. The Berlin strain is on the other hand more susceptible to this protocol and develops diabetes at close to 100% incidence in 8–12 days [25].

In the RIP-NP model, the NP antigen is also expressed in the thymus, why deletion of high-affinity anti-self $CD8^+$ T lymphocytes occurs, resulting in a slow-onset type 1 diabetes (1–6 months) following LCMV infection. Onset of diabetes in this model is dependent on both $CD4^+$ and $CD8^+$ T lymphocytes, whereas the $CD4^+$ T lymphocytes are dispensable in the RIP-GP model [26].

In this chapter, we describe the basic method for CVB-induced accelerated disease onset in the NOD mouse model. Furthermore, we provide protocols for how to induce diabetes in transgenic mice expressing LCVM antigens under the RIP promoter. Additionally, the induction of diabetes in RIP-LCMV transgenic mice by adoptive transfer and peptide immunization is described. Variation in susceptibility to infection and disease development as well as other detailed technical aspects such as the need for an animal biosafety level 2 (ABSL2) facility is discussed in Subheading 4 (Notes).

## 2 Materials

### 2.1 Induction of Diabetes by CVB Infection

1. Female NOD mice aged 8–12 weeks (e.g., from the Jackson lab or Taconic Biosciences).
2. Coxsackie B virus (ATCC).
3. Plain media (RPMI or similar).
4. 1 mL syringes
5. 25 gauge needle
6. Blood glucose measurement equipment.
7. Surgical equipment (scissors and forceps).

### 2.2 Induction of Diabetes by LCMV Infection

1. Transgenic mice, RIP-GP Armstrong or RIP-NP (Jackson Laboratories), RIP-GP Berlin [11].
2. Virus should be prepared beforehand by a single passage of LCMV Armstrong clone 53b [12] on BHK-21 cells. Determine virus titer through plaque assay, and freeze aliquots at −80 °C.
3. Sterile phosphate-buffered saline (PBS).
4. Insulin syringes (31G).
5. Blood glucose measurement equipment.

### 2.3 Diabetes Induction in RIP-GP Mice Through Adoptive Transfer/Peptide Immunization (Nonviral)

#### 2.3.1 Preparation of P14 Splenocytes

1. P14 T cell receptor (TCR) transgenic mice on C57Bl/6 background (Jackson Laboratories) (*see* **Note 1**).
2. $CO_2$ or isoflurane anesthetic equipment for euthanasia of mice.
3. Surgical equipment (scissors and forceps).
4. Sterile Hank's buffered saline solution (HBSS) without $Ca^{2+}$ and $Mg^{2+}$.
5. Sterile Hank's buffered saline solution (HBSS) with $Ca^{2+}$ and $Mg^{2+}$ and 2% fetal calf serum (FCS).
6. 1 mL sterile syringe
7. Sterile plastic Petri dish.
8. Sterile cell strainers (70 μm and 100 μm pores).
9. 15 and 50 mL sterile conical plastic tubes
10. Ammonium-chloride-potassium (ACK) lysing buffer.
11. Refrigerated centrifuge.

#### 2.3.2 Isolation of P14 $CD8^+$ T Cells

1. P14 TCR transgenic mice on C57Bl/6 background (Jackson Laboratories) (*see* **Note 1**).
2. $CO_2$ or isoflurane anesthetic equipment for euthanasia of mice.
3. Surgical equipment (scissors and forceps).

4. Sterile Hank's buffered saline solution (HBSS) without $Ca^{2+}$ and $Mg^{2+}$.
5. Sterile Hank's buffered saline solution (HBSS) with $Ca^{2+}$ and $Mg^{2+}$ and 2% FCS.
6. 1 mL sterile syringe
7. Sterile cell strainers (70 μm and 100 μm pores).
8. 15 and 50 mL sterile conical plastic tubes
9. Ammonium-chloride-potassium (ACK) lysing buffer.
10. Refrigerated centrifuge.
11. Kit for isolating $CD8^{+}$ T cells (e.g., MACS $CD8a^{+}$ T cell isolation kit, Miltenyi Biotec).
12. Buffer(s) associated with isolation kit of choice.

*2.3.3 Peptide Immunization with LCMV-GP*

1. RIP-GP Berlin [11] or RIP-GP Armstrong (Jackson Laboratories) transgenic mice.
2. P14 TCR transgenic splenocytes (P14 TCR transgenic mice available from Jackson Laboratories) (*see* **Note 1**).
3. Synthetic LCMV-GP peptide ($GP_{33–41}$, NH2-KAVYNFATM-COOH, Genscript) (*see* **Note 2**).
4. Adjuvant CpG (IDT).
5. Sterile ultrapure water.
6. Sterile phosphate-buffered saline (PBS).
7. Polyinosinic:polycytidylic acid (poly(I:C), Sigma).
8. Insulin syringes (31G).
9. Isoflurane anesthetic equipment (if required by institutional animal care and use committee).
10. Blood glucose measurement equipment.

## 3 Methods

### *3.1 Acceleration of Diabetes Development in NOD Mice by Coxsackie B Virus Infection (See Note 3)*

1. Thaw CVB stock on ice (*see* **Note 4**). Vortex the tube to mix the tube contents.
2. Dilute the virus in plain media to desired concentration (*see* **Note 5**).
3. Bring syringes and glucose measurement equipment to the animal facility (*see* **Note 6**).
4. Once at the animal BSL2 unit, work in a laminar flow hood to prevent contamination.
5. Measure weight and blood glucose levels of all mice to be infected with CVB or infected in mock.

6. Inject 150 μL virus (e.g., $10^4$ PFU) intraperitoneally in each mouse (*see* **Note 7**). Mock (media alone) infected animals serve as control.
7. Return mice to their cages.
8. Properly dispose of or decontaminate any material that has been in contact with virus.
9. Monitor animal health including weights daily for the first 7 days, every second day for day 7–14 p.i., and after 14 days weekly unless more frequently required due to local ethical guidelines. Blood glucose measurements are recommended biweekly during the two first weeks and after that on a weekly basis. If blood glucose is high (>13.8 mmol/l or >250 mg/dl), measure blood glucose also on the next day, and if it is still high, sacrifice mice and collect pancreata for histological evaluation (*see* **Note 8**).

### 3.2 Induction of Diabetes by LCMV Infection

There are two basic approaches for induction of diabetes in the RIP-GP/NP mice: infection with LCMV or (for RIP-GP mice only) transfer of GP-specific P14 $CD8^+$ T cells followed by a peptide immunization protocol. P14 $CD8^+$ T cells are specific for the $GP_{33–41}$ antigen presented on MHC class I molecules H-$2D^b$ [27]. As presented in the introduction, the RIP-GP Armstrong strain is not very susceptible to the nonviral induction protocol. Additionally, we have recently further developed the immunization protocol, decreasing the amount of $GP_{33–41}$ peptide administered, and using isolated $CD8^+$ P14 T cells [28], why two different approaches will be mentioned in the methods section below. The different induction protocols are also shown schematically in Fig. 1:

1. Bring out LCMV viral stock aliquots from freezer. Keep on dry ice as long as possible.
2. Bring PBS, insulin syringes, pipette and pipette tips, and glucose measurement equipment to the animal facility (*see* **Note 6**).
3. Once at the ABSL2 unit, work in a laminar flow hood to prevent contamination.
4. Measure and note the blood glucose values of all mice to be infected.
5. Add PBS to the thawed viral stock to a dilution of $10^5$ plaque-forming units (PFU)/mL.
6. Inject 100 μL virus (i.e., $10^4$ PFU) intraperitoneally in each mouse.
7. Return mice to their cages.
8. Properly dispose of or decontaminate any material that has been in contact with virus.

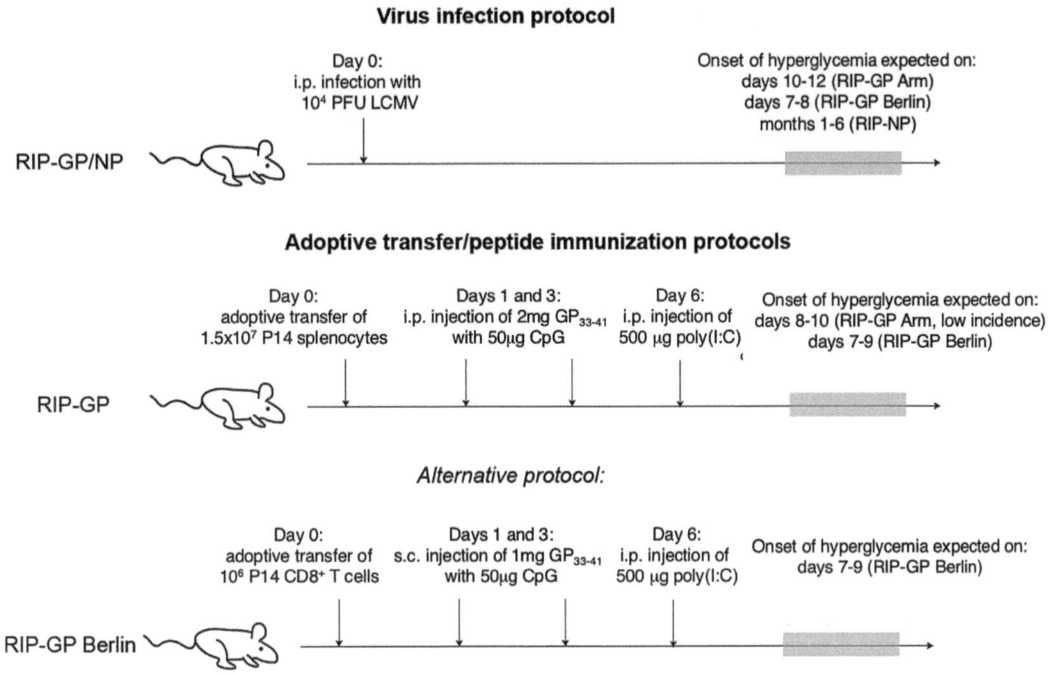


**Fig. 1** Schematic illustrations of diabetes induction protocols in LCMV-RIP-GP/NP mice. The figure outlines how diabetes can be induced in LCMV-RIP-GP/NP mice using LCMV or via the transfer of splenocytes/T cells specific for LCMV antigen followed by peptide immunization and stimulation with a viral mimic (poly(I:C)). The time to diabetes development is indicated. *Arm* Armstrong strain of LCMV; *Berlin* Berlin strain of LCMV; *GP* glycoprotein of LCMV; *i.p.* intraperitoneal; *NP* nucleoprotein of LCMV; *PFU* plaque-forming units; *poly(I:C)* polyinosinic:polycytidylic acid; *RIP* rat insulin promoter; *s.c.* subcutaneous

9. For the RIP-GP Berlin strain, subsequent daily blood glucose measurements need to commence at 4 days post-infection (first animals diabetic ~day 5–6). For the RIP-GP Armstrong strain, daily blood glucose measurements can commence at 6 days post-infection (first animals diabetic ~day 7–8). In slow-onset RIP-NP mice, blood glucose levels need to be followed at a frequency suitable for the experiment protocol and local animal ethics guidelines (*see* **Notes 8** and **9**).

### 3.3 Diabetes Induction in RIP-GP Mice Through Adoptive Transfer/Peptide Immunization (Nonviral)

#### 3.3.1 Preparation of P14 Splenocytes

1. Humanely euthanize P14 mice according to experiment protocol and local ethics guidelines.
2. Use scissors to make an incision in the left flank of the mouse to expose the spleen.
3. Remove the spleen and place in a tube containing 5 mL HBSS without $Ca^{2+}$ and $Mg^{2+}$ and place on ice.
4. Work in a sterile environment; place a cell strainer in a Petri dish.
5. Pour the spleen and HBSS from the tube onto the cell strainer.

6. Use the plunger from a 1 mL sterile syringe to mince the spleen against the strainer. Mince until only the fibrous capsule of the spleen remains.
7. Take 5 mL of fresh, cold HBSS without $Ca^{2+}$ and $Mg^{2+}$ and wash the cell strainer so that remaining cells end up in the Petri dish.
8. Discard the cell strainer.
9. Transfer the cell suspension (~10 mL) to a 15 mL plastic conical tube and centrifuge at ~400 × *g*, 6 min, 4 °C.
10. Discard the supernatant.
11. Add 2 mL ACK lysis buffer to the tube and resuspend the pellet. Allow to sit at room temperature for 2 min.
12. Fill up the tube with HBSS (with $Ca^{2+}$ and $Mg^{2+}$) containing 2% FBS.
13. Centrifuge at ~400 × *g*, 6 min, 4 °C.
14. Discard the supernatant.
15. Resuspend in 5 mL HBSS (with $Ca^{2+}$ and $Mg^{2+}$) containing 2% FBS.
16. Pour the cell suspension over a cell strainer placed on top of a 50 mL conical tube. Rinse the 15 mL tube with an additional 5 mL of HBSS (with $Ca^{2+}$ and $Mg^{2+}$) containing 2% FBS and pour this over the cell strainer to rinse off any cells that may remain in the tube and strainer.
17. Perform cell count.
18. Centrifuge at ~400 × *g*, 6 min, 4 °C.
19. Resuspend cells in sterile PBS at desired dilution. Keep on ice.

#### *3.3.2 Isolation of P14 $CD8^+$ T Cells (Alternative Protocol)*

1. Humanely euthanize P14 mice according to experiment protocol and local ethics guidelines.
2. Use scissors to make an incision in the left flank of the mouse to expose the spleen.
3. Remove the spleen and place in a tube containing HBSS and put on ice.
4. Work in a sterile environment; place a cell strainer in a Petri dish.
5. Pour the spleen and HBSS from the tube onto the cell strainer.
6. Use the plunger from a 1 mL sterile syringe to mince the spleen against the strainer. Mince until only the fibrous capsule of the spleen remains.
7. Take 5 mL of fresh, cold HBSS and wash the cell strainer so that remaining cells end up in the Petri dish.
8. Discard the cell strainer.

9. Transfer the cell suspension (~10 mL) to a 15 mL plastic conical tube and centrifuge at ~400 × *g*, 6 min, 4 °C.
10. Discard the supernatant.
11. Add 2 mL ACK lysis buffer to the tube and resuspend the pellet. Allow to sit at room temperature for 2 min.
12. Fill up the tube with HBSS (with $Ca^{2+}$ and $Mg^{2+}$) containing 2% FBS.
13. Centrifuge at ~400 × *g*, 6 min, 4 °C.
14. Discard the supernatant.
15. Resuspend in 5 mL HBSS (with $Ca^{2+}$ and $Mg^{2+}$) containing 2% FBS.
16. Pour the cell suspension over a cell strainer placed on top of a 50 mL conical tube. Rinse the 15 mL tube with an additional 5 mL of HBSS (with $Ca^{2+}$ and $Mg^{2+}$) containing 2% FBS and pour this over the cell strainer to rinse off any cells that may remain in the tube or strainer.
17. Perform cell count.
18. Centrifuge at ~400 × *g*, 6 min, 4 °C.
19. Resuspend cells in a buffer and dilution stated by your $CD8^+$ T cell isolation protocol.
20. Perform $CD8^+$ T cell isolation.
21. Perform cell count.
22. Centrifuge at ~400 × *g*, 6 min, 4 °C.
23. Resuspend isolated P14 $CD8^+$ T cells in sterile PBS at desired dilution. Keep on ice.

#### *3.3.3 Induction of Diabetes*

1. On day 0, inject $1.5 \times 10^7$ P14 splenocytes intravenously in each RIP-GP mouse.
2. The following day (day 1), inject 2 mg $GP_{33-41}$ peptide in ultrapure water containing 50 μg of CpG adjuvant intraperitoneally.
3. On day 3, inject 2 mg $GP_{33-41}$ peptide in ultrapure water containing 50 μg of CpG adjuvant intraperitoneally.
4. On day 6 after P14 transfer, inject 500 μg poly(I:C) intraperitoneally.
5. On following days, monitor blood glucose levels. If blood glucose is high (>13.8 mmol/L or >250 mg/dL), measure blood glucose also on next day and, if it is still high, sacrifice mice and collect pancreata for histological evaluation (*see* **Notes 8–12**).

*3.3.4 Induction of Diabetes (Alternative Protocol)*

1. On day 0, inject $1 \times 10^6$ P14 $CD8^+$ T cells intravenously in each RIP-GP mouse.
2. The following day (day 1), inject 1 mg $GP_{33-41}$ peptide in ultrapure water containing 50 μg of CpG adjuvant subcutaneously (over the interscapular space).
3. On day 3, inject 1 mg $GP_{33-41}$ peptide in ultrapure water containing 50 μg of CpG adjuvant subcutaneously (over the interscapular space).
4. On day 6 after P14 $CD8^+$ T cell transfer, inject 500 μg poly(I: C) intraperitoneally.
5. On following days, monitor blood glucose levels. If blood glucose is high (>13.8 mmol/L or >250 mg/dL), measure blood glucose also on next day and, if it is still high, sacrifice mice and collect pancreata for histological evaluation (*see* **Notes 8–12**).

## 4 Notes

1. In intravital imaging experiments, transferred P14 cells can be taken from mice expressing fluorescent reporters (e.g., GFP, dsRed) [29].
2. Peptide should be sterile filtered following resuspension prior to storage. PBS, syringes, and needles must be sterile as well.
3. The protocols described here have been developed by several laboratories, and some smaller variations in the protocols may exist between different laboratories. It is instrumental for each laboratory to optimize the protocols for its environment and material.
4. CVB stocks can be prepared in, for example, HeLa or Vero cells [30]. It is recommended that virus is extracted by repeated freeze-thawing followed by purification using a sucrose gradient and ultracentrifugation [31]. Titrations of virus stocks can be performed either by the TCID50 method or by standard plaque assay for EVs on HeLa cells [30].
5. At high infectious doses, a CVB infection is lethal. The lethal dose varies depending on virus stock, mouse strain, and animal facility. It is recommended that pilot experiments are conducted to identify the dose of CVB that induces systemic virus infection without causing significant weight loss and/or death. Systemic virus spread can be monitored, for example, by a histological assessment of the pancreas after termination of the experiment. If infected, the acinar cells of the exocrine part of the pancreas will be destroyed over time [8, 10, 17, 32, 33]. A small blood sample (e.g., 20 μL, diluted 1:1 in 12 mmol

EDTA) can be retrieved on day 3 or 4 p.i. for viremia measurements using standard plaque assay [30].

6. Since both CVB and LCMV are classified as BSL2 agents, mice need to be housed in an ABSL2 facility following infection with CVB or LCMV. All samples from infected mice need to be handled with regard to biohazard regulations or be treated to inactivate the virus (e.g., by paraformaldehyde, bleach or similar). Note that EtOH is not effective in inactivating CVB.
7. The age at which prediabetic NOD mice will develop diabetes with a faster kinetic after CVB infection may vary between laboratories. It is advised that animals are infected around 1–2 weeks prior to the age at which the first female NOD mice in the colony spontaneously develop diabetes.
8. Commonly used protocols deem animals diabetic when presenting with a blood glucose level >13.8 mmol/l or >250 mg/l on two consecutive days. Local animal ethics guidelines may however vary and should be applied when required.
9. Islet-specific T cells can be tracked in blood, tissue homogenates, or tissue sections using GP-specific tetramers.
10. If the adoptive transfer protocol is used, P14 T cells can be tracked when using donors and recipients with different congenic markers such as Ly5.1/5.2.
11. Incidence of diabetes can be increased in the RIP-GP Armstrong strain when using the adoptive transfer/peptide immunization protocol. By introducing $CD4^+$ T cell help by transfer of $1.5 \times 10^7$ in vitro-activated Smarta TCR transgenic splenocytes on day 4 following P14 splenocyte transfer, the incidence rises to ~60% from 0% to 10% [25]. Smarta TCR transgenic $CD4^+$ T cells are specific for the LCMV-$GP_{64-80}$ peptide presented on MHC class II molecules I-$A^b$ [34].
12. The effect of regulatory T cells (Tregs) can be studied in the RIP-GP Berlin model using the adoptive transfer/peptide immunization protocol. Transfer of $1 \times 10^6$ in vitro-generated Smarta $CD4^+$ Tregs on day 8 following P14 splenocyte transfer reduces incidence to ~20% from ~100% [25].

## References


1. DiMeglio L, Evans-Molina C, Oram R (2018) Type 1 diabetes. Lancet 16(391):2449–2462
2. Hyoty H (2016) Viruses in type 1 diabetes. Pediatr Diabetes 17(Suppl 22):56–64. https://doi.org/10.1111/pedi.12370
3. Rewers M, Ludvigsson J (2016) Environmental risk factors for type 1 diabetes. Lancet 387 (10035):2340–2348. https://doi.org/10.1016/S0140-6736(16)30507-4
4. Yeung WC, Rawlinson WD, Craig ME (2011) Enterovirus infection and type 1 diabetes mellitus: systematic review and meta-analysis of observational molecular studies. BMJ 342: d35. https://doi.org/10.1136/bmj.d35

5. Lenzen S (2017) Animal models of human type 1 diabetes for evaluating combination therapies and successful translation to the patient with type 1 diabetes. Diabetes Metab Res Rev 33(7). https://doi.org/10.1002/dmrr.2915
6. Mullen Y (2017) Development of the nonobese diabetic mouse and contribution of animal models for understanding type 1 diabetes. Pancreas 46(4):455–466. https://doi.org/10.1097/MPA.0000000000000828
7. You S, Chatenoud L (2016) Autoimmune diabetes: an overview of experimental models and novel therapeutics. Methods Mol Biol 1371:117–142. https://doi.org/10.1007/978-1-4939-3139-2_8
8. Flodstrom M, Maday A, Balakrishna D, Cleary MM, Yoshimura A, Sarvetnick N (2002) Target cell defense prevents the development of diabetes after viral infection. Nat Immunol 3 (4):373–382
9. Horwitz MS, Bradley LM, Harbertson J, Krahl T, Lee J, Sarvetnick N (1998) Diabetes induced by Coxsackie virus: initiation by bystander damage and not molecular mimicry. Nat Med 4(7):781–785
10. Larsson PG, Lakshmikanth T, Svedin E, King C, Flodstrom-Tullberg M (2013) Previous maternal infection protects offspring from enterovirus infection and prevents experimental diabetes development in mice. Diabetologia 56(4):867–874. https://doi.org/10.1007/s00125-013-2834-z
11. Ohashi PS, Oehen S, Buerki K, Pircher H, Ohashi CT, Odermatt B, Malissen B, Zinkernagel RM, Hengartner H (1991) Ablation of "tolerance" and induction of diabetes by virus infection in viral antigen transgenic mice. Cell 65(2):305–317
12. Oldstone MB, Nerenberg M, Southern P, Price J, Lewicki H (1991) Virus infection triggers insulin-dependent diabetes mellitus in a transgenic model: role of anti-self (virus) immune response. Cell 65(2):319–331
13. Serreze DV, Ottendorfer EW, Ellis TM, Gauntt CJ, Atkinson MA (2000) Acceleration of type 1 diabetes by a coxsackievirus infection requires a preexisting critical mass of autoreactive T-cells in pancreatic islets. Diabetes 49 (5):708–711
14. Hara N, Alkanani AK, Ir D, Robertson CE, Wagner BD, Frank DN, Zipris D (2012) Prevention of virus-induced type 1 diabetes with antibiotic therapy. J Immunol 189 (8):3805–3814. https://doi.org/10.4049/jimmunol.1201257
15. Homann D, Jahreis A, Wolfe T, Hughes A, Coon B, van Stipdonk MJ, Prilliman KR, Schoenberger SP, von Herrath MG (2002) CD40L blockade prevents autoimmune diabetes by induction of bitypic NK/DC regulatory cells. Immunity 16(3):403–415
16. Larsson PG, Lakshmikanth T, Laitinen OH, Utorova R, Jacobson S, Oikarinen M, Domsgen E, Koivunen MR, Chaux P, Devard N, Lecouturier V, Almond J, Knip M, Hyoty H, Flodstrom-Tullberg M (2015) A preclinical study on the efficacy and safety of a new vaccine against Coxsackievirus B1 reveals no risk for accelerated diabetes development in mouse models. Diabetologia 58(2):346–354. https://doi.org/10.1007/s00125-014-3436-0
17. Stone VM, Hankaniemi MM, Svedin E, Sioofy-Khojine A, Oikarinen S, Hyoty H, Laitinen OH, Hytonen VP, Flodstrom-Tullberg M (2018) A Coxsackievirus B vaccine protects against virus-induced diabetes in an experimental mouse model of type 1 diabetes. Diabetologia 61(2):476–481. https://doi.org/10.1007/s00125-017-4492-z
18. Weets I, Van Autreve J, Van der Auwera BJ, Schuit FC, Du Caju MV, Decochez K, De Leeuw IH, Keymeulen B, Mathieu C, Rottiers R, Dorchy H, Quartier E, Gorus FK, Belgian Diabetes R (2001) Male-to-female excess in diabetes diagnosed in early adulthood is not specific for the immune-mediated form nor is it HLA-DQ restricted: possible relation to increased body mass index. Diabetologia 44 (1):40–47
19. Kanno T, Kim K, Kono K, Drescher KM, Chapman NM, Tracy S (2006) Group B coxsackievirus diabetogenic phenotype correlates with replication efficiency. J Virol 80 (11):5637–5643
20. Horwitz MS, Ilic A, Fine C, Balasa B, Sarvetnick N (2004) Coxsackieviral-mediated diabetes: induction requires antigen-presenting cells and is accompanied by phagocytosis of beta cells. Clin Immunol 110(2):134–144
21. Stene LC, Oikarinen S, Hyoty H, Barriga KJ, Norris JM, Klingensmith G, Hutton JC, Erlich HA, Eisenbarth GS, Rewers M (2010) Enterovirus infection and progression from islet autoimmunity to type 1 diabetes: the Diabetes and Autoimmunity Study in the Young (DAISY). Diabetes 59(12):3174–3180
22. Kurts C, Heath WR, Carbone FR, Allison J, Miller JF, Kosaka H (1996) Constitutive class I-restricted exogenous presentation of self antigens in vivo. J Exp Med 184(3):923–930
23. Blanas E, Carbone FR, Allison J, Miller JF, Heath WR (1996) Induction of autoimmune diabetes by oral administration of autoantigen. Science 274(5293):1707–1709
24. Byersdorfer CA, Schweitzer GG, Unanue ER (2005) Diabetes is predicted by the beta cell

level of autoantigen. J Immunol 175 (7):4347–4354
25. Martinic MM, Huber C, Coppieters K, Oldham JE, Gavin AL, von Herrath MG (2010) Expression level of a pancreatic neo-antigen in beta cells determines degree of diabetes pathogenesis. J Autoimmun 35(4):404–413. https://doi.org/10.1016/j.jaut.2010.08.006
26. von Herrath MG, Dockter J, Oldstone MB (1994) How virus induces a rapid or slow onset insulin-dependent diabetes mellitus in a transgenic model. Immunity 1(3):231–242
27. Pircher H, Burki K, Lang R, Hengartner H, Zinkernagel RM (1989) Tolerance induction in double specific T-cell receptor transgenic mice varies with antigen. Nature 342 (6249):559–561. https://doi.org/10.1038/342559a0
28. Christoffersson G, Chodaczek G, Ratliff SS, Coppieters K, von Herrath MG (2018) Suppression of diabetes by accumulation of non--islet-specific CD8+ effector T cells in pancreatic islets. Sci Immunol 3(21):eaam6533
29. Christoffersson G, von Herrath MG (2016) A deeper look into type 1 diabetes - imaging immune responses during onset of disease. Front Immunol 7:313. https://doi.org/10.3389/fimmu.2016.00313
30. Burrill CP, Strings VR, Andino R (2013) Poliovirus: generation, quantification, propagation, purification, and storage. Curr Protoc Microbiol Chapter 15:Unit 15H 11. https://doi.org/10.1002/9780471729259.mc15h01s29
31. Hankaniemi MM, Laitinen OH, Stone VM, Sioofy-Khojine A, Maatta JAE, Larsson PG, Marjomaki V, Hyoty H, Flodstrom-Tullberg-M, Hytonen VP (2017) Optimized production and purification of Coxsackievirus B1 vaccine and its preclinical evaluation in a mouse model. Vaccine 35(30):3718–3725. https://doi.org/10.1016/j.vaccine.2017.05.057
32. Flodstrom M, Horwitz MS, Maday A, Balakrishna D, Rodriguez E, Sarvetnick N (2001) A critical role for inducible nitric oxide synthase in host survival following coxsackievirus B4 infection. Virology 281(2):205–215
33. Tracy S, Drescher KM (2007) Coxsackievirus infections and NOD mice: relevant models of protection from, and induction of, type 1 diabetes. Ann N Y Acad Sci 1103:143–151
34. Oxenius A, Bachmann MF, Zinkernagel RM, Hengartner H (1998) Virus-specific MHC-class II-restricted TCR-transgenic mice: effects on humoral and cellular immune responses after viral infection. Eur J Immunol 28(1):390–400

# Chapter 8

# Rat Models of Virus-Induced Type 1 Diabetes

**James C. Needell and Danny Zipris**

## Abstract

Studies performed in humans and animal models have implicated the environment in the etiology of type 1 diabetes (T1D), but the nature and timing of the interactions triggering β cell autoimmunity are poorly understood. Virus infections have been postulated to be involved in disease mechanisms, but the underlying mechanisms are not known. It is exceedingly difficult to establish a cause-and-effect relationship between viral infection and diabetes in humans. Thus, we have used the BioBreeding Diabetes-Resistant (BBDR) and the LEW1.WR1 rat models of virus-induced disease to elucidate how virus infection leads to T1D. The immunophenotype of these strains is normal, and spontaneous diabetes does not occur in a specific pathogen-free environment. However, β cell inflammation and diabetes with many similarities to the human disease are induced by infection with the parvovirus Kilham rat virus (KRV). KRV-induced diabetes in the BBDR and LEW1.WR1 rat models is limited to young animals and can be induced in both male and female rats. Thus, these animals provide a powerful experimental tool to identify mechanisms underlying virus-induced T1D development.

**Key words** BioBreeding Diabetes-Resistant (BBDR) rat, Chang human liver cell line, H-1 parvovirus, Kilham rat virus, LEW1.WR1 rat, Normal rat kidney (NRK) cell line

## 1 Introduction

Type 1 diabetes (T1D) is a multistage proinflammatory disorder that results from specific islet beta cell destruction and loss of insulin secretion [1–3]. Although there is compelling evidence that genetic factors determine susceptibility to T1D, many lines of evidence, including the dramatic rise of diabetes incidence in recent decades, indicate an important role for nongenetic factor [4–8]. Indeed, a large body of evidence supports the notion that environmental factors are linked with mechanisms that trigger β cell autoimmunity [9–16]. However, no single environmental factor has as yet been definitively associated with disease pathogenesis [15].

Viruses have been hypothesized to play a role in disease mechanisms; however, the nature of these viruses and the mechanism by which they trigger diabetes are currently unknown [13, 14,

Aileen J. F. King (ed.), *Animal Models of Diabetes: Methods and Protocols*, Methods in Molecular Biology, vol. 2128,
https://doi.org/10.1007/978-1-0716-0385-7_8, 

17–25]. Ethical considerations prohibit us from performing human experimentation that involves infectious agents, and therefore establishing a cause-and-effect relationship between virus infection and diabetes in genetically susceptible individuals and identifying mechanisms of virus-induced disease are extremely difficult. We have therefore used the diabetes-prone BioBreeding Diabetes-Resistant (BBDR) and LEW1.WR1 rat models of Kilham rat virus (KRV)-induced T1D to better understand how virus infection leads to β cell inflammation and destruction.

### 1.1 Kilham Rat Virus and H-1 Parvovirus

KRV and the homologous H-1 parvovirus belong to the Parvoviridae virus group, a family of single-stranded DNA viruses that infect various species [26–29]. Parvoviruses are among the smallest eucaryotic viruses, and they have been isolated from a wide variety of hosts including rodents and humans [26–29]. The virus genomic DNA consists of 5 Kb nucleotides and is encapsidated by protein in an icosahedral nonenveloped particle [26, 28]. Vertebrate parvoviruses are divided into two subgroups based on their ability to replicate autonomously in infected cells. The adeno-associated virus subgroup is defective and requires coinfection with adenovirus for its replication, whereas the autonomous virus subgroup does not require helper viruses for replication, albeit they require the host cell to be in the late S or G2 phase for successful infection [30]. KRV encodes three structural proteins (VP1, VP2, and VP3) and two nonstructural proteins (NS1 and NS2) [31, 32]. The replication of KRV takes place in the nucleus of infected cells, and the genome is not integrated with that of the cell [33]. KRV infects lymphoid organs such as the spleen, thymus, lymph nodes [34], and Peyer's patches [35]. We recently obtained evidence that KRV proteins can be detected in β cells on day 5 post-infection; however, islet destruction and diabetes are observed only after insulitis is detected at 2–4 weeks following infection [36]. KRV shares a high degree of homology with H-1; KRV and H-1 VP proteins are ~80% homologous, and their NS proteins are 100% homologous [27].

### 1.2 BioBreeding Diabetes-Resistant (BBDR) and LEW1.WR1 Rats

There are two inbred strains of BB rats: the diabetes-prone BB (BBDP) rat that develops spontaneous diabetes and the diabetes-resistant BB (BBDR) rat. BBDR rats were derived from BBDP forebears bred for resistance to autoimmunity [37, 38]. The BBDR and LEW1.WR1 rats share the high-risk class II major histocompatibility complex (MHC) allele designated RT1B/Du. In contrast to BBDP rats that develop severe lymphopenia, both the BBDR and LEW1.WR1 rats exhibit a normal immunophenotype [38, 39] and both develop T1D following infection with KRV [40]. Diabetes does not develop spontaneously in these rat strains in a specific pathogen-free environment. Hyperglycemia is typically observed in BBDR and LEW1.WR1 rats infected with $1 \times 10^7$

PFUs of KRV on days 14–28 post-virus inoculation at disease incidences of ~25% and ~60%, respectively [6, 7, 35, 41–44] via mechanisms associated with alterations in the intestinal microbiome [35] and the upregulation of both adaptive [41, 45] and innate immunity [16, 46–48]. Activation of the innate immune system prior to virus infection with Toll-like receptor (TLR) agonists such as the virus mimic polyinosinic:polycytidylic acid (poly (I:C)) considerably exacerbates disease development [35, 44, 46–48]. The animal disease closely resembles the human disorder with respect to histopathology, pathogenesis, lack of sex bias, and MHC class II association [49].

Infecting animals with the KRV homolog H-1 parvovirus leads to virus-specific humoral and cellular immune responses as does KRV in BBDR and LEW1.WR1 rats; however, H-1 induces innate immune activation to a lesser degree than KRV and does not induce T1D and as such has been used as a control in our studies [44]. KRV-induced β cell autoimmunity does not develop in rat strains such Wistar Furth, PVG.R8, and BBDP rats [41–44]. The ability to induce virus-induced T1D is not KRV-specific, since viruses from other viral groups, i.e., enterovirus, poxvirus, and herpesvirus, can also lead to disease onset [50, 51]. Virus-induced diabetes in the BBDR and LEW1.WR1 rat models is MHC-dependent, rat- and strain-specific, and limited to young animals immediately following weaning and can be equally induced in male and female rats [50]. Finally, infection with KRV results in the development of β cell inflammation (insulitis) but not exocrine pancreatitis [47]. Diabetes is diagnosed when a persistent elevation in blood glucose levels is observed.

## 2 Materials

1. KRV stock (ATCC® VR-1790™).
2. Normal rat kidney cell line (NRK, ATCC® CRL-6509™).
3. Chang liver cell line (ATCC® CCL-13™).
4. 150 $cm^2$ tissue culture flasks.
5. 6-well plates.
6. Complete DMEM (cDMEM) *supplemented with* 10% FBS, 100 U/mL penicillin/streptomycin, and 2 mM L-glutamine.
7. 50 μM 2-mercaptoethanol
8. Agarose, UltraPure.
9. Crystal violet.
10. Formaldehyde.
11. Sterile PBS.

12. Water bath.
13. BBDR and LEW1.WR1 rats of either sex at 21–25 days of age.
14. Poly (I:C), sodium salt (Sigma).
15. 1 mL syringes
16. Diastix Reagent Strips for Urinalysis (Bayer).
17. Scalpel blade.
18. Clean gauze.
19. Blood glucose meter (we use the OneTouch Ultra2 meter).

## 3 Methods

### 3.1 KRV Production

1. Plate 150 $cm^2$ tissue culture flasks with NRK cells (ATCC® CRL-6509™) in cDMEM.
2. Incubate NRK cultures at 37 °C for 24 h to ~50% confluency.
3. Remove supernatants, cover cells with 15 mL of 1:10 dilution of KRV stock, and incubate for 1.5 h (*see* **Note 1**).
4. Add 30 mL of cDMEM to culture flasks and incubate for 4 days.
5. Remove medium to 50 mL sterile test tubes and centrifuge at 800 × *g* to remove cell debris.
6. Aliquot and keep at −80 °C until use.

### 3.2 KRV Plaque Assay

1. The day prior to the assay, plate NRK cells in 6-well plates with cDMEM and incubate at 37 ° C.
2. The day after plating, visually check the confluency and viability of the cells prior to starting the assay to ensure standard cellular morphology and a ~50% confluent monolayer is present.
3. Remove media and add 1 mL of fresh cDMEM.
4. Prepare tenfold serial dilutions of KRV.
5. Add 0.1 mL of virus dilutions to each well and gently rock to mix, and incubate for 1.5 h with rocking every 20 min.
6. Prepare an agarose overlay mixture. First, equilibrate water bath to 56 ° C. Determine the total volume of agar solution needed: multiply 4 mL/plate by the number of plaque assays. Of this total volume, one half is autoclaved $dH_2O$ + agarose to make a 2% solution, and the other half is cDMEM.
7. Add the appropriate amount of autoclaved $dH_2O$ into an appropriately sized sterile glass bottle and slowly add the agarose while gently swirling to mix the $dH_2O$ and agarose. Microwave the mixture until just boiling until the agarose has completely dissolved and then transfer to the 56 ° C water bath. Allow mixture to cool to 56 ° C.

8. Place the DMEM medium in a 56 ° C water bath.
9. Add an equal volume of the DMEM to the agarose solution to complete the final agar solution and mix well.
10. Carefully aspirate off media from the cells and slowly add 4 mL of the agar solution to each plate and allow agar to solidify. Ensure the solution is warm, but not hot to the touch to prevent cell death and reduced viral titers (*see* **Note 2**).
11. Incubate cultures at 37 °C.
12. After adding the overlay, incubate plates for 5–6 days to produce distinct plaques. Once liquid is overlaid, plates are incubated at 37 °C (*see* **Note 3**).
13. At the end of the incubation, carefully remove agar plug using a spatula.
14. Add 1 mL of 0.1% crystal violet solution (0.1% w/v crystal violet, 1% formaldehyde in PBS).
15. The solution is allowed to sit for several minutes and then the crystal violet solution is removed.
16. Rinse cells with 1 mL of PBS.
17. Count the plaques in each well, taking the average for any technical replicates of the same dilution. The negative control should have a uniform monolayer and can be used as a reference control.
18. Determine the viral titer of the stock sample by taking the average number of plaques for a dilution and the inverse of the total dilution factor (see formula below):

$$\text{PFU/mL} = \frac{\text{Average number of plaques}}{\text{Dilution} \times \text{Volume of diluted virus volume added}}$$

### 3.3 H-1 Parvovirus Production and Plaque Assay

For H-1 parvovirus production and titration. follow the above protocols using the Chang liver cell line.

### 3.4 KRV and Poly (I:C) plus KRV-Induced Diabetes

1. For KRV inoculation, inject 21- to 25-day-old BBDR or LEW1.WR1 rats intraperitoneally with $1 \times 10^7$ PFUs of KRV in a total volume of 0.5 mL of PBS.
2. For KRV plus poly (I:C) treatment, make a 1 mg/mL solution of poly (I:C) in sterile PBS.
3. Administer 21–25 days of age rats with 3 μg of poly (I:C) per 1 g of body weight in a total volume of 0.5 mL PBS intraperitoneally on 3 consecutive days. On the fourth day, inject the poly (I:C)-treated animals with $1 \times 10^7$ PFUs of KRV in a total volume of 0.5 mL of PBS.

### 3.5 Diabetes Monitoring

1. Beginning 10 days following treatment, test for glycosuria three times weekly for up to 40 days following treatment. For the urine test, rats are manually restrained and the abdomen is gently depressed until the animal urinates (only one drop is needed on the test strip). Apply the indicator end of the glucose strip on the urine and evaluate the color change on the test strip to assess the glucose level (*see* **Note 4**).
2. Once glucosuria is observed, diabetes is confirmed by monitoring glucose levels in a drop of blood by tail nicks. To perform tail nicks, bedding material and feces are removed and the tail is disinfected with 70% ethanol and rinsed with water. Using a new scalpel blade, cut approximately 1–2 mm of the distal tail at an angle perpendicular to the work surface is done. Gentle pressure proximal to the collection site is applied to occlude venous return and ease collection, and a volume of about two blood drops is applied on the glucose strip. Gentle pressure for 30–45 s on the wound using clean gauze is used, to stop any hemorrhaging. Disease is diagnosed on the basis of a plasma glucose >250 mg/dL.

## 4 Notes

1. Periodic rocking to ensure even coverage and prevent the cellular monolayer from drying.
2. Most agarose overlays will begin to solidify below 42 °C; work quickly and/or prepare small batches to prevent solidification while handling.
3. Plates should not be moved as movement of the liquid overlays during the incubation period will result in smeared plaques. Agarose overlays are semisolid and can be moved or checked periodically under a light microscope to monitor plaque development.
4. Urine test strips cannot detect glucose until the blood glucose level is ~180 mg/dL. It is a less invasive method to determine whether the animal has developed hyperglycemia.

### References

1. Schranz DB, Lernmark A (1998) Immunology in diabetes: an update. Diabetes Metab Rev 14:3–29
2. Seissler J, De Sonnaville JJ, Morgenthaler NG et al (1998) Immunological heterogeneity in type I diabetes: presence of distinct autoantibody patterns in patients with acute onset and slowly progressive disease. Diabetologia 41:891–897
3. Srikanta S, Ganda OP, Jackson RA et al (1983) Type I diabetes mellitus in monozygotic twins: chronic progressive beta cell dysfunction. Ann Intern Med 99:320–326
4. Jun HS, Yoon JW (2003) A new look at viruses in type 1 diabetes. Diabetes Metab Res 19:8–31
5. Jun HS, Yoon JW (2001) The role of viruses in Type I diabetes: two distinct cellular and

molecular pathogenic mechanisms of virus-induced diabetes in animals. Diabetologia 44:271–285
6. Knip M, Simell O (2012) Environmental triggers of type 1 diabetes. Cold Spring Harb Perspect Med 2:a007690
7. Knip M, Veijola R, Virtanen SM et al (2005) Environmental triggers and determinants of type 1 diabetes. Diabetes 54:S125–S136
8. Soltesz G, Patterson CC, Dahlquist G (2007) Worldwide childhood type 1 diabetes incidence--what can we learn from epidemiology? Pediatr Diabetes 8(Suppl 6):6–14
9. Filippi CM, Von Herrath MG (2008) Viral trigger for type 1 diabetes: pros and cons. Diabetes 57:2863–2871
10. Forlenza GP, Rewers M (2011) The epidemic of type 1 diabetes: what is it telling us? Curr Opin Endocrinol Diabetes Obes 18:248–251
11. Gale EM (2002) The rise of childhood type 1 diabetes in the 20th century. Diabetes 51:3353–3361
12. Hu Y, Wong FS, Wen L (2017) Antibiotics, gut microbiota, environment in early life and type 1 diabetes. Pharmacol Res 119:219–226
13. Kondrashova A, Hyöty H (2014) Role of viruses and other microbes in the pathogenesis of type 1 diabetes. Int Rev Immunol 33:284–295
14. Op De Beeck A, Eizirik DL (2016) Viral infections in type 1 diabetes mellitus — why the β cells? Nat Rev Endocrinol 12:263
15. Stene LC, Gale EM (2013) The prenatal environment and type 1 diabetes. Diabetologia 56:1888–1897
16. Zipris D (2009) Epidemiology of type 1 diabetes and what animal models teach us about the role of viruses in disease mechanisms. Clin Immunol 131:11–23
17. Drescher KM, Von Herrath M, Tracy S (2015) Enteroviruses, hygiene and type 1 diabetes: toward a preventive vaccine. Rev Med Virol 25:19–32
18. Gamble DR, Taylor KW (1969) Seasonal incidence of diabetes mellitus. Br Med J 3:631–633
19. Ghazarian L, Diana J, Simoni Y et al (2012) Prevention or acceleration of type 1 diabetes by viruses. Cell Mol Life Sci 70:239–255
20. Hober D, Sauter P (2010) Pathogenesis of type 1 diabetes mellitus: interplay between enterovirus and host. Nat Rev Endocrinol 6:279
21. Imagawa A, Hanafusa T (2011) Fulminant type 1 diabetes—an important subtype in East Asia. Diabetes Metab Res Rev 27:959–964
22. Lonnrot M, Korpela K, Knip M et al (2000) Enterovirus infection as a risk factor for beta-cell autoimmunity in a prospectively observed birth cohort: the Finnish Diabetes Prediction and Prevention Study. Diabetes 49:1314–1318
23. Lönnrot M, Lynch KF, Elding Larsson H et al (2017) Respiratory infections are temporally associated with initiation of type 1 diabetes autoimmunity: the TEDDY study. Diabetologia 60:1931–1940
24. Richardson SJ, Morgan NG, Foulis AK (2014) Pancreatic pathology in type 1 diabetes mellitus. Endocr Pathol 25:80–92
25. Schneider DA, Von Herrath MG (2014) Potential viral pathogenic mechanism in human type 1 diabetes. Diabetologia 57:2009–2018
26. Brown DD, Salzman LA (1984) Sequence homology between the structural proteins of Kilham rat virus. J Virol 49:1018–1020
27. Jacoby RO, Ball-Goodrich LJ, Besselsen DG et al (1996) Rodent parvovirus infections. Lab Anim Sci 46:370–380
28. Salzman LA (1978) The parvoviruses. In: Nayak DP (ed) Molecular biology of animal viruses. Marcel Dekker, New York, NY
29. Takasawa N, Munakata Y, Ishii KK et al (2004) Human parvovirus B19 transgenic mice become susceptible to polyarthritis. J Immunol 173:4675–4683
30. Tattersall P, Dc W (1978) The parvoviruses-an introduction. Cold Spring Habor Laboratories, Cold Spring Harbor, NY
31. Berns KI (1990) Parvovirus replication. Microbiol Rev 54:316–329
32. Gunther M, May P (1976) Isolation and structural characterization of monomeric and dimeric forms of replicative intermediates of Kilham rat virus DNA. J Virol 20:86–95
33. Tennant RW, Hand RE (1970) Requirement of cellular synthesis for Kilham rat virus replication. Virology 42:1054–1063
34. Brown DW, Welsh RM, Like AA (1993) Infection of peripancreatic lymph nodes but not islets precedes Kilham rat virus-induced diabetes in BB/Wor rats. J Virol 67:5873–5878
35. Hara N, Alkanani AK, Ir D et al (2012) Prevention of virus-induced type 1 diabetes with antibiotic therapy. J Immunol 189:3805–3814
36. Alkanani AK, Hara N, Gianani R et al (2014) Kilham rat virus-induced type 1 diabetes involves beta cell infection and intra-islet JAK--STAT activation prior to insulitis. Virology 468–470:19–27
37. Crisa L, Greiner DL, Mordes JP et al (1990) Biochemical studies of RT6 alloantigens in

BB/Wor and normal rats. Evidence for intact unexpressed RT6a structural gene in diabetes-prone BB rats. Diabetes 39:1279–1288

38. Rossini AA, Handler ES, Mordes JP et al (1995) Human autoimmune diabetes mellitus: lessons from BB rats and NOD mice--Caveat emptor. Clin Immunol Immunopathol 74:2–9
39. Greiner DL, Rossini AA, Mordes JP (2001) Translating data from animal models into methods for preventing human autoimmune diabetes mellitus: caveat emptor and primum non nocere. Clin Immunol 100:134–143
40. Mordes JP, Bortell R, Blankenhorn EP et al (2004) Rat models of type 1 diabetes: genetics, environment, and autoimmunity. ILAR J 45:278–291
41. Ellerman KE, Richards CA, Guberski DL et al (1996) Kilham rat triggers T-cell-dependent autoimmune diabetes in multiple strains of rat. Diabetes 45:557–562
42. Guberski DL, Thomas VA, Shek WR et al (1991) Induction of type I diabetes by Kilham's rat virus in diabetes-resistant BB/Wor rats. Science 254:1010–1013
43. Zipris D, Hillebrands JL, Welsh RM et al (2003) Infections that induce autoimmune diabetes in BBDR rats modulate CD4+CD25 + T cell populations. J Immunol 170:3592–3602
44. Zipris D, Lien E, Xie JX et al (2005) TLR activation synergizes with Kilham rat virus infection to induce diabetes in BBDR rats. J Immunol 174:131–142
45. Chung YH, Jun HS, Son M et al (2000) Cellular and molecular mechanism for Kilham rat virus-induced autoimmune diabetes in DR-BB rats. J Immunol 165:2866–2876
46. Zipris D (2008) Innate immunity and its role in type 1 diabetes. Curr Opin Endocrinol Diabetes Obes 15:326–331
47. Zipris D (2011) Innate immunity in type 1 diabetes. Diabetes Metab Res Rev 27:824–829
48. Zipris D (2010) Toll-like receptors and type 1 diabetes. Adv Exp Med Biol 654:585–610
49. Qaisar N, Lin S, Ryan G et al (2017) A critical role for the type i interferon receptor in virus-induced autoimmune diabetes in rats. Diabetes 66:145–157
50. Mordes JP, Zipris D, Liu Z et al (2013) Viruses and autoimmune diabetes in rats. In: Taylor K, Hyöty H, Toniolo A, Zuckerman JA (eds) Diabetes and viruses. Springer, New York, NY, pp 57–70
51. Tirabassi RS, Guberski DL, Blankenhorn EP et al (2009) Infection with viruses from several families triggers autoimmune diabetes in LEW.1WR1 rats. Diabetes 59:110

# Chapter 9

# Large Animal Models of Diabetes


## Barbara Ludwig, Eckhard Wolf, Uwe Schönmann, and Stefan Ludwig



## Abstract

Safe and reliable large animal diabetes models are a key prerequisite for advanced preclinical studies on diabetes. Chemical induction is the standard model of diabetes in rodents but is often critiqued in higher animals due to reduced efficacy, relevant side effects, and inadequate mortality rate. In this chapter, we aim to describe both pharmacological and surgical approaches for reproducible and safe diabetes models in minipigs and primates. In addition, genetically modified pig models for diabetes research are described.



**Key words** Diabetes model, Diabetes induction, Pig model, Primate model, Streptozotocin, Pancreatectomy


## Abbreviations

| | |
|---|---|
| BG | Blood glucose |
| BW | Body weight |
| im | Intramuscular |
| iv | Intravenous |
| NHP | Nonhuman primate |
| sc | Subcutaneous |
| STZ | Streptozotocin |

## 1 Introduction

A reliable and safe model of insulin-deficient diabetes in large animals is a key prerequisite in preclinical research of various therapeutic approaches, particularly islet transplantation including different islet sources such as human, xenogeneic or iPS-/ES-derived insulin-producing cells and (bio-)artificial devices. The pig is especially eligible due to the resemblance of the gastrointestinum, the morphology of the pancreas, metabolic responses, and the pharmacokinetics to humans [1]. Miniature pigs such as the Goettingen minipigs are particularly suitable for long-term studies because of


Aileen J. F. King (ed.), *Animal Models of Diabetes: Methods and Protocols*, Methods in Molecular Biology, vol. 2128,
https://doi.org/10.1007/978-1-0716-0385-7_9, © Springer Science+Business Media, LLC, part of Springer Nature 2020

their small size, comparably low costs (e.g., versus primates), and benign personality. Although the use of nonhuman primates in diabetes research is still required especially for safety evaluation of novel cell-based diabetes therapies in order to fulfill the requirements of competent authorities and recommendations of the respective professional societies [2], these models are critically discussed due to ethical considerations. Also, species-specific characteristics with regard to glucose metabolism and insulin action must be considered [1, 3, 4]. Generally, there are two different methods for diabetes induction available, chemically with the beta cell-specific toxin streptozotocin (STZ) or alloxan and the surgical approach of (sub)total pancreatectomy [5]. The usage of chemical ablation is considered the standard method in rodents and is also recommended by many authors for large animals [6, 7]. However, the efficacy in large animals can be unreliable due to the very narrow window of dosing and a lot-to-lot variability of streptozotocin. Moreover, after induction of hyperglycemia, a relevant recovery of endogenous beta cells may occur during long-term studies and challenge the obtained results [8, 9]. Therefore, the surgical approach of complete pancreatectomy for induction of insulin-deficient diabetes is a valuable alternative. Following a reliable diabetes induction, the establishment of an optimal regimen for postoperative diabetes care particularly for prolonged follow-up periods is of critical importance to generate valuable experimental data. Generally, conventional blood glucose monitoring (comparable to self-monitoring of blood glucose in patients) is an appropriate way. However, the recent advances in diabetes technology and continuous glucose monitoring systems (CGMS), diabetes aids that are successfully used in clinical practice are also applicable in large animal diabetes models. Furthermore, insulin application methods of multiple daily injections can ideally be replaced by continuous subcutaneous insulin infusion pumps. The protocols are applicable in both pig and (with limitations) primates for pharmacological testing and medical or biological device implantation studies.

## 2 Materials

### 2.1 Anesthesia/Premedication

1. Ventilation and respiratory monitoring equipment (we use the EVITA XL).
2. Endotracheal tubes.
3. Central venous line.
4. Benzodiazepine (short acting such as midazolam).
5. Antimuscarinic/anticholinergic agent (we use atropine).
6. NMDA receptor antagonist.
7. Crystalloid solution (such as saline or Ringer's lactate).

**2.2 Postoperative Care**

1. Analgesic (non-opioid, such as metamizole).
2. Antibiotic.

**2.3 Pancreatectomy**

1. Surgical abdomen instrument kit.
2. Absorbable suture.

**2.4 Glucose Monitoring**

1. Standard glucometer.
2. Flash glucose monitoring system.
3. Continuous glucose monitoring system.

**2.5 Insulin Application**

1. Insulin pen system.
2. Insulin pump system.
3. Long-acting insulin (such as insulin glargine).
4. Short-acting insulin (such as insulin glulisine).

**2.6 Treatment for Exocrine Pancreas Insufficiency**

1. Pancrelipase.

## 3 Methods

### 3.1 Animals

All studies in animals must be approved and in accordance with the guidelines established by the respective institutions and competent authorities.

When a large animal diabetes model is planned, one has to thoroughly consider various practical issues that determine the most suitable breed, the optimal age, and weight. In our experience, miniature pigs, such as the Göttingen minipig, are most suitable because of their small size and weight, easy handling, and benign personality. These factors are of high relevance especially for studies with a prolonged follow-up.

For a nonhuman primate (NHP) diabetes model with long-term follow-up, rhesus macaques or cynomolgus monkeys are appropriate species. Studies must be performed at professionalized primate centers with specialized infrastructure and personnel. The NHP model has several limitations and specificities with regard to glucose metabolism (insulin action, kinetics, etc.) [20, 21] as well as practical issues such as potential risks during handling and intolerance against foreign materials (catheters, sensors, etc.). All these factors may significantly influence the study outcome and must be considered at an early stage of planning an experiment.

Adapted to the respective species, the animal house infrastructure should facilitate individual housing and standard physical environment and operating conditions (19–23 °C; 40–70% relative humidity; 12:12 h day/night cycle). Moreover, depending on the study subject and protocol, individual feeding schedules must be implementable, and water must be provided ad libitum. Furthermore, one should guarantee a daily clinical monitoring including assessment of activity level, body condition, body weight (once weekly), skin quality, hydration status, posture, respiration, and social behavior. In addition, and depending on the study protocol, control of metabolic parameters and therapeutic interventions (e.g., insulin application) must be performed potentially several times per day.

### 3.2 Blood Sampling

Depending on the type of study that is planned, regular and potentially frequent blood sampling, iv medication, and performance of functional tests such as intravenous glucose tolerance tests might be required.

#### 3.2.1 Pig Model

An easy, permanent, and non-traumatizing venous access is of critical importance. In our experience, the establishment of a central venous line prior or in parallel to the study initiation is of major advantage. Adopting the following protocol, a durable access can be ensured even in long term without infectious or other complications:

1. A triple lumen catheter is prepared for placement into the right v. jugularis.
2. Midazolam (1 mg/kg) and ketamine (10 mg/kg) are administered intramuscularly (im) for sedation (*see* **Note 1**).
3. The right neck region is cleaned, shaved, disinfected, and covered with surgical drapes.
4. Via a 5 cm paratracheal incision, the internal jugular vein is dissected, encircled, and cranially ligated.
5. Through a longitudinal incision, the triple lumen catheter is inserted and fixed.
6. To prevent manipulation and dislocation, the catheter is tunneled subcutaneously and exposed on the back of the animal.
7. A circular bandage is used for protection (Fig. 1).

For long-term studies, the central venous line can be replaced as needed simply by the Seldinger technique (*see* **Note 2**).

#### 3.2.2 NHP Model

For long-term studies in primates, sufficient sedation is required for any invasive procedure such as functional tests or iv medication. However, an extensive training program, adapted to the specific demands of the study (glucose monitoring, exogenous insulin treatment) is obligatory depending on the standards of the respective primate center.

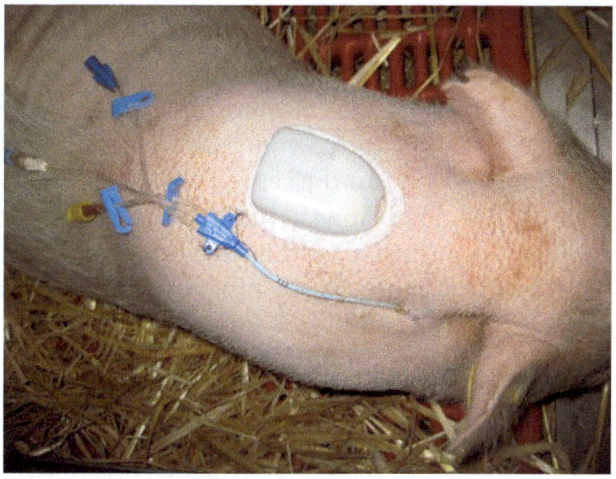

**Fig. 1** Göttingen minipig with central venous line subcutaneously tunneled to the neck region for physical protection. An insulin pump (mylife OmniPod) is placed in the shaved neck region for continuous insulin delivery [32]

### 3.3 Anesthesia

The procedure is generally equally applicable to pigs and NHP. However, local expertise and center-specific standards should be respected. The following protocol describes the detailed procedure exemplary for pigs.

Prior to intervention, the animals must be fasted overnight with free access to water. The premedication is performed as described above (Subheading 3.2.1). If vascular access is not yet available, it can easily be established by puncture of an ear vein. When sufficient sedation is achieved, animals are placed on the operating table and kept in supine position during the entire procedure. Intramuscular premedication can be augmented by intravenous injection of ketamine and/or midazolam if necessary, but the animal should always show sufficient spontaneous breathing activity, even in supine position. For maintenance anesthesia, the following steps should be followed:

1. The trachea of the animals is intubated transorally with a cuffed endotracheal tube.
2. The lungs are ventilated with a standard ventilator (ideally with options for children) set in volume-controlled mode using the following settings: fraction of inspired oxygen of 0.5, VT of 8 mL/kg, positive end-expiratory pressure of 5 cm $H_2O$, inspiratory:expiratory ratio in the range of 1:1 to 1:2, and respiratory rate adjusted to achieve an end-tidal $CO_2$ between 35 and 45 mmHg (we recommend use of capnometry or capillary blood gas monitoring).
3. Anesthesia is performed as total intravenous anesthesia with midazolam (0.5 mg/kg bolus and 1 mg/kg/h) and ketamine (2 mg/kg bolus and 15 mg/kg/h). If neuromuscular blockade is desired, paralysis is achieved with atracurium (1 mg/kg bolus and 3 mg/kg/h).

4. Intravascular volume is maintained with a crystalloid solution at a rate of 10 mL/kg/h.

For monitoring, oxygen saturation (*see* **Note 3**) should be measured continuously as well as a standard electrocardiogram.

In the case of continuous neuromuscular blockade, atracurium infusion should be stopped approximately 20 min before the end of surgery. When surgery has reached its end, midazolam and ketamine can be stopped, and slow infusion of metamizole (*see* Subheading 3.5.1) should be started. When sufficient spontaneous breathing activity is detectable, the pig may be disconnected from the ventilator and brought to the recovery room with the uncuffed endotracheal tube still in place, breathing room air. The pig should be placed in a lateral position, the pulse oximetry still in place. Extubation can be performed when the animal tries to stand up on its own feet or shows recurring coughing. Oxygen saturation should be >90% at all times.

### 3.4 Diabetes Induction Using Streptozotocin

Chemical induction of diabetes using streptozotocin is the standard model in rodents. Various authors and research groups recommend this method also for the pig and NHP model [7]. By all means, special attention and care should be exercised in order to minimize side effects and even mortality rate in this model. We recommend the following protocol for safe and most reliable pharmacological diabetes induction:

1. Animals should be fasted overnight prior to STZ injection and for at least 2 h thereafter (*see* **Note 4**).
2. Filter-sterilized streptozotocin is diluted in 100 mM sodium citrate to 25 mg/mL (pH 4.5) and administrated intravenously (ideally via a central venous line) at 50 mg/kg and 150 mg/kg body weight (BW) for primates or pigs, respectively.
3. For safety monitoring, hepatic function and renal function should be assessed prior to and 1 and 4 weeks after STZ injection by assessment of serum aspartate aminotransferase (AST), alanine aminotransferase (ALT), and plasma creatinine (*see* **Note 5**).

Using this protocol, animals should develop a significant increase in blood glucose within 1 week, but especially in pigs, this effect is partially reversible. A number of animals might recover with regard to blood glucose control and endogenous insulin secretion within the following weeks. Therefore, a uniformly induced irreversible diabetes cannot be achieved in our experience. Increasing the dosage of STZ (e.g., to 200 mg/kg BW in pigs) is not the solution since in our (and others) experience the renal and hepatic toxicity significantly increases.

### 3.5 Diabetes Induction by Total Pancreatectomy

#### 3.5.1 Pig Model

In order to achieve irreversible insulin-deficient diabetes, surgical pancreatectomy is the recommended procedure according to our experience.

After anesthesia is established, the abdominal region is cleaned and shaved prior to disinfection and covered with surgical drapes. The surgical procedure should be carried out according to the following procedure:

1. After median laparotomy, a routine splenectomy is performed (*see* **Note 6**).
2. The tail of the pancreas is carefully dissected from the surrounding tissue, starting laterally. The upper and lower margins of the pancreas tail are detached from the retroperitoneal attachments.
3. The tail and the distal parts of the body are mobilized to the junction of the splenic vein, superior mesenteric vein, and portal vein. The posterior attachments of the pancreas and all tissues between the splenic artery and the junction of the splenic vein and portal vein are divided. The body of the pancreas is dissected and freed from the portal vein, tied off the vascular branches between them.
4. The dissection is continued along the superior margin of the pancreas, and the left gastric artery is dissected and separated from its origin at the splenic artery.
5. Next, the body of the pancreas that wraps the portal vein with a large anterior and thin posterior ring is dissected and liberated.
6. The head of the pancreas is then dissected and mobilized to the right of the aorta. The duodenum is pulled up and the pancreas dissected and liberated from the right portion of the portal vein and the intrahepatic vena cava.
7. The pancreas is then carefully separated from the duodenum, preserving the duodenal vascular arcade, and the pancreatic branches can be individually tied and divided, separating the head of the pancreas from the duodenum.
8. The pancreas is then excised by dissection, ligation, and division of the small vessels from the superior mesenteric artery to the pancreas.
9. The portal ring around the portal vein is dissected from the surrounding structures. The inferior and anterior portion of the pancreas head attached to the mesocolon is transected to remove the organ, finishing the total pancreatectomy.
10. The pancreas is ready to be removed after sectioning of the portal ring.

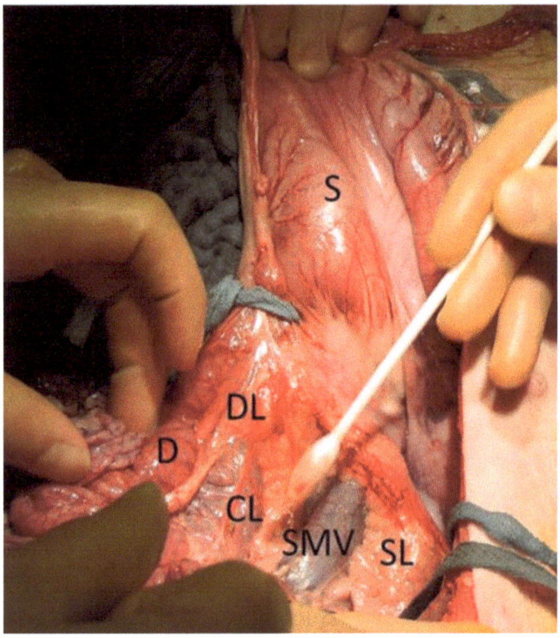


**Fig. 2** Intraoperative situs with duodenum (D), duodenal lobe of the pancreas (DL), connecting lobe of the pancreas (CL), stomach (S), V. mesenterica superior (SMV), and splenic lobe of the pancreas (SL) [32]

11. In the end, both the duodenum and the common bile duct were preserved. The bile duct can be preserved because, in contrast to humans, in swine the bile duct ends in the duodenum apart from the pancreatic duct (Fig. 2).
12. To prevent herniation, the relatively loose duodenum is fixed to the retroperitoneal space.
13. The abdomen and the skin were sutured with absorbable material.

For 2–5 days postoperatively, the animals should receive analgesia with metamizole (single dose of 500 mg iv, maximum daily dose of 1500 mg iv) according to clinical appearance (*see* **Note 7**) [10]. Prophylactic antimicrobial therapy should be administered with single shot of an antibiotic such as tazobactam.

Enteral feeding can be allowed 24 h post-pancreatectomy, and substitution of exogenous pancreatic enzymes should be started (*see* **Note 8**).

#### *3.5.2 NHP Model*

In order to reach optimal efficiency and reduce the surgical risk associated with total pancreatectomy, we recommend for the NHP model a combination of subtotal pancreatectomy followed by low-dose STZ application.

Rhesus macaques (*Macaca mulatta*) of ~10 kg body weight are suitable animals for this procedure.

1. The animals are fasted overnight with free access to water.
2. After premedication (as described above), the anesthesia can be performed as total iv anesthesia (see above).
3. The abdominal region is cleaned and shaved before disinfection and covered with surgical drapes.
4. After median laparotomy, the pancreas is carefully dissected from the surrounding tissue and liberated.
5. Particular precaution must be exercised in the duodenal part of the pancreas, preserving the duodenal vascular arcade and thereby preventing ischemic complications. Therefore, a very small part of pancreatic tissue is left in situ (*see* **Note 9**).
6. The abdomen and the skin are sutured with absorbable material.

Postoperatively, insulin therapy must be started immediately by either multiple daily injections of long- and short-acting insulin or when primate jackets or vests are used, continuous glucose systems can be installed and operated remote-controlled (*see* also Subheading 3.7).

### *3.6 Metabolic Monitoring*

In case of an insulin dependent, particularly a completely insulin-deficient diabetes, regular blood glucose monitoring should be performed according to clinical practice. We suggest a three to four times daily blood glucose testing in order to avoid hypoglycemia and hyperglycemic excursions. For conventional capillary blood glucose (BG) testing (using a standard commercial glucometer) in pigs, the ear tip is the most appropriate site for testing. However, even trained pigs are not much in favor of this procedure particularly for long-term studies. Therefore, we highly recommend to utilize either a flash glucose monitoring system (FreeStyle Libre®, Abbott) or a continuous glucose monitoring system (CGMS). With the former, BG levels can be obtained from a sensor that is placed subcutaneously (no calibration, life span of 2 weeks) simply by scanning. This painless, even contact-free procedure allows to display the current glucose reading, also the latest 8 h of continuous glucose data and a trend arrow that shows if glucose levels are going up or down or changing slowly. In our experience, this system is the most suitable method for BG monitoring and with regard to error-rate and ease of use in pigs also superior to CGM systems (*see* **Note 10**). In primates, artificial devices can only be applied if primate vests or jackets are used. For long-term studies it seems a valuable alternative to training the animals in finger prick testing.

For additional functional beta cell testing in animals undergoing diabetes induction (residual endogenous insulin secretion) or after beta cell replacement therapy (reconstituted endogenous insulin secretion), an intravenous glucose tolerance test is a well-tolerated, viable, and reliable method. The key prerequisite is a stable venous line, ideally a permanent central venous line (see above) for easy blood sampling as described for the pig model. In this setting, no anesthesia or sedation is needed, and the test can be performed stress-free for both the animal and the experimenter. In primates, general anesthesia is usually necessary to perform these tests.

To obtain solid results, we recommend the following protocol (to be adapted to the respective research question):

1. Depending on the research question and the diabetic status of the animal, exogenous insulin treatment should be suspended prior to the test.
2. After overnight fasting and retrieving baseline samples, a sterile glucose solution at a dose of 0.5 g/kg bodyweight is quickly injected intravenously (iv).
3. Serum samples for measurement of glucose and potentially hormone levels are taken at the desired time points (e.g., 5, 10, 20, 30, 60, 90, 120, and 180 min post glucose injection).
4. Hormone levels can be determined using species-specific enzyme-linked immunosorbent assay (ELISA) or radioimmunoassay (RIA).

If the animals are completely insulin deficient, we recommend continuing insulin treatment with long-acting insulin or basal rate of insulin pump prior and during the test to prevent ketoacidosis.

### 3.7 Insulin Substitution

Insulin treatment can generally be performed either by multiple daily injections (MDI) by sc injection of long-acting insulin and repeated application of rapid-acting insulin or by insulin pump therapy (continuous subcutaneous insulin infusion; CSII). From our experience, in pigs, the latter method is strongly recommended due to the possibility of subtle adjustment of baseline insulin doses and the atraumatic application of bolus insulin. The insulin catheter (or patch pump) should be placed in the shaved neck region of the animal and replaced every 2 days. This procedure is easily manageable even without sedation in trained animals (*see* **Note 11**). The insulin dose should be determined by stepwise adjustments in order to reach near-normal glycemic control with avoidance of hypoglycemic events. In primates, this approach again depends on the utilization of primate vests or jackets.

### 3.8 Genetically Tailored Pig Models for Diabetes Research

In addition to partial or total pancreatectomy and chemical destruction of the insulin-producing beta cells, dietary interventions (e.g., feeding a high-calorie, high-fat diet to induce obesity) and genetic modifications have been used to generate pig models for diabetes research. Obesity in pigs is routinely induced by high-fat and/or high-carbohydrate diets (reviewed in ref. 11). Minipig lines such as Ossabaw, Yucatan, or Göttingen minipigs are most widely used since they can be maintained to adulthood at reasonable costs. Dietary supplementation of cholesterol causes additional dyslipidemia (elevated cholesterol, low-density and high-density lipoprotein). While impaired glucose tolerance was inconsistently observed, diet-induced obesity in domestic or minipigs did not lead to an overt diabetic phenotype (reviewed in ref. 11). This requires additional manipulations, such as damage of beta cells by streptozotocin or alloxan, or impairment of beta cell function by specific genetic modifications. Genetic engineering of pigs has been remarkably refined (reviewed in ref. 12) and is the most recent approach of generating tailored large animal models for diabetes research. A broad spectrum of techniques for targeted genetic modification of pigs facilitates the generation of large animal models that mimic disease mechanisms of various types of diabetes mellitus on a molecular level (reviewed in ref. 13). The combination of dietary intervention and genetic modification appears to be particularly suitable to mimic several aspects of the multifactorial type 2 diabetes mellitus (reviewed in ref. 11, 14.

#### 3.8.1 $GIPR^{dn}$ Transgenic Pigs: A Prediabetic Large Animal Model

Many type 2 diabetic patients show a reduced incretin effect that is explained by a reduced function of the incretin hormone GIP (glucose-dependent insulinotropic polypeptide). The incretin hormones GIP and GLP1 (glucagon-like peptide-1) are secreted upon nutrient ingestion by specific endocrine cells in the small intestine. Among other functions, incretins bind to specific receptors of beta cells and potentiate insulin secretion. To mimic the reduced function of GIP in a large animal model, we generated transgenic pigs expressing a dominant-negative GIP-receptor ($GIPR^{dn}$) under the control of a rat *Ins2* promoter sequence [15]. $GIPR^{dn}$ binds GIP with similar affinity as the intact GIPR (a classical seven transmembrane domain G-protein coupled receptor); however—due to the deletion of eight amino acids and an additional amino acid exchange in the third intracellular domain of the receptor—it does not transduce the signal [16]. The $GIPR^{dn}$ transgenic pig model resembles important aspects of prediabetes, including a reduced incretin effect, impaired glucose tolerance, initially delayed and in later stages quantitatively reduced insulin secretion (Fig. 3) and a progressive reduction of beta cell mass. The reproducible and progressive phenotype of this model was used in a targeted metabolomics approach to identify biomarker candidates that change in their plasma concentration with progression of the phenotype in the prediabetic period. In particular, metabolomic signatures of amino acids and lipids were identified that showed a high correlation with beta cell mass [17]. Moreover, the $GIPR^{dn}$ transgenic pig model was used to characterize effects of the GLP1 receptor agonist liraglutide, which is clinically approved for treatment of adult type 2 diabetics, in juvenile organisms [18].

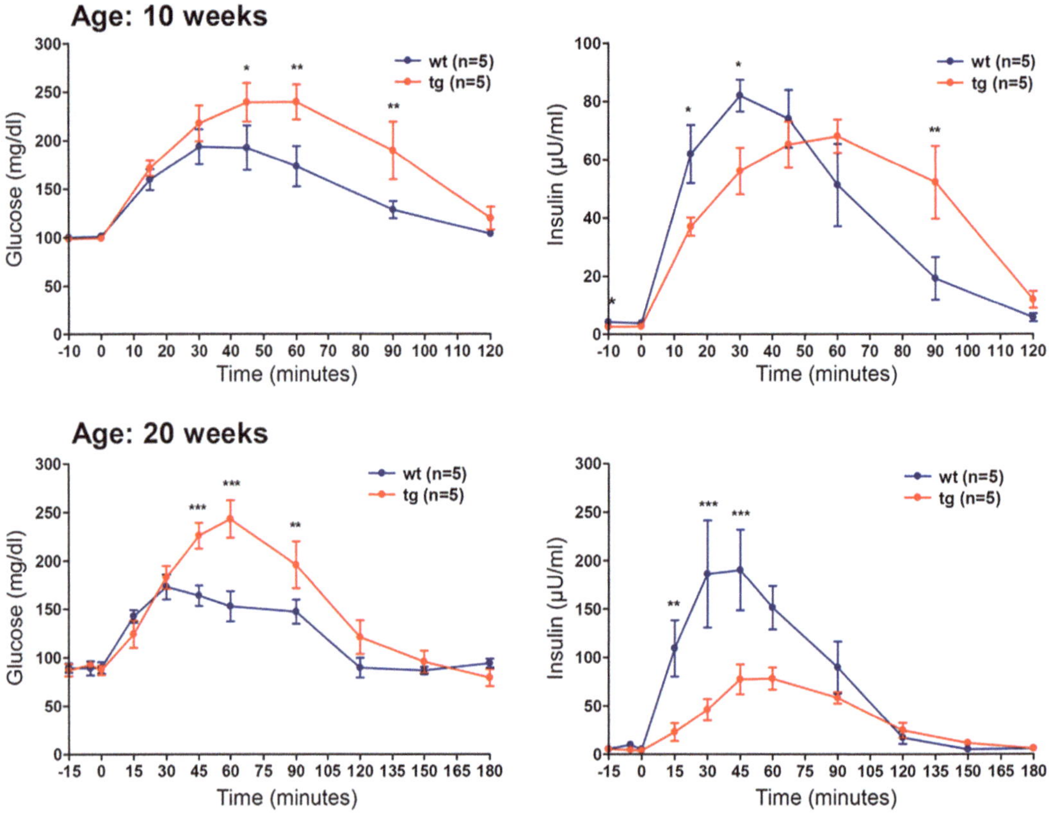


**Fig. 3** Progressive deterioration of oral glucose tolerance in prediabetic GIPR$^{dn}$ transgenic pigs (from [15])

#### *3.8.2 INS$^{C94Y}$ Transgenic Pigs: A Clinically Diabetic Large Animal Model*

The expression of mutant insulin may—depending on the type of mutation and the expression level—lead to permanent neonatal diabetes mellitus (now termed mutant insulin gene-induced diabetes of youth—MIDY) (reviewed in [19]). In humans, more than 50 different mutations of the *INS* gene are known. We generated transgenic pigs that express mutant insulin C94Y [19]. A corresponding mutation was also found in MIDY patients. The C94Y mutation in our pig model disrupts one of the two disulfide bonds between the A and B chain of the insulin molecule, resulting in misfolded insulin, accumulation of proinsulin in the endoplasmic reticulum (ER), and chronic ER stress that cannot be solved by intrinsic repair mechanisms, the so-called unfolded protein response (UPR). This finally results in beta cell apoptosis. MIDY piglets get diabetic within the first week after birth. Since beta cell mass is unaltered at this stage, a deficit in insulin secretion seems to be the primary cause. With increasing age, a loss of beta cell mass is observed. At age 4.5 months, the beta cell mass of MIDY pigs is 70% reduced compared to their wild-type littermates, and the beta cells show morphological hallmarks of ER stress (Fig. 4). We thus developed an insulin substitution therapy that restored normoglycemia and almost normal growth.

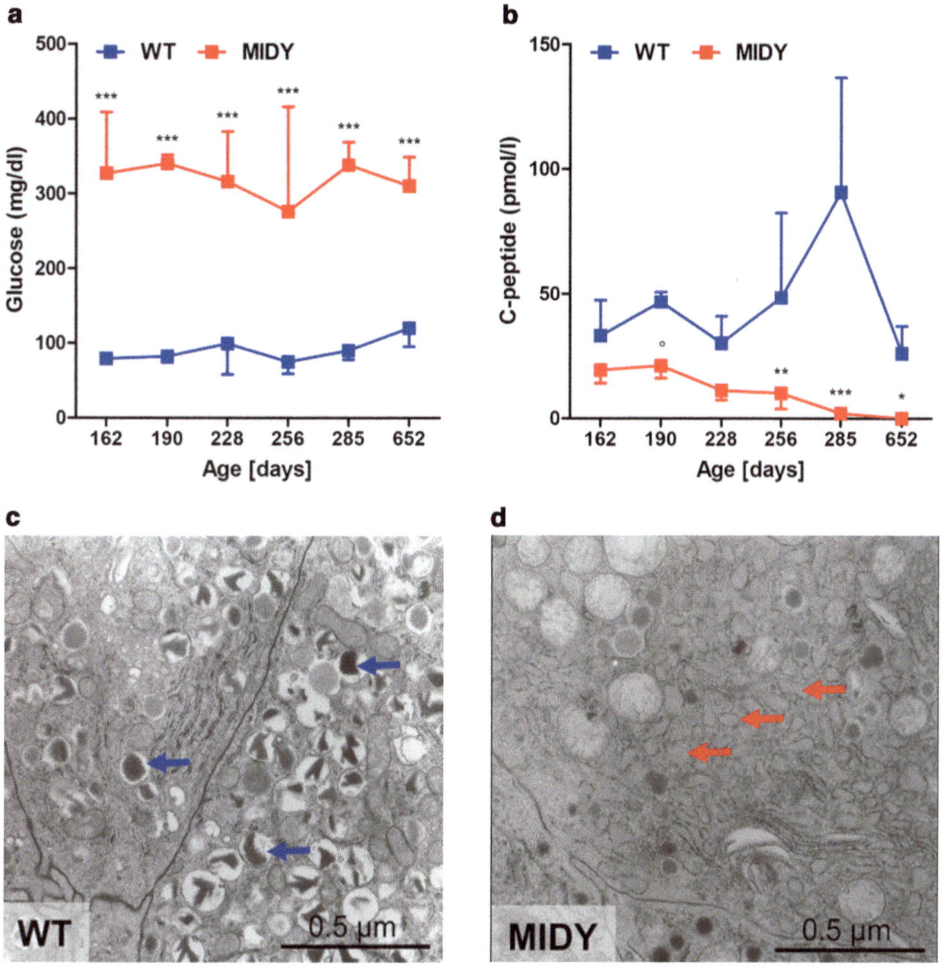


**Fig. 4** Consequences of expression of mutant insulin C94Y in MIDY pigs. (**a**) Permanently elevated fasting blood glucose levels. (**b**) Decreasing plasma C-peptide concentrations indicating perturbed insulin secretion and a decrease in beta cell mass. (**c**, **d**) Ultrastructural changes of beta cells from MIDY pigs (age, 4.5 months), which are indicative of ER stress. Beta cells of wild-type (WT) pigs (**c**) show multiple insulin granules (blue arrows). In beta cells from MIDY pigs, the number of insulin granules is markedly reduced, and characteristic dilations of the endoplasmic reticulum are visible (**d**; red arrows; from [19])

MIDY pigs are an interesting model for a broad range of applications, such as the preclinical testing of novel treatments or diagnostics (e.g., new insulin formulations, continuous glucose monitoring systems, insulin pumps, artificial pancreas) or the evaluation of early stages of diabetic complications in the kidneys, eyes, or microvasculature. Already at age 5 months, reduced capillarization and pericyte investment was observed in the myocardium of MIDY pigs compared to age-matched controls. After experimental induction of an ischemic lesion, the myocardium responded with increased fibrosis. Local gene therapy with thymosin B4 markedly improved capillarization and pericyte investment in wild-type pigs, but only to a lesser extent in MIDY pigs [20]. These findings are clinically relevant since reduced capillarization and pericyte investment is also observed in the myocardium from diabetic patients.

#### 3.8.3 The Munich MIDY-Pig Biobank: A Unique Resource for Studying Systemic Consequences of Diabetes Mellitus

To study the consequences of insulin insufficiency and chromic hyperglycemia in a multi-organ, multi-omics approach, we established a comprehensive biobank from four 2-year-old MIDY pigs and five age-matched controls ([21]; see report in [22]). In this context, the first standardized protocol for systematic sampling and processing of a broad spectrum of organs and tissues from porcine biomedical models was established [13]. The Munich MIDY-Pig Biobank harbors more than 20,000 redundant samples of different body fluids and of ~50 different organs and tissues. Tissue samples were preserved to facilitate holistic molecular profiling studies (e.g., of transcriptome, proteome, lipidome, metabolome) and transcript and protein localization studies as well as qualitative and quantitative pathohistological investigations. The retina of MIDY pigs showed interesting diabetes-associated alterations with similarities to diabetic retinopathy in human patients [23].

#### 3.8.4 Transgenic Pigs Expressing Mutant HNF1A: A Model for MODY3

Different autosomal dominant mutations in the hepatocyte nuclear factor-1α (*HNF1A*) gene, located on human chromosome 12q, are responsible for defects in insulin secretion and derailed glucose homeostasis and have been categorized as maturity onset diabetes of the young 3 (MODY 3) (reviewed in ref. 24). To generate a large animal model for this condition, Umeyama et al. [24] produced transgenic pigs expressing a dominant-negative human HNF1A (P291fsinsC). The expression vector included the mutant *HNF1A* cDNA under the transcriptional control of cytomegalovirus immediate-early gene enhancer/porcine insulin promoter sequences and flanked by chicken β-globin insulators. Expression of transgene-derived mRNA was shown in the brain, heart, lung, liver, pancreas, spleen, and kidney, and the 315 amino acid HNF1A (P291fsinsC) protein was detected in the brain, lung, heart, pancreas, spleen, and kidney. Persistent diabetes with non-fasting blood glucose levels above 200 mg/dl was observed in four longer-living transgenic pigs. Histological analysis revealed abnormal pancreatic islet morphogenesis, immature renal development, and pathological alterations of the kidneys, such as glomerular hypertrophy and sclerosis [25] as well as liver alterations [24]. Since expression of the transgene was not limited to pancreatic beta cells, it is not clear whether the kidney lesions were caused by the diabetic condition or toxic effects of the locally expressed HNF-1α (P291fsinsC) protein. The latter is likely the case, since $INS^{C94Y}$ transgenic pigs (see above) develop similar hyperglycemia but show no signs of diabetic nephropathy, at least before 12 months.

#### 3.8.5 INS-eGFP Transgenic Pigs: A Model for Ex Vivo Studies of Beta Cell Maturation

Functional studies of the endocrine cells in the islets of Langerhans are a prerequisite for better understanding the various forms of diabetes. Since the access to human islets from healthy or diabetic subjects is limited, islets from rodent models are often used for in vitro studies, revealing important knowledge on beta cell function. Several therapeutic concepts for diabetes mellitus are based on the idea to activate the regeneration of beta cells or the transdifferentiation of pancreatic progenitor cells or other endocrine cell types into beta cells. Due to structural (e.g., distribution of the various endocrine cell types) and molecular differences (e.g., transcription factors of endocrine cells) between rodent and human islets of Langerhans, findings in rodent islets may not adequately resemble the situation in human islets. Alternatively, porcine islets of Langerhans can be used as a model. Islets of adult pigs are structurally similar to human islets, but their isolation is difficult and expensive. In contrast, the isolation of neonatal islet-like cell clusters (NICCs) from piglets is less difficult, but NICCs are immature and require maturation in vitro. To facilitate monitoring of this process in living cells, we generated transgenic pigs expressing enhanced green fluorescent protein (eGFP) under the control of the porcine INS promoter [26]. The use of this model facilitates in vitro and in vivo maturation studies of NICCs (Fig. 5) and molecular analyses of FACS-sorted beta cells.

#### 3.8.6 Pancreatogenesis-Disabled Pigs: Potential Hosts for Human Pancreas Development?

The group of Hiromitsu Nakauchi (Tokyo University) provided the first proof of concept for developing an allo- or xenogeneic pancreas in interspecific chimeras [27]. They used mouse blastocysts with defective copies of the pancreatic and duodenal homeobox 1 (*Pdx1*) gene, which is essential for pancreas development. Embryos and fetuses developing from these blastocysts have an empty pancreas niche, and apancreatic pups die shortly after birth from severe hyperglycemia. This lethal phenotype could be rescued by injecting *Pdx1*-intact mouse embryonic stem cells (ESC) or induced pluripotent stem cells (iPSC) or rat iPSC into the PDX1-defective blastocysts. The resulting chimeras had a pancreas entirely derived from injected pluripotent stem cells, demonstrating that intra- and interspecific blastocyst complementation with pluripotent stem cells can form an entire organ in a host engineered to have a free developmental niche.

As a first step toward generation of a human pancreas in an animal host, Matsunari et al. [28] generated pancreatogenesis-disabled transgenic pigs that express *Hes1* (Hairy enhancer split-1) under *PDX1* promoter control. Since no fully functional porcine pluripotent stem cells are available, blastocyst complementation was performed with embryonic blastomeres expressing the fluorescent marker Kusabira-Orange (K-O). The resulting chimeric fetuses and offspring had pancreata derived entirely from the K-O-labeled cells.

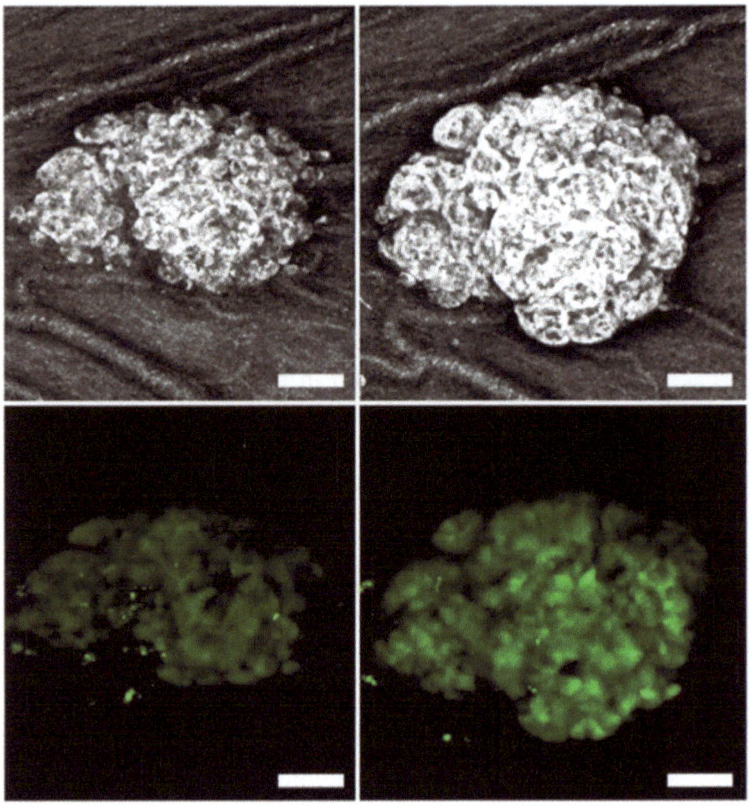


**Fig. 5** Neonatal islet-like cell clusters (NICCs) from INS-eGFP transgenic pigs can be used to study cell proliferation and maturation of NICCs in vitro and in vivo (from [26])

The availability of specific organogenesis-impaired pig hosts opens the possibility of targeted organ generation from human pluripotent stem cells (PSC), but whether this is feasible between phylogenetically distant species remains to be seen. A recent study evaluated the ability of different types of human PSC to contribute to chimeras after injection into porcine and bovine blastocysts. While naïve PSC were reported to engraft in blastocysts of both host species, contribution to postimplantation pig embryos was limited. A higher (although still very low) degree of chimerism was observed when using an intermediate PSC type [29]. These observations suggest that species barriers prevent extensive postimplantation chimerism.

### 3.9 Summary

In summary, in our experience, the large animal diabetes model using minipigs and primates by (sub)total pancreatectomy is a feasible approach for diabetes studies when complete endogenous insulin deficiency is the primary prerequisite. A skilled and committed team of surgeon, anesthesiologist, veterinarian, diabetologist, and students/technicians as well as an adequate environment for the intervention and the follow-up are essential prerequisites for the establishment of such model. With appropriate postoperative care, pancreatectomized minipigs can be kept without complications and are well suitable for, e.g., transplantation experiments. Genetic engineering of pigs is now well established and facilitates the generation of tailored (pre)diabetic pig models and improved donors for islet xenotransplantation ([30]; reviewed in ref. 31).

Although still required in some settings and regulatory situations, the imperative of the primate model should be evaluated very critical for practical reasons, ethical considerations, and suitability for metabolic studies.

## 4 Notes

1. For premedication, always attach the syringe with the medication mixture to a long extension line type Heidelberger. Thereby, you can follow the potentially moving animal after placing the needle and continue injecting.
2. Use the sterile plastic cover, which the venous line is usually protected within the packing and pull it through the "tunnel." Thereby, it is much easier to establish a safe and strainless permanent central venous line access.
3. In pigs, the oxygen probe (standard probe placed on fingertip in humans) can be pulled across the tail for secure and sensitive measurement.
4. Fasting period prior to and after STZ administration increases the susceptibility of beta cells for STZ-induced damage and therefore the efficacy of the toxin.
5. In our experience, pigs tolerated doses 150 mg/kg STZ without significant increase of AST, ALT, or creatinine.
6. Splenectomy preceding manipulation on the pancreas significantly simplifies the procedure by creating space through removal of this relatively large organ and elimination of the risk of distortion of the splenic artery following pancreatectomy with subsequent infarctions. CAVEAT: Splenectomy compromises the immune competence of the animal (potentially critical in transplantation studies involving immunosuppression), and impact of splenectomy on immunological readouts must be considered.
7. We recommend applying analgesics slowly (e.g., during 1 h) in combination with fluid substitution (e.g., sodium chloride 0.9%, 200 mL).

8. We recommend starting with low-dose substitution (e.g., Creon 25,000 IU/day). However, one should test temporarily stopping the medication, since in our experience minipigs did not show any clinical symptoms of exocrine pancreas insufficiency, and therefore we could discontinue substitution.
9. Importantly, this strategy of subtotal pancreatectomy also allows for preserving the anatomical drainage of the bile duct.
10. Regardless of placing the CGM probe or the flash glucose sensor, we recommend the thoroughly shaved and cleaned neck region of the pig for insertion. The skin is considerably less compact and error rate is significantly lower.
11. Due to the high learning aptitude of pigs, one should associate the procedure from the very beginning with a favorite treat (e.g., strawberry yoghurt). The animals then even learn to lie down and tolerate exchange of catheter without any sedation needed.

## References

1. Graham ML, Schuurman HJ (2015) Validity of animal models of type 1 diabetes, and strategies to enhance their utility in translational research. Eur J Pharmacol 759:221–230. https://doi.org/10.1016/j.ejphar.2015.02.054
2. Hering BJ, Cozzi E, Spizzo T, Cowan PJ, Rayat GR, Cooper DK, Denner J (2016) First update of the International Xenotransplantation Association consensus statement on conditions for undertaking clinical trials of porcine islet products in type 1 diabetes--Executive summary. Xenotransplantation 23(1):3–13. https://doi.org/10.1111/xen.12231
3. Swindle MM, Makin A, Herron AJ, Clubb FJ Jr, Frazier KS (2012) Swine as models in biomedical research and toxicology testing. Vet Pathol 49(2):344–356. https://doi.org/10.1177/0300985811402846
4. Swindle MM, Smith AC (1998) Comparative anatomy and physiology of the pig. Scand J Lab Anim Sci 25(Supplement 1):11–21
5. King AJ (2012) The use of animal models in diabetes research. Br J Pharmacol 166 (3):877–894. https://doi.org/10.1111/j.1476-5381.2012.01911.x
6. Grussner R, Nakhleh R, Grussner A, Tomadze G, Diem P, Sutherland D (1993) Streptozotocin-induced diabetes mellitus in pigs. Horm Metab Res 25(4):199–203
7. Wilson JD, Dhall DP, Simeonovic CJ, Lafferty KJ (1986) Induction and management of diabetes mellitus in the pig. Aust J Exp Biol Med Sci 64(Pt 6):489–500
8. Bottino R, Criscimanna A, Casu A, He J, Van der Windt DJ, Rudert WA, Giordano C, Trucco M (2009) Recovery of endogenous beta-cell function in nonhuman primates after chemical diabetes induction and islet transplantation. Diabetes 58(2):442–447. https://doi.org/10.2337/db08-1127
9. de la Garza-Rodea AS, Knaan-Shanzer S, den Hartigh JD, Verhaegen AP, van Bekkum DW (2010) Anomer-equilibrated streptozotocin solution for the induction of experimental diabetes in mice (Mus musculus). J Am Assoc Lab Anim Sci 49(1):40–44
10. Schmidt W, Humke R (1985) Ueber die Anwendung von Novalgin in der Schweinepraxis. Tierarztl Umsch 40:584–589
11. Renner S, Dobenecker B, Blutke A, Zols S, Wanke R, Ritzmann M, Wolf E (2016) Comparative aspects of rodent and nonrodent animal models for mechanistic and translational diabetes research. Theriogenology 86 (1):406–421. https://doi.org/10.1016/j.theriogenology.2016.04.055
12. Whitelaw CB, Sheets TP, Lillico SG, Telugu BP (2016) Engineering large animal models of human disease. J Pathol 238(2):247–256. https://doi.org/10.1002/path.4648
13. Wolf E, Braun-Reichhart C, Streckel E, Renner S (2014) Genetically engineered pig models for diabetes research. Transgenic Res 23

(1):27–38. https://doi.org/10.1007/s11248-013-9755-y

14. Kleinert M, Clemmensen C, Hofmann SM, Moore MC, Renner S, Woods SC, Huypens P, Beckers J, de Angelis MH, Schurmann A, Bakhti M, Klingenspor M, Heiman M, Cherrington AD, Ristow M, Lickert H, Wolf E, Havel PJ, Muller TD, Tschop MH (2018) Animal models of obesity and diabetes mellitus. Nat Rev Endocrinol 14(3):140–162. https://doi.org/10.1038/nrendo.2017.161
15. Renner S, Fehlings C, Herbach N, Hofmann A, von Waldthausen DC, Kessler B, Ulrichs K, Chodnevskaja I, Moskalenko V, Amselgruber W, Goke B, Pfeifer A, Wanke R, Wolf E (2010) Glucose intolerance and reduced proliferation of pancreatic beta-cells in transgenic pigs with impaired glucose-dependent insulinotropic polypeptide function. Diabetes 59(5):1228–1238. https://doi.org/10.2337/db09-0519
16. Renner S, Blutke A, Streckel E, Wanke R, Wolf E (2016) Incretin actions and consequences of incretin-based therapies: lessons from complementary animal models. J Pathol 238(2):345–358. https://doi.org/10.1002/path.4655
17. Renner S, Romisch-Margl W, Prehn C, Krebs S, Adamski J, Goke B, Blum H, Suhre K, Roscher AA, Wolf E (2012) Changing metabolic signatures of amino acids and lipids during the prediabetic period in a pig model with impaired incretin function and reduced beta-cell mass. Diabetes 61(8):2166–2175. https://doi.org/10.2337/db11-1133
18. Streckel E, Braun-Reichhart C, Herbach N, Dahlhoff M, Kessler B, Blutke A, Bahr A, Ubel N, Eddicks M, Ritzmann M, Krebs S, Goke B, Blum H, Wanke R, Wolf E, Renner S (2015) Effects of the glucagon-like peptide-1 receptor agonist liraglutide in juvenile transgenic pigs modeling a pre-diabetic condition. J Transl Med 13:73. https://doi.org/10.1186/s12967-015-0431-2
19. Renner S, Braun-Reichhart C, Blutke A, Herbach N, Emrich D, Streckel E, Wunsch A, Kessler B, Kurome M, Bahr A, Klymiuk N, Krebs S, Puk O, Nagashima H, Graw J, Blum H, Wanke R, Wolf E (2013) Permanent neonatal diabetes in INS(C94Y) transgenic pigs. Diabetes 62(5):1505–1511. https://doi.org/10.2337/db12-1065
20. Hinkel R, Howe A, Renner S, Ng J, Lee S, Klett K, Kaczmarek V, Moretti A, Laugwitz KL, Skroblin P, Mayr M, Milting H, Dendorfer A, Reichart B, Wolf E, Kupatt C (2017) Diabetes mellitus-induced microvascular destabilization in the myocardium. J Am Coll Cardiol 69(2):131–143. https://doi.org/10.1016/j.jacc.2016.10.058
21. Blutke A, Renner S, Flenkenthaler F, Backman M, Haesner S, Kemter E, Landstrom E, Braun-Reichhart C, Albl B, Streckel E, Rathkolb B, Prehn C, Palladini A, Grzybek M, Krebs S, Bauersachs S, Bahr A, Bruhschwein A, Deeg CA, De Monte E, Dmochewitz M, Eberle C, Emrich D, Fux R, Groth F, Gumbert S, Heitmann A, Hinrichs A, Kessler B, Kurome M, Leipig-Rudolph M, Matiasek K, Ozturk H, Otzdorff C, Reichenbach M, Reichenbach HD, Rieger A, Rieseberg B, Rosati M, Saucedo MN, Schleicher A, Schneider MR, Simmet K, Steinmetz J, Ubel N, Zehetmaier P, Jung A, Adamski J, Coskun U, Hrabe de Angelis M, Simmet C, Ritzmann M, Meyer-Lindenberg A, Blum H, Arnold GJ, Frohlich T, Wanke R, Wolf E (2017) The Munich MIDY Pig Biobank - A unique resource for studying organ crosstalk in diabetes. Mol Metab 6(8):931–940. https://doi.org/10.1016/j.molmet.2017.06.004
22. Abbott A (2015) Inside the first pig biobank. Nature 519(7544):397–398. https://doi.org/10.1038/519397a
23. Kleinwort KJH, Amann B, Hauck SM, Hirmer S, Blutke A, Renner S, Uhl PB, Lutterberg K, Sekundo W, Wolf E, Deeg CA (2017) Retinopathy with central oedema in an INS (C94Y) transgenic pig model of long-term diabetes. Diabetologia 60(8):1541–1549. https://doi.org/10.1007/s00125-017-4290-7
24. Umeyama K, Watanabe M, Saito H, Kurome M, Tohi S, Matsunari H, Miki K, Nagashima H (2009) Dominant-negative mutant hepatocyte nuclear factor 1alpha induces diabetes in transgenic-cloned pigs. Transgenic Res 18(5):697–706. https://doi.org/10.1007/s11248-009-9262-3
25. Hara S, Umeyama K, Yokoo T, Nagashima H, Nagata M (2014) Diffuse glomerular nodular lesions in diabetic pigs carrying a dominant-negative mutant hepatocyte nuclear factor 1-alpha, an inheritant diabetic gene in humans. PLoS One 9(3):e92219. https://doi.org/10.1371/journal.pone.0092219
26. Kemter E, Cohrs CM, Schafer M, Schuster M, Steinmeyer K, Wolf-van Buerck L, Wolf A, Wuensch A, Kurome M, Kessler B, Zakhartchenko V, Loehn M, Ivashchenko Y, Seissler J, Schulte AM, Speier S, Wolf E (2017) INS-eGFP transgenic pigs: a novel reporter system for studying maturation, growth and vascularisation of neonatal islet-

like cell clusters. Diabetologia 60 (6):1152–1156. https://doi.org/10.1007/s00125-017-4250-2

27. Kobayashi T, Yamaguchi T, Hamanaka S, Kato-Itoh M, Yamazaki Y, Ibata M, Sato H, Lee YS, Usui J, Knisely AS, Hirabayashi M, Nakauchi H (2010) Generation of rat pancreas in mouse by interspecific blastocyst injection of pluripotent stem cells. Cell 142(5):787–799. https://doi.org/10.1016/j.cell.2010.07.039

28. Matsunari H, Nagashima H, Watanabe M, Umeyama K, Nakano K, Nagaya M, Kobayashi T, Yamaguchi T, Sumazaki R, Herzenberg LA, Nakauchi H (2013) Blastocyst complementation generates exogenic pancreas in vivo in apancreatic cloned pigs. Proc Natl Acad Sci U S A 110(12):4557–4562. https://doi.org/10.1073/pnas.1222902110

29. Wu J, Platero-Luengo A, Sakurai M, Sugawara A, Gil MA, Yamauchi T, Suzuki K, Bogliotti YS, Cuello C, Morales Valencia M, Okumura D, Luo J, Vilarino M, Parrilla I, Soto DA, Martinez CA, Hishida T, Sanchez-Bautista S, Martinez-Martinez ML, Wang H, Nohalez A, Aizawa E, Martinez-Redondo P, Ocampo A, Reddy P, Roca J, Maga EA, Esteban CR, Berggren WT, Nunez Delicado E, Lajara J, Guillen I, Guillen P, Campistol JM, Martinez EA, Ross PJ, Izpisua Belmonte JC (2017) Interspecies chimerism with mammalian pluripotent stem cells. Cell 168 (3):473–486.e415. https://doi.org/10.1016/j.cell.2016.12.036

30. Klymiuk N, van Buerck L, Bahr A, Offers M, Kessler B, Wuensch A, Kurome M, Thormann M, Lochner K, Nagashima H, Herbach N, Wanke R, Seissler J, Wolf E (2012) Xenografted islet cell clusters from INSLEA29Y transgenic pigs rescue diabetes and prevent immune rejection in humanized mice. Diabetes 61(6):1527–1532. https://doi.org/10.2337/db11-1325

31. Klymiuk N, Ludwig B, Seissler J, Reichart B, Wolf E (2016) Current concepts of using pigs as a source for beta-cell replacement therapy of type 1 diabetes. Curr Mol Bio Rep 2(2):73–82. https://doi.org/10.1007/s40610-016-0039-1

32. Heinke S, Ludwig B, Schubert U, Schmid J, Kiss T, Steffen A, Bornstein S, Ludwig S (2016) Diabetes induction by total pancreatectomy in minipigs with simultaneous splenectomy: a feasible approach for advanced diabetes research. Xenotransplantation 23 (5):405–413. https://doi.org/10.1111/xen.12255

# Chapter 10

# Use of Streptozotocin in Rodent Models of Islet Transplantation

**Aileen J. F. King, Elisabet Estil·les, and Eduard Montanya**


## Abstract

Streptozotocin (STZ) selectively destroys beta cells and is widely used to induce experimental diabetes in rodents. Rodent beta cells are very sensitive to the toxic effects of STZ, while human beta cells are highly resistant to STZ. Taking advantage of this characteristic, here, we describe two protocols for the induction of STZ-diabetes. In the first model, hyperglycemia is induced prior to islet transplantation, whereas in the second model, STZ is injected after islet transplantation. The former model has many applications and thus is the most commonly used method. However, when implanting human islets into mice, there are clear benefits to administering STZ after the transplantation. It reduces the cost and burden of experiments and the number of human islets needed for transplantation and improves the welfare and survival of animals used in the experiments. In both methods, a key step in the experimental protocol is to remove the graft-bearing kidney at the end of the experiment and monitor onset of hyperglycemia. This can be used to demonstrate that the glycemic control of the animal is due to the engrafted islets and not regeneration of endogenous beta cells. This chapter outlines protocols of administering streptozotocin pre- and post-islet transplantation in mice as well as nephrectomy to remove the graft-bearing kidney.


**Key words** Streptozotocin, Human islet transplantation, Pancreatic beta cell, Nephrectomy

## 1 Introduction

Inducing diabetes by chemical means is a relatively cheap and reliable method of inducing hyperglycemia "on demand." It can be used in a wide variety of species and experimental settings, but for this chapter, we will concentrate on murine recipients of either human or mouse islet transplants. Alloxan and streptozotocin (STZ) are compounds commonly used to induce beta cell death and thereby hyperglycemia through lack of insulin production. Their structural similarity to glucose allows uptake by GLUT2 transporters, and thus these compounds selectively accumulate within the beta cells. Streptozotocin alkylates DNA and subsequent activation of poly-(ADP-ribose) polymerase (PARP) leads to energy depletion. Alloxan induces the formation of free radicals,

Aileen J. F. King (ed.), *Animal Models of Diabetes: Methods and Protocols*, Methods in Molecular Biology, vol. 2128, https://doi.org/10.1007/978-1-0716-0385-7_10, 

which beta cells have a poor defense mechanism against. It also causes oxidation of essential-SH groups, inhibiting key proteins such as glucokinase. In both cases, the actions of the compounds lead to rapid beta cell death. This chapter will focus on the use of streptozotocin. In high doses, diabetes can typically be induced within days. However, if the aim is to transplant rodent islets into a stably hyperglycemic animal, it is advisable to wait at least 5–7 days before transplantation.

It should be noted that the susceptibility to streptozotocin is species, strain, and gender dependent. Thus, doses may have to be optimized for the model in which it is to be used. In the traditional model of STZ-induced diabetes used in experimental islet transplantation, STZ is injected first, and when diabetes is confirmed, transplantation is performed. In rodent islet transplantation, the availability of donors and recipients is at the discretion of the investigator, and islet isolation and transplantation can be planned well in advance and performed at the most convenient time after diabetes is confirmed in the recipient. In contrast, the availability of human islets depends on organ donation, and prospective STZ-diabetic recipients, usually fragile immunodeficient mice, must be kept in the Animal Care Facility for unpredictable lapses of time, requiring more intensive care, including daily treatment with insulin, until a human islet preparation is available. After STZ injection, and despite insulin treatment, their condition often deteriorates, and mortality is increased [1]. Based on the well-known resistance of human islets to STZ [2], we have described a new model in which STZ-diabetes is induced *after* human islet transplantation [3]. We have shown that STZ has no detectable toxic effects on β-cell death, mass, and function of human transplanted islets. This model has several advantages. First, it reduces the burden and cost of the experiments, since it is not necessary to maintain a pool of chronically hyperglycemic immunodeficient recipients for undetermined periods of time waiting for a human islet preparation to become available. Second, it reduces the number of human islets that must be transplanted, since it is well established that restoration of normoglycemia in STZ-diabetic recipients requires the transplantation of a higher islet number than that needed to maintain normoglycemia [4–7]. Third, it reduces the number of animals used in the experiments, increasing the survival, and also the welfare, of the animals. Moreover, the model is close to the usual clinical practice, where normoglycemia is maintained before and after islet transplantation to avoid the deleterious effects of hyperglycemia on beta cells.

At the end of any experiment using streptozotocin for beta cell depletion, it is essential to show that the glycemic control of the animal is due to the implanted islets rather than regeneration of the beta cells of the endogenous pancreas. This can be investigated by carrying out a nephrectomy of the graft-bearing kidney. After the removal of the implanted islets by this method, the animal should develop hyperglycemia if the graft was responsible for the glycemic control of the animal.

This chapter will therefore outline the methods of injecting STZ before islet transplantation, injecting STZ after islet transplantation, and the removal of the graft-bearing kidney. These methods are suitable for use in mice and rats, but this chapter will focus on the use of mice as recipients (for methods regarding the use of STZ in larger animals, *see* Chapter 9).

## 2 Materials

### 2.1 Animals

1. In the model where STZ is administered before islet transplantation, the recipient strain will depend on the experimental design. If immune rejection is of interest (or conversely to be avoided), the source of the islets should be taken into consideration when choosing the recipient (*see* **Note 1**).
2. In the context of using STZ after human islet transplantation, immunodeficient mice should be used as recipients, e.g., Athymic Nude-Foxn1nu mice (*see* **Note 2**).

### 2.2 Streptozotocin Solution and Administration

1. Citric acid.
2. Sterile saline solution.
3. Streptozotocin (STZ) (*see* **Note 3**).
4. 22-μm pore filter.
5. 1.5-mL sterile microcentrifuge tubes.
6. 1-mL sterile syringes with 25- or 27-gauge needle.
7. Aluminum foil.

### 2.3 Islet Transplantation

1. Islets (the source will depend on the objective of the experiment) in an appropriate media or saline in 1.5-mL sterile microcentrifuge tubes (*see* **Note 4**).
2. Media appropriate for the islets.
3. Sterile saline.
4. 5-mL syringe and 19G needle.
5. Hamilton syringe (500 μL).
6. PE50 tubing (*see* **Note 5**).
7. Connector tubing (plastic tubing of around 3–5 mm in length, suitable to connect PE50 tubing to the end of a P200 pipette tip).
8. Centrifuge and suitable inserts (*see* **Note 6**).

9. P200 pipette and tips (with some tips cut to fit the Hamilton syringe).
10. Anesthetic (*see* **Note** 7).
11. Analgesic (*see* **Note** 7).
12. Surgical instruments (scalpel, two pairs of forceps, scissors, watchmaker forceps, needle holder).
13. Razor.
14. Antiseptic skin scrub.
15. Heating pad.
16. Gauze.
17. 23G needle.
18. Cauteriser.
19. Suture appropriate for the peritoneum and skin.

### 2.4 Equipment to Monitor Glycemic Control

1. Scales.
2. Sterile needles (we suggest 25–30 gauge; *see* **Note 8**).
3. Blood glucose meter (any model, e.g., Accu-Chek Performa; *see* **Note 9**).
4. Blood glucose test strips to fit meter.

### 2.5 Nephrectomy

1. Anesthetic (*see* **Note** 7).
2. Analgesic (*see* **Note** 7).
3. Razor.
4. Antiseptic skin scrub.
5. Heating pad.
6. Gauze.
7. Surgical instruments (scalpel, two pairs of forceps, scissors, hemostatic clamp, needle holder).

## 3 Methods

We will describe two methods of administering STZ below. In the first, hyperglycemia is induced with a single high dose of STZ prior to islet transplantation. In the latter, we will describe a model where STZ is administered after islet transplantation. This is useful in models where the implanted islets have specific resistance to STZ (e.g., human islets), and the goal is to specifically destroy endogenous beta cells in a species that is STZ susceptible (e.g., mouse) after the implanted islets have engrafted.

### 3.1 Prepare Citrate Buffer

1. Make an 11 mM citric acid solution in saline.
2. Mix and adjust the pH of the solution to pH 4.5 (*see* **Note 10**).

### 3.2 Streptozotocin Injection Before Islet Transplantation

#### 3.2.1 Inject STZ

1. Weigh mouse to determine the amount of STZ needed.
2. Weigh appropriate amount of STZ and cover it with aluminum foil.
3. Dissolve the weighed STZ in citrate buffer (pH 4.5) in a sterile microfuge tube (*see* **Note 11**). Vortex until STZ is completely in solution. This must be done immediately prior to injection, since STZ degrades rapidly once in solution (*see* **Note 12**).
4. Filter-sterilize the STZ solution using a 0.22-μm pore filter.
5. Inject the appropriate volume of STZ solution intraperitoneally using 1-ml sterile syringes with 25–27G needles according to local guidelines. To calculate the volume to inject, use the following formula (*see* **Note 13**):

$$\text{Volume (mL)} = \frac{180\ \text{mg/kg} \ \times \text{weight of mouse (kg)}}{\text{Concentration of STZ (mg/mL)}}$$

   Return the mice to their home cages. Provide free access to normal food and water.
6. Monitor the mice over the next few days by weighing them and measuring blood glucose at least three to four times a week (*see* **Note 14**).
7. Five to 7 days after the STZ injection, check if the mice are overtly hyperglycemic before commencing the study (*see* **Note 15**).

#### 3.2.2 Prepare Islets for Transplantation

**This step should be carried out as surgery is starting (*see* Note 16).**

1. Attach a 200-μL pipette tip to the Hamilton syringe (*see* **Note 17**).
2. Fill the Hamilton syringe and tip with media or saline as appropriate, with no air bubbles.
3. Aspirate the islets into the tip (*see* **Note 18**) and allow to pellet by gravity (*see* **Note 19**).
4. Attach the PE50 tubing to the pipette tip which is attached to the Hamilton syringe and carefully inject the liquid and islet pellet into the tubing, around 5 cm down. Bend the PE50 tubing back on itself just after the pellet and hold in place using a piece of connector tubing.
5. Remove the tip with PE50 tubing from the Hamilton syringe and place in a centrifuge for 2 min at 200 RCF (*see* **Notes 6** and **20**).
6. Reattach the tip with PE50 tubing to the Hamilton syringe, ensuring no air bubbles (*see* **Note 21**).

#### 3.2.3 Islet Transplantation to the Subcapsular Kidney

1. Anesthetize the mouse with isoflurane or another appropriate anesthetic.
2. Shave the fur on the left flank and treat the skin with surgical scrub, followed by 70% alcohol.
3. Administer analgesic and lay the mouse on a heat pad.
4. Lay drapes over the mouse, with the operation area exposed, and check if the mouse is adequately anesthetized (e.g., by lack of reaction to a paw pinch).
5. Use a scalpel to make a 1.5–2-cm incision in the skin over the left kidney.
6. Use scissors to cut a small incision on the peritoneum above the left kidney.
7. Externalize the kidney by gently pressing under each side of the kidney to allow it to pop out.
8. Keep the kidney moist using sterile saline.
9. Make an incision in the kidney capsule by using either a scalpel or by dragging a 23G needle (*see* **Note 22**).
10. Using the watchmaker forceps to lift the kidney capsule, place the PE50 tubing (prepared in Subheading 3.2.2) under the kidney capsule and push the tube toward the top quarter of the kidney.
11. Use the Hamilton syringe screw function to slowly inject the islet pellet under the kidney capsule. Carefully pull the PE50 tubing out.
12. Cauterize the kidney capsule at the incision.
13. Suture the peritoneum.
14. Suture the skin.
15. Let the mouse recover (*see* **Note 23**).
16. Monitor weight and blood glucose concentrations for the remainder of the study.

### 3.3 STZ Injection After Islet Transplantation

In this protocol, human islets are transplanted to inmunodeficient recipients and STZ is injected several days after islet transplantation to destroy the beta cells of the endogenous pancreas. This capitalizes on the differing susceptibilities between human and mouse beta cells to the effects of STZ and negates the welfare and cost implications of keeping diabetic animals while waiting for human islets to become available. Depending on the transplanted human beta cell mass, hyperglycemia will or will not be detected after STZ injection (*see* **Note 24**). Frequency of blood glucose and body weight measurements should be planned according to the aim and objectives of the study.

##### 3.3.1 Prepare Islets for Transplantation

**This step should be carried out as surgery is starting (*see* Note 16).**

1. Attach a 200-μL pipette tip to the Hamilton syringe (*see* **Note 17**).
2. Fill the Hamilton syringe and tip with media or saline as appropriate, with no air bubbles.
3. Aspirate the islets into the tip (*see* **Note 18**) and allow to pellet by gravity (*see* **Note 19**).
4. Attach the PE50 tubing to the pipette tip which is attached to the Hamilton syringe and carefully inject the liquid and islet pellet into the tubing, around 5 cm down. Bend the PE50 tubing back on itself just after the pellet and hold in place using a piece of connector tubing.
5. Remove the tip with PE50 tubing from the Hamilton syringe and place in a centrifuge for 2 min at 200 RCF (*see* **Notes 6** and **20**).
6. Reattach the PE50 tubing to the Hamilton syringe, ensuring no air bubbles (*see* **Note 21**).

##### 3.3.2 Islet Transplantation to the Subcapsular Kidney

1. Anesthetize the mouse with isoflurane or another appropriate anesthetic.
2. Shave the fur on the left flank and treat the skin with surgical scrub, followed by 70% alcohol.
3. Administer analgesic and lay the mouse on a heat pad.
4. Lay drapes over the mouse, with the operation area exposed, and check if the mouse is adequately anesthetized (e.g., by lack of reaction to a paw pinch).
5. Use a scalpel to make a 1.5–2-cm incision in the skin over the left kidney.
6. Use scissors to cut a small incision on the peritoneum above the left kidney.
7. Externalize the kidney by gently pressing under each side of the kidney to allow it to pop out.
8. Keep the kidney moist using sterile saline.
9. Make an incision in the kidney capsule by using either a scalpel or by dragging a 23G needle (*see* **Note 22**).
10. Using the watchmaker forceps to lift the kidney capsule, place the PE50 tubing (prepared in Subheading 3.2.2) under the kidney capsule and push the tube toward the top quarter of the kidney.
11. Use the Hamilton syringe screw function to slowly inject the islet pellet under the kidney capsule. Carefully pull the PE50 tubing out.

12. Cauterize the kidney capsule at the incision.
13. Suture the peritoneum.
14. Suture the skin.
15. Let the mouse recover (*see* **Note 23**).

Return the mice to their home cages for 10–14 days during which time the islets should engraft.

#### *3.3.3 Streptozotocin Injections*

**Ten to 14 days after transplantation, treat mice with five consecutive intraperitoneal daily doses of 70 mg/kg of body weight of STZ.**

1. Weigh the mouse to determine the amount of STZ needed.
2. Weigh appropriate amount of STZ and cover it with aluminum foil.
3. Dissolve the weighed STZ in citrate buffer (pH 4.5) in a sterile microfuge tube (*see* **Note 3**). Vortex until STZ is completely in solution. This must be done immediately prior to injection, since STZ degrades rapidly once in solution (*see* **Note 4**).
4. Filter-sterilize the STZ solution using a 0.22-μm pore filter.
5. Inject the appropriate volume of STZ solution intraperitoneally using 1-ml sterile syringes with 25-gauge 5/8 in. needle. To calculate the volume to inject, use the following formula:

$$\text{Volume (mL)} = \frac{70\ \text{mg/kg} \ \times \text{weight of mouse (kg)}}{\text{Concentration of STZ (mg/mL)}}$$

6. Return the mice to their home cages. Provide free access to normal food and water.
7. Repeat **steps 2–6** on the next 4 consecutive days to complete five daily injections.
8. Monitor weight and blood glucose concentrations as appropriate, depending on study design.

### *3.4 Monitoring Animals*

1. Monitor body weight and blood glucose concentrations regularly depending on the study design (*see* **Note 25**). In the first week after treatments, this should be every 1–2 days to check animal welfare. This can be reduced to one to two times a week in stable animals. Always check local guidelines and veterinary advice.
2. To measure blood glucose concentrations: obtain a drop of blood, in non-fasting conditions at the same time of the day, through a needle prick to the tip of the tail, and measure blood glucose level with a glucometer (*see* **Notes 9** and **26**).

### 3.5 Nephrectomy

Nephrectomizing to confirm that the animal reverts to hyperglycemia (in the case of STZ before transplantation models) or becomes hyperglycemic (in the case of STZ after transplantation) is an important control to ensure that the normoglycemic status is indeed dependent on the graft rather than regeneration of endogenous beta cells.

1. Anesthetize the mouse with isoflurane or another appropriate anesthetic.
2. Shave the fur on the left flank and treat the skin with surgical scrub, followed by 70% alcohol.
3. Administer analgesic and lay the mouse on a heat pad.
4. Lay drapes over the mouse, with the operation area exposed, and check if the mouse is adequately anesthetized (e.g., by lack of reaction to a paw pinch).
5. Use a scalpel to make a 1.5–2-cm incision in the skin over the left kidney.
6. Use scissors to cut a small incision on the peritoneum above the left kidney.
7. Use scissors to carefully free the kidney from surrounding tissue (being careful not to cut the renal artery or vein).
8. Externalize the kidney.
9. Clamp the renal artery, renal vein, and ureter using a hemostatic clamp.
10. Use a scalpel to cut the kidney free, above the point of clamping.
11. Ligate the artery, vein, and ureter using nonabsorbable suture. Repeat this two to three times, each time ensuring that a tight ligation is achieved that will prevent bleeding.
12. On the last ligation, leave longer ends of the suture which you can hold with forceps when you release the clamp. If there is any sign of bleeding, re-clamp immediately below this suture and repeat **step 11**.
13. When satisfied that there is no bleeding from the vessels, cut the suture from the last ligation to a shorter length and then suture the peritoneum and skin.
14. Monitor the recovery of the mouse (*see* **Note 23**).
15. Measure blood glucose concentrations every day. When stable hyperglycemia has been detected (>20 mM for 3 consecutive days), the mouse should be killed. This typically occurs in the first week.
16. The pancreas can also be removed when the animal is killed which can serve as additional evidence of a lack of endogenous beta cell regeneration by histological processing and staining for insulin.

## 4 Notes

1. Male mice are often used as recipients to avoid effects of the estrus cycle. If female mice are used as recipients, it should be noted that female mice may be more resistant to the effects of streptozotocin and the dose may need to be adjusted. It should be noted that different strains can also have different susceptibilities to STZ. It is most common to use adult mice as recipients at around 8–12 weeks. If the experimental design requires that the islets are not rejected, then either immune-deficient animals should be used or inbred recipients should be used in conjunction with donor islets from the same strain. All experimental protocols using live animals must be reviewed and approved by the Ethical Committee for Animal Experimentation of the Institution.
2. Manipulations of immunodeficient animals must be performed in a sterile environment (laminar air-flow hood) with sterile materials. Although immune competent animals could be used when transplanting human islets to mice, this would be specifically in studies where the xenogeneic immune response is of interest.
3. Wear appropriate personal protective equipment for handling STZ (gloves, mask, and lab coat), and diligently follow all waste disposal regulations when disposing waste materials.
4. The microcentrifuge tube containing the islets can be stored on ice prior to transplantation, but prolonged periods (>2 h) should be avoided.
5. To avoid islets sticking to the tube, it is recommended that the tubing is pre-treated with a silicon coating such as Sigmacote. The tube should then be sterilized (we recommend gas sterilization).
6. The centrifuge is needed at the time of transplantation and thus should be either permanently in the animal facility or portable. As inserts, we use 1-mL syringes with the plunger removed and tip cut off. We place these modified 1-mL syringes within a 15-mL tube with the barrel flange resting on the top. This is used to contain the PE50 tubing for centrifugation. Once the islets have been loaded, the PE50 tubing attached to the cut pipette is inserted through the syringe with the cut pipette tip resting in the top of the syringe.
7. Veterinary advice should be sought and local guidelines followed.
8. The smallest needle possible should be used to generate a drop suitable for the chosen blood glucose meter. For meters that require a larger volume of blood, a larger needle can be used.

9. Most laboratories use blood glucose meters suitable for human use. It has been argued that these are not the best solution for rodents, but nonetheless they are the most widely used due to their ease of use and reasonable price. The upper limit of the meter should be considered (which typically varies between 27 mM and 33 mM), especially in studies using high single doses of streptozotocin. Some specialized meters are available that are suitable for rodent use and can measure blood glucose concentrations of up to 50 mM. However, such meters and their strips are substantially more expensive and may require a bigger blood droplet.
10. Citrate buffer can be stored at 4 °C for 1 month.
11. If an STZ concentration of 15 mg/mL of citrate buffer is used, the content of a 1.5-mL sterile microcentrifuge tube will allow the injection of 6–8 mice with a body weight of around 35 g.
12. STZ degrades within 15–20 min after dissolving in citrate buffer. The solution should be inspected visually to ensure the total dissolution of STZ prior to administration. The reconstituted solution is clear and colorless. It has to be used directly after preparation, and it cannot be stored.
13. The efficacy of streptozotocin in inducing hyperglycemia depends on the strain and gender of the mouse. We use 180 mg/kg for male C57BL/6 mice. When starting a study with a new strain, it is suggested that a literature search is carried out, and once a dose has been chosen, the mice are closely monitored. Mice losing excess weight (>15% in 72 h) should be killed. If mice consistently fail to become diabetic, an increased dose could be considered. If an individual mouse does not become diabetic, local guidelines may allow a second injection of STZ, but it is not recommended to exceed that. Some laboratories recommend fasting the mice prior to the injection, but we have not found that necessary.
14. It is expected that hyperglycemia develops within 1–2 days and there is a modest reduction in weight. If animals lose excessive weight (>15% in 72 h or 20% overall) or show signs of markedly decreased condition, they should be killed. Depending on local guidelines and the study design, insulin can be used to prevent excess weight. In our hands, this is usually not necessary, and while hyperglycemic, the mice lose around 10% body weight. STZ can cause liver and kidney toxicity, and if this is suspected, the animal should be killed and a lower dose of STZ considered.
15. There is no consensus of when mice should be considered diabetic. Normally used thresholds of designating an animal hyperglycemic include blood glucose values above 13.4 mM, 16.7 mM, or 20 mM.

16. Ideally, a second person should prepare the islets for transplantation to coincide with the surgeon having the animal ready to implant the islets. This avoids the islets being stored in pelleted form within the tubing for prolonged periods.
17. The top of the pipette tip will need to be cut prior to attaching the Hamilton syringe. It is suggested that this is done prior to autoclaving the pipette tips to keep them as sterile as possible.
18. If the islets are in a microcentrifuge tube, they will most likely have sedimented to the bottom by gravity and can be picked from the bottom. A short centrifuge can also be used to pellet the islets at the bottom of the tube, which may be necessary for larger numbers of islets. At this stage, the Hamilton syringe plunger can be used unthreaded to allow the uptake of the islets.
19. Take up a little air after the islets and hold the Hamilton syringe upright to allow the islets to settle at the bottom of the tip.
20. During the centrifugation stage, thread the plunger in the Hamilton syringe.
21. The liquid level at the top of the pipette tip will be lowered after centrifugation which will need topping up to ensure no air is introduced when reassembling the Hamilton syringe.
22. The incision should be made toward the lower pole of the kidney, which allows a larger area for islets to be implanted into.
23. The mouse should make a rapid recovery, showing normal movement within a few hours. The recovery is more rapid after islet transplantation compared with after nephrectomy.
24. Typically, larger numbers of human islets are needed to achieve normoglycemia compared to rodent islets.
25. Body weight is an essential measurement in animals that have been treated with STZ. While a small reduction in body weight after STZ treatment is normal (~10%), excessive loss of weight could either be due to toxic effects of STZ or due to excessive hyperglycemia. Animals showing excess body weight loss (>20%) should be killed. To show improved blood glucose homeostasis, a reduction in blood glucose must also be accompanied by stabilization or a gain in body weight.
26. Blood glucose should be measured while inducing as a little stress as possible in the mouse. This includes using the smallest needle possible to gain an appropriate amount of blood for the chosen blood glucose meter. Also the mouse should not be heavily restrained; holding the end of the tail lightly while allowing the animal to move is recommended.

## References


1. Graham ML, Janecek JL, Kittredge JA, Hering BJ, Schuurman HJ (2011) The streptozotocin-induced diabetic nude mouse model: differences between animals from different sources. Comp Med 61(4):356–360
2. Yang H, Wright JR Jr (2002) Human beta cells are exceedingly resistant to streptozotocin in vivo. Endocrinology 143(7):2491–2495
3. Estil Les E, Tellez N, Nacher M, Montanya E (2018) A model for human islet transplantation to immunodeficient streptozotocin-induced diabetic mice. Cell Transplant:1684–1691. https://doi.org/10.1177/0963689718801006
4. Merino JF, Nacher V, Raurell M, Aranda O, Soler J, Montanya E (1997) Improved outcome of islet transplantation in insulin-treated diabetic mice: effects on beta-cell mass and function. Diabetologia 40(9):1004–1010
5. Merino JF, Nacher V, Raurell M, Biarnes M, Soler J, Montanya E (2000) Optimal insulin treatment in syngeneic islet transplantation. Cell Transplant 9(1):11–18
6. Montana E, Bonner-Weir S, Weir GC (1993) Beta cell mass and growth after syngeneic islet cell transplantation in normal and streptozocin diabetic C57BL/6 mice. J Clin Invest 91 (3):780–787. https://doi.org/10.1172/JCI116297
7. Matsumoto I, Sawada T, Nakano M, Sakai T, Liu B, Ansite JD, Zhang HJ, Kandaswamy R, Sutherland DE, Hering BJ (2004) Improvement in islet yield from obese donors for human islet transplants. Transplantation 78(6):880–885. 00007890-200409270-00014 [pii]

# Chapter 11

# Transplantation of Islets of Langerhans into the Anterior Chamber of the Eye for Longitudinal In Vivo Imaging

**Christian M. Cohrs, Chunguang Chen, and Stephan Speier**

## Abstract

Noninvasive in vivo imaging techniques are attractive tools to longitudinally study various aspects of islet of Langerhans physiology and pathophysiology. Unfortunately, most imaging modalities currently applicable for clinical use do not allow the comprehensive investigation of islet cell biology due to limitations in resolution and/or sensitivity, while high-resolution imaging technologies like laser scanning microscopy (LSM) lack the penetration depth to assess islets of Langerhans within the pancreas. Significant progress in this area was made by the combination of LSM with the anterior chamber of the mouse eye platform, utilizing the cornea as a natural body window to study cell physiology of transplanted islets of Langerhans. We here describe the transplantation and longitudinal in vivo imaging of islets of Langerhans in the anterior chamber of the mouse eye as a versatile tool to study different features of islet physiology in health and disease.

**Key words** Islets of Langerhans, In vivo imaging, Anterior chamber of the eye, Transplantation, Diabetes

## 1 Introduction

In vitro studies have significantly contributed to our current understanding of physiological and pathological processes in the islets of Langerhans. However, in vitro experimental settings often do not fully reflect the in vivo situation, thus making it necessary to perform experiments that better resemble the physiological systemic conditions. Apart from new in situ approaches (*see* Chapter 20—Panzer et al.) where the natural environment of islets is preserved, various noninvasive in vivo imaging techniques including computer tomography (CT), magnetic resonance imaging (MRI), or positron emission tomography (PET) are being used to investigate islet mass and function. However, while these imaging modalities have the penetration depth to assess islets within the pancreas in the abdominal cavity, they are limited in their potential to assess islet cell physiology by low spatial and/or temporal resolution, reduced

Aileen J. F. King (ed.), *Animal Models of Diabetes: Methods and Protocols*, Methods in Molecular Biology, vol. 2128,
https://doi.org/10.1007/978-1-0716-0385-7_11, 

signal sensitivity, or both [1]. Contrariwise, high-resolution imaging technologies like laser scanning microscopy (LSM) lack the penetration depth to assess islet biology within the pancreas in vivo [2].

Utilization of the anterior chamber (AC) of the mouse eye as a natural body window for longitudinal LSM imaging of transplanted islets of Langerhans [3, 4] overcomes the limitations of penetration depth, enabling the longitudinal study of islet cell biology in vivo. The initial experiments described by Speier et al. in 2008 have shown that islets transplanted into the AC develop the typical endogenous fenestrated endothelium, become re-innervated, and regain function after engraftment, resembling the endogenous in vivo situation within the pancreas [3, 4]. By choosing appropriate recipient mouse strains in combination with donor islets from mouse, pig, and human, syngeneic as well as allo- and xenotransplantations have been performed to assess in vivo islet biology of different species [5–17]. In addition to address general islet physiology, including cell morphology and function [3, 11, 13, 14], various studies investigating type 1 and type 2 diabetes pathogenesis have been reported using this innovative approach, providing novel insight into disease development [5, 9, 10, 15]. Thereby, the availability of genetically encoded reporters expressed in donor islets facilitated and corroborated the versatility of this powerful noninvasive imaging technique to study islet biology in health and disease.

Here, we provide a detailed description of islet transplantation into the anterior chamber of the mouse eye and their longitudinal confocal or two-photon LSM imaging using the example of islet engraftment and revascularization.

## 2 Materials

### 2.1 Animals

Donor as well as recipient mice have to be carefully chosen according to the aim of the study. While the donor and recipient strain should be identical for syngeneic islet transplantation, allo- or xenotransplantation requires the use of an immune-deficient recipient mouse strain. Additionally, using albino mouse strains as recipients will facilitate imaging of the grafts due to the non-pigmented iris (*see* **Note 1**).

### 2.2 Reagents

1. PBS sterile.
2. Isoflurane.
3. 100% $O_2$.
4. Pilocarpine: prepare a 0.2–0.4% pilocarpine solution diluted in 0.9% NaCl.

5. Vidisic or similar eye liquid gel.
6. Vascular labels: Fluorescently labeled Dextran (e.g., FITC-Dextran), Qtracker Vascular Label (655 or 705).
7. Carprofen or similar (*see* **Note 2**).

### 2.3 Equipment

#### 2.3.1 General Equipment

1. Syringes: 1 mL, 5 mL.
2. Cannulas: 25G, 27G, 30G.
3. Pipettes: 1000 μL, 200 μL, 20 μL, 10 μL.
4. Cell culture dishes: 100 mm, 60 mm, 35 mm.
5. Cotton swabs.
6. 80% Ethanol.

#### 2.3.2 Transplantation

1. Anesthesia unit for rodents.
2. 0.5-mL micro syringe (Hamilton).
3. Polyethylene tubing: 0.4 mm inner diameter (i.d.) and 0.8 mm outer diameter (o.d.).
4. Stereomicroscope.
5. Head-holding unit with gas mask and eye stabilizer connected to a UST-2 Solid Universal Joint and Dumont no. 5 forceps covered with polyethylene tubing (i.d. = 0.8 mm, o.d. = 1.2 mm) (Fig. 1a).
6. Custom-made beveled glass cannula (i.d. 0.32 mm, o.d. 0.4 mm).

#### 2.3.3 Imaging

1. Upright laser scanning microscope equipped with Argon and HeNe lasers as well as a two-photon laser with respective non-descanned detectors. In order to image single islets, a 10–40× water immersion objective with a long working distance should be used.
2. Custom-made stage with head holder adapter, nose piece, and eye stabilizer connected to a UST-2 Solid Universal Joint and a heating pad (Fig. 2a).
3. For mouse intubation: Mouse intubation kit with an IV catheter, 20 GA, 1.1 × 25 mm (we use the MiniVent Ventilator for Mice, Model 845).

## 3 Methods

### 3.1 Transplantation of Pancreatic Islets into the Anterior Chamber of the Eye

1. Isolated pancreatic islets should be cultured in suitable media and should be recovered after isolation at least overnight.
2. Prepare all solutions and have all equipment ready before starting the transplantation.

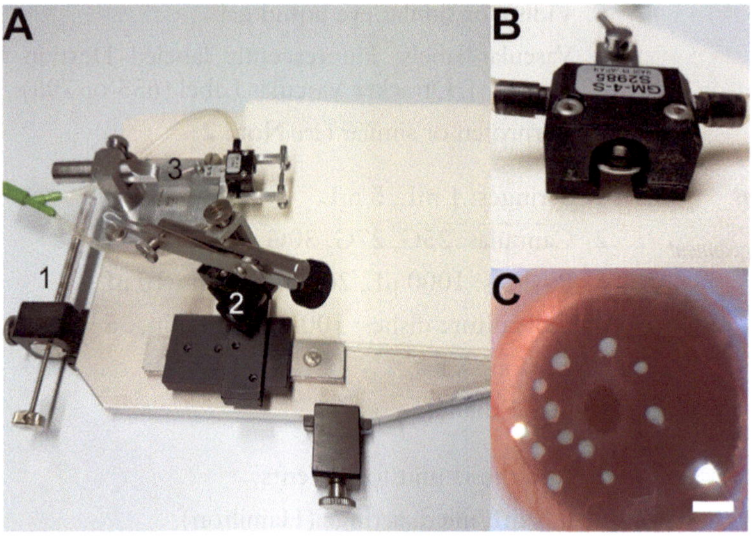


**Fig. 1** Islet of Langerhans transplantation into the anterior chamber of the mouse eye. Custom-made stage with mouse head holder for islet transplantation (**a**) with fixed micro syringe (1), universal joint holding curved forceps for eye exposure (2), and the head holder unit with tubing connections for isoflurane anesthesia for the nose piece (magnified view in (**b**)) (3). Picture of a mouse eye after successful mouse islet transplantation (**c**). Scale bar = 500 μm

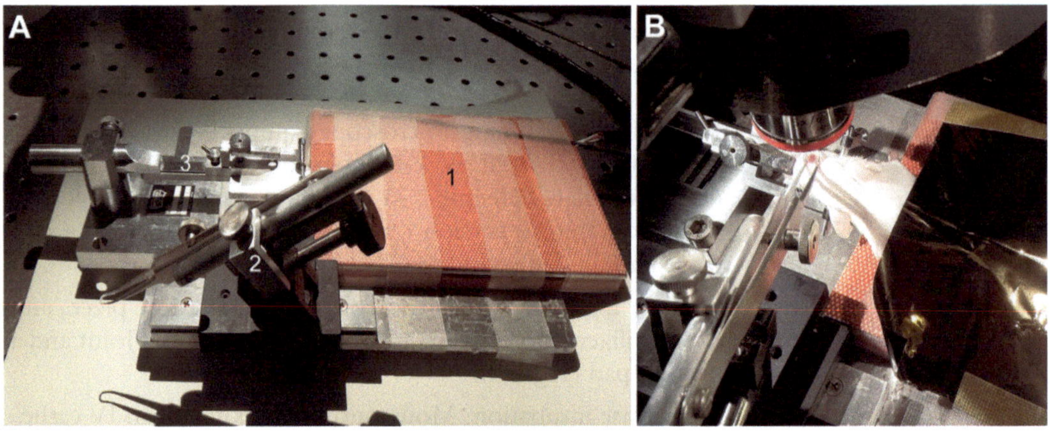


**Fig. 2** Setup for in vivo imaging. Custom-made stage with mouse head holder for in vivo imaging (**a**) with a heating pad to control the body temperature of the anesthetized mouse (1), universal joint holding curved forceps for eye exposure (2), and the head holder unit (3). Picture of a mouse before starting the imaging session with the head fixed by the head holder and the eye being exposed by the eye stabilizer (**b**)

(a) Fill the 0.5-mL micro syringe with sterile PBS.

(b) Connect the polyethylene tubing with a 30G cannula and connect it to the micro syringe.

(c) Insert the glass cannula (approximately half of its length) into the other end of the polyethylene tubing and glue at the tubing/cannula junction using superglue.

(d) Briefly prime the 35-mm dishes with culture media, remove excessive solution, and wash and fill the dish with sterile PBS (*see* **Note 3**).

(e) Prepare Carprofen and prefill a 1-mL syringe equipped with a 30G cannula.

(f) Prefill another 1-mL syringe with sterile PBS and connect a 27G cannula.

(g) Briefly disinfect all surfaces using 80% ethanol.

3. Transfer up to 30 islets from the culture medium to a 35-mm cell culture dish filled with sterile PBS. Rotating the dish in small circles will gather the islets in the middle of the dish to facilitate aspiration.
4. Slowly aspirate islets to the glass cannula attached to the polyethylene tubing and the micro syringe in a minimal volume (up to 20 μL) as larger volumes will complicate injection into the anterior chamber of the eye. For further reduction of the injection volume, fix the cannula in an upside down position until infusion as this will allow the islets to settle at the end of the cannula.
5. Place the mouse in an incubation chamber pre-flooded with 2–3% isoflurane (v/v) in $O_2$ to induce anesthesia. Subsequently transfer the mouse to the head holder equipped with a gas mask (nose piece) connected to the isoflurane/$O_2$ mixture (*see* Fig. 1a, b). Depending on the mouse strain, the isoflurane concentration can be reduced to 1.5–2%. For postoperative pain relief, administer adequate amounts of analgesic drugs (*see* **Note 1**).
6. Fix the mouse in the head holder and turn the mouse eye used for transplantation upward. Use the eye stabilizer covered with a polyethylene tubing at the end that is attached to the UST-2 Solid Universal Joint to slightly expose the eye until the corneoscleral junction is visible to achieve proper stabilization for transplantation (*see* **Note 4**).
7. Use the 1-mL syringe filled with sterile PBS and cover the eye with a drop of PBS. Subsequently, use the 27G needle to puncture the cornea close to the sclera. This is best performed when the needle is inserted in an upright angle that allows fast closure of the incision afterward and minimizes the possibility to damage the iris that causes bleedings (*see* **Note 5**).
8. Carefully insert the blunt cannula with the previously aspirated islets into the incision making sure that the iris is again not being damaged by the cannula. Slowly infuse the islets into the anterior chamber with the minimal amount of solution possible (*see* **Note 6**). After infusion, slowly withdraw the cannula and close the incision by gently pushing the cornea at the site of the incision.

9. If islets are clustering together due to the infusion process, careful movement of the single islets can be achieved by applying minimal pressure to the cornea using the blunt cannula. If possible, acquire a picture of the transplanted eye (Fig. 1c).
10. Carefully remove the eye stabilizer and allow the islets to settle onto the iris for an additional 10–15 min.
11. Place a drop of Vidisic onto the transplanted eye to ensure moistening.
12. Stop the isoflurane anesthesia, turn the mouse back into a horizontal position, and leave the mouse in the nose piece providing the mouse with additional $O_2$. Before the mouse wakes up, remove the mouse from the holder and place it back into a new cage placed onto a heating plate and ensure full recovery of the mouse.
13. To ensure a successful transplantation, observe the transplanted eye the following days for potential irritations.
14. Before imaging the first time, allow the cornea to heal for at least 2 days.

### 3.2 Imaging of Transplanted Islets in the Anterior Chamber of the Eye

1. Place the mouse in an incubation chamber pre-flooded with 2–3% isoflurane (v/v) in $O_2$ to induce anesthesia for 7–10 min.
2. In order to reduce movement artifacts during imaging caused by gasping under isoflurane anesthesia, intubate the mouse by using an iv catheter (20G) and the mouse intubation kit. Connect the iv catheter to the MiniVent ventilator for mice using a stroke volume of 270 μL and a frequency of 240 strokes/min.
3. Place the mouse onto the custom-made stage equipped with a heating pad set to 37 °C (Fig. 2a).
4. Fix the mouse into the head holder. Further fix the mouse at the nose piece. Use the eye stabilizer covered with a polyethylene tubing at the end that is attached to the UST-2 Solid Universal Joint to slightly expose the eye until the corneoscleral junction is visible to achieve proper stabilization (Fig. 2b) (*see* **Note 5**).
5. Place the mouse under an upright microscope and use a low-magnification objective (if applicable) to get an overview of the islet graft localization on the iris (*see* **Note 7**).
6. For muscle dilation, cover the eye with 0.2–0.4% pilocarpine for 10 min.
7. Based on the transplanted islets and the potential fluorescent markers being expressed in the endocrine cells, choose a suitable vascular label that will be delivered by iv tail vein injection. Injection of the dye can be performed during pilocarpine incubation time.

8. Remove pilocarpine using a cotton swab and place a drop of Vidisic onto the eye. Vidisic serves as immersion liquid for imaging with the higher magnification water immersion objectives.
9. Change to a water immersion dipping objective with a long working distance (10–40× objectives).
10. For morphology and volume, islets can be imaged by assessing islet backscatter. This can be achieved by using, e.g., the 633-nm laser (also other laser lines can be used; *see* [13]) to acquire reflection at a narrow emission range ± 5 nm of the laser wavelength (*see* **Note 8**).
11. Depending on the used LSM system, additional fluorescent reporters might be simultaneously or independently imaged from the backscatter acquisition (*see* **Note 8**).
12. To ensure imaging the vasculature and potential other fluorescent reporters within the donor islets, two-photon imaging can be performed after acquiring the confocal images of the islet. This will allow assessing fluorescent signals throughout the entire islet (*see* **Note 8**).
13. Repeat **steps 10–12** for all islets that can be imaged within the anterior chamber of one mouse (*see* **Note 9**).
14. After imaging is finished, remove the mouse from the microscope, free the eye, bring the mouse into a horizontal position, and unfix the mouse head from the head holder.
15. If using isoflurane, stop anesthesia and only deliver $O_2$ through the MiniVent ventilator to facilitate fast recovery of the mouse (approximately 2–3 min). When whiskers start moving, remove the iv catheter and place the mouse back into a cage on a heating plate to ensure stable body temperature during the recovery phase.
16. Make sure that the mouse fully recovers from anesthesia.
17. Repeat **steps 1–16** for every desired time point (*see* **Note 10**).
18. For image processing and analysis, use the microscope software, ImageJ, or any suitable program for 3D rendering and quantification (e.g., Imaris).

## 4 Notes

1. All experiments on living animals have to be permitted in advance by the local ethical authorities.
2. For pain relief, use drugs that are in concert with suggestions by the local ethical authorities. Make sure that the injection performed before transplantation of the islets will last at least

for 12 h. If animals show any sign of pain after 12 h, inject an additional dose of the drug. Usually, animals show no sign of pain after 24 h. Check carefully the transplanted eye and the general appearance of the animal daily within the first week after transplantation.

3. Quickly rinse the dish with culture media before adding sterile PBS as the FBS within the media will prevent the islets sticking to the surface of the cell culture dish.
4. Fix the eye carefully. Always make sure that blood flow is still visible in blood vessels of the iris as complete clamping for the entire transplantation procedure will damage the eye.
5. As sharp needles are necessary to perform an accurate incision, use a fresh cannula for each mouse.
6. Avoid infusing excessive amounts of solution as this would result in too high pressure within the anterior chamber. This may cause irritation of the eye and may result in poor transplantation outcome as islets tend to get pushed out of the incision as soon as the cannula is being removed. Additionally, avoid pressurized infusion of the islets as this will alter morphology and can also lead to disintegration of single islets.
7. Before imaging the islets within the anterior chamber for the first time, it is advantageous to draw an "islet map" of the anterior chamber of the eye. This facilitates relocation of the imaged islets at every following imaging time point.
8. Do not exceed a laser power of over 20 mW onto the sample as this may cause irritation of the islets and the iris if performed multiple times.
9. In order to minimize the burden on the mouse, limit the entire anesthesia time (induction of anesthesia—waking up after imaging) to 1.5 h per imaging time point. This usually allows imaging of seven to ten islets with both reflection and two-photon acquisition.
10. Although engraftment and revascularization is initiated shortly after transplantation and develops fast, avoid imaging single mice too often. It is beneficial to keep at least 2–3 days recovery time between two imaging sessions.

## Acknowledgments

This work was supported with funds from the Paul Langerhans Institute Dresden (PLID) of Helmholtz Zentrum München at the University Clinic Carl Gustav Carus of Technische Universität Dresden, the German Ministry for Education and Research (BMBF) to the German Centre for Diabetes Research (DZD), the DFG SFB/Transregio 127, and the European Foundation for the Study of Diabetes (EFSD)/Boehringer Ingelheim Basic Research.

## References


1. Laurent D, Vinet L, Lamprianou S, Daval M, Filhoulaud G, Ktorza A et al (2016) Pancreatic beta-cell imaging in humans: fiction or option? Diabetes Obes Metab 18(1):6–15
2. Ahlgren U, Gotthardt M (2010) Approaches for imaging islets: recent advances and future prospects. Adv Exp Med Biol 654:39–57
3. Speier S, Nyqvist D, Cabrera O, Yu J, Molano RD, Pileggi A et al (2008) Noninvasive in vivo imaging of pancreatic islet cell biology. Nat Med 14(5):574–578
4. Speier S, Nyqvist D, Kohler M, Caicedo A, Leibiger IB, Berggren PO (2008) Noninvasive high-resolution in vivo imaging of cell biology in the anterior chamber of the mouse eye. Nat Protoc 3(8):1278–1286
5. Abdulreda MH, Rodriguez-Diaz R, Caicedo A, Berggren PO (2016) Liraglutide compromises pancreatic beta cell function in a humanized mouse model. Cell Metab 23 (3):541–546
6. Ali Y, Diez J, Selander L, Zheng X, Edlund H, Berggren PO (2016) The anterior chamber of the eye is a transplantation site that supports and enables visualisation of beta cell development in mice. Diabetologia 59(5):1007–1011
7. Almaca J, Molina J, Arrojo EDR, Abdulreda MH, Jeon WB, Berggren PO et al (2014) Young capillary vessels rejuvenate aged pancreatic islets. Proc Natl Acad Sci U S A 111 (49):17612–17617
8. Bader E, Migliorini A, Gegg M, Moruzzi N, Gerdes J, Roscioni SS et al (2016) Identification of proliferative and mature beta-cells in the islets of Langerhans. Nature 535 (7612):430–434
9. Chen C, Chmelova H, Cohrs CM, Chouinard JA, Jahn SR, Stertmann J et al (2016) Alterations in beta-cell calcium dynamics and efficacy outweigh islet mass adaptation in compensation of insulin resistance and prediabetes onset. Diabetes 65(9):2676–2685
10. Chmelova H, Cohrs CM, Chouinard JA, Petzold C, Kuhn M, Chen C et al (2015) Distinct roles of beta-cell mass and function during type 1 diabetes onset and remission. Diabetes 64(6):2148–2160
11. Cohrs CM, Chen C, Jahn SR, Stertmann J, Chmelova H, Weitz J et al (2017) Vessel network architecture of adult human islets promotes distinct cell-cell interactions in situ and is altered after transplantation. Endocrinology 158(5):1373–1385
12. Ilegems E, Dicker A, Speier S, Sharma A, Bahow A, Edlund PK et al (2013) Reporter islets in the eye reveal the plasticity of the endocrine pancreas. Proc Natl Acad Sci U S A 110 (51):20581–20586
13. Ilegems E, van Krieken PP, Edlund PK, Dicker A, Alanentalo T, Eriksson M et al (2015) Light scattering as an intrinsic indicator for pancreatic islet cell mass and secretion. Sci Rep 5:10740
14. Kemter E, Cohrs CM, Schafer M, Schuster M, Steinmeyer K, Wolf-van Buerck L et al (2017) INS-eGFP transgenic pigs: a novel reporter system for studying maturation, growth and vascularisation of neonatal islet-like cell clusters. Diabetologia 60(6):1152–1156
15. Li G, Wu B, Ward MG, Chong AC, Mukherjee S, Chen S et al (2016) Multifunctional in vivo imaging of pancreatic islets during diabetes development. J Cell Sci 129 (14):2865–2875
16. van Krieken PP, Dicker A, Eriksson M, Herrera PL, Ahlgren U, Berggren PO et al (2017) Kinetics of functional beta cell mass decay in a diphtheria toxin receptor mouse model of diabetes. Sci Rep 7(1):12440
17. Weitz JR, Makhmutova M, Almaca J, Stertmann J, Aamodt K, Brissova M et al (2018) Mouse pancreatic islet macrophages use locally released ATP to monitor beta cell activity. Diabetologia 61(1):182–192

# Chapter 12

# Multicolor Labeling and Tracing of Pancreatic Beta-Cell Proliferation in Zebrafish

**Sumeet Pal Singh and Nikolay Ninov**

## Abstract

During embryogenesis, beta-cells arise from the dorsal and ventral bud originating in the endoderm germ layer. As the animal develops to adulthood, the beta-cell mass dramatically increases. The expansion of the beta-cell population is driven by cell division among the embryonic beta-cells and supplanted by neogenesis from post-embryonic progenitors. Here, we describe a protocol for multicolor clonal analysis in zebrafish to define the contribution of individual embryonic beta-cells to the increase in cell numbers. This technique provides insights into the proliferative history of individual beta-cells in an islet. This insight helps in defining the replicative heterogeneity among individual beta-cells during development. Additionally, the ability to discriminate individual cells based on unique color signatures helps quantify the volume occupied by beta-cells and define the contribution of cellular size to the beta-cell mass.

**Key words** Single cell, Lineage tracing, Brainbow, Quiescence, Heterogeneity

## 1 Introduction

### 1.1 Pancreatic Beta-Cell Development in Zebrafish

The zebrafish pancreas shares physiological similarities with the human pancreas. Similar to the mammalian counterpart, the adult zebrafish pancreas consists of the endocrine lineage organized into multiple islets of Langerhans surrounded by the exocrine lineage [1]. The exocrine lineage is composed of acinar and ductal cells and aids in the digestive process. The endocrine islets are composed of insulin-producing beta-cells, glucagon-producing alpha-cells, somatostatin-producing delta-cells, ghrelin-producing epsilon-cells, and pancreatic polypeptide (PP)-producing cells. Cell ablation experiments and direct visualization of the glucose-stimulated calcium influx have confirmed the role of the pancreatic beta-cells in regulating blood glucose levels in zebrafish [2, 3].

Similar to the mammalian counterpart, the zebrafish pancreas develops from the endoderm germ layer [4]. The endoderm germ layer forms during gastrulation between 6 and 9 h post-fertilization (hpf). The first appearance of the beta-cells in the zebrafish embryo

Aileen J. F. King (ed.), *Animal Models of Diabetes: Methods and Protocols*, Methods in Molecular Biology, vol. 2128,
https://doi.org/10.1007/978-1-0716-0385-7_12, 

occurs as dispersed cells in the dorsal bud anlage within the endoderm around 14 hpf [5]. The dorsal bud-derived beta-cells (DBCs) express the insulin gene and coalesce together by 24 hpf to form the embryonic islet (Fig. 1a). Additional beta-cells differentiate from the ventral pancreatic bud and are added to the embryonic islet around 48 hpf. The dorsal and ventral bud-derived beta-cells (D + VBCs) constitute the primary islet of the zebrafish (Fig. 1a). As development proceeds, more beta-cells differentiating from the Notch-positive ductal cells are added to the primary islet [6, 7]. In addition, the Notch-positive ductal cells differentiate to form smaller clusters of endocrine cells within the pancreas, called secondary islets. The secondary islets lie posterior to the primary islet (Fig. 1a). The simple and stereotypical architecture of the zebrafish islets helps identify and isolate the zebrafish primary islet at any developmental stage.

### 1.2 Lineage Tracing of Beta-Cells Using a Multicolor Barcode Technique

During development, the number of beta-cells increases as the animal develops to adulthood. This increase in cell numbers is driven in part by nutrient-dependent replication of differentiated beta-cells [8]. However, not all beta-cells harbor equal replication potential. Specifically, in zebrafish, the dorsal bud DBCs were shown to be largely quiescent until 12 dpf [9]. In contrast, the ventral bud-derived beta-cells (VBCs) proliferate to support the increase in beta-cell numbers. However, it remained unclear if the DBCs were quiescent throughout the lifetime of the zebrafish or if their quiescence was flexible in nature.

To illuminate the dynamics of cell proliferation in beta-cells during development, we adapted the multicolor barcode system, called Brainbow [10]. The Brainbow system elegantly modifies the Cre-Lox recombination system to produce a palette of multiple colors. The modification consists of a Cre-reporter construct in which different colored fluorescent proteins are flanked with specific Lox sites (Fig. 1b). In the Brainbow1.0L construct, the red fluorescent protein (RFP) is expressed in the non-recombined (or "default") state. Upon exposure to Cre recombinase, the red fluorescent protein is replaced by either yellow or blue fluorescent proteins. The replacement depends on the type of Lox site excised by the Cre recombinase and is stochastic. If multiple copies of the construct are integrated into the genome, each cassette can independently undergo changes in the fluorescent protein upon induction of Cre recombinase (Fig. 1c). The combination of different levels of red, green, and blue fluorescent proteins being expressed from each cassette after recombination determines the color signature of the cell. As the recombination is irreversible, the color signature is passed onto the daughter cells upon cell division. Individual cells with unique color signatures can divide and form multicellular clones, or remain as single cells, indicating quiescence (Fig. 1d). This allows recording of the proliferative histories of individual cells.

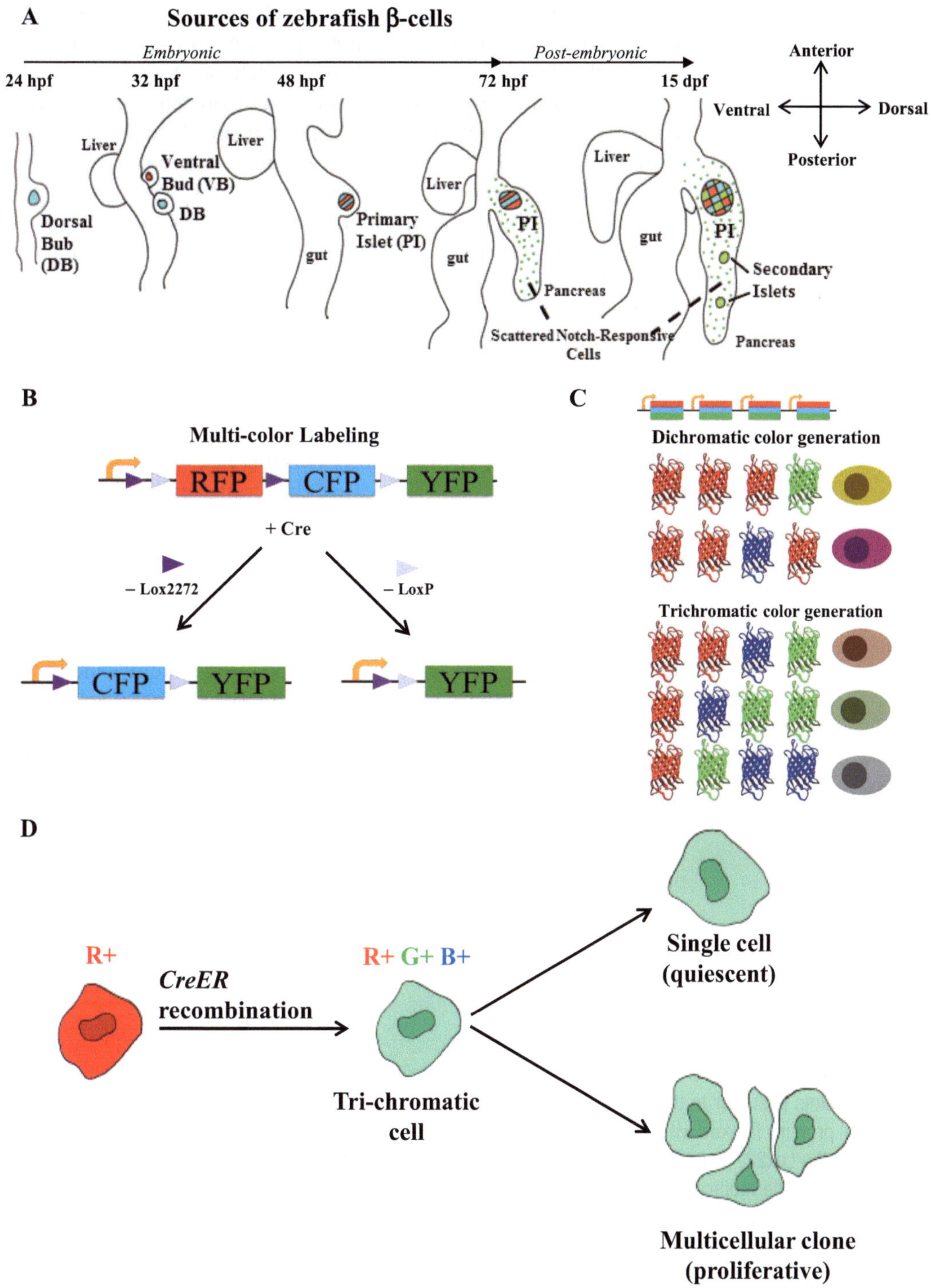


**Fig. 1** Multicolor lineage tracing of zebrafish beta-cells. (**a**) Cartoon depicting the development of the zebrafish pancreas. Around 24 hpf, the dorsal bud-derived beta-cells coalesce together to form the embryonic islet. Ventral bud-derived beta-cells are added to the islet by 48 hpf, giving rise to the primary islet. As development proceeds, Notch-responsive cells (NRCs) can differentiate into beta-cells, which contribute to the beta-cell

The capacity to follow individual cell lineages depends on the discriminatory power of the color signature. Color signatures can be broadly divided into two categories: dichromatic (those that are formed by combination of two fluorescent proteins) and trichromatic (those that are formed by combination of three fluorescent proteins) (Fig. 1c). A single recombination event induces the expression of green or blue fluorescent proteins, which along with the default red fluorescent protein generates a dichromatic cell. On the other hand, two independent recombination events are required to add green and blue fluorescent proteins to the default red fluorescent protein, thereby generating a trichromatic cell. As Cre-Lox recombination is a stochastic event, the proportion of trichromatic cells induced within the beta-cell population will be lower than the proportion of dichromatic cells. The lower proportion of trichromatic cells reduces the probability to label two cells with the same color combination. This increases the uniqueness of the color signature and increases the confidence when tracing individual cells [10, 11]. The computational pipeline presented in the chapter automatically restricts the clonal analysis to trichromatic cells.

In this protocol, we will outline the steps needed to adapt the system to zebrafish beta-cells and use the adapted version to analyze the proliferation of the DBCs [3]. The adapted system, called Beta-bow, drives the Brainbow1.0L construct from the beta-cell specific insulin promoter. Using Beta-bow, we showed that most DBCs remain quiescent until 15 days post-fertilization (dpf). However, between 15 and 30 dpf, a subset of DBCs generated multicellular clones, suggesting the initiation of cell division within this quiescent beta-cell population [3]. Using the steps outlined here, the protocol can be adapted to other cells undergoing proliferation.

**Fig. 1** (continued) numbers in the primary islet. Additionally, the NRCs contribute to the generation of secondary islets posterior to the primary islet. (**b**) The multicolor labeling cassette, Brainbow-1.0L, contains fluorescent proteins flanked with specific Lox sites. Upon induction of the Cre recombinase, the DNA is excised using either the Lox2272 or LoxP site, which leads to a switch from RFP expression to CFP or YFP expression, respectively. (**c**) The multicolor labeling cassette can be inserted as multiple repeats within the genome. This gives rise to discrete recombination events within each cell. Each recombination event can independently switch RFP expression to CFP or YFP, which results in the development of a color palette. The palette is broadly classified into two categories: dichromatic (composed of the default RFP and either CFP or YFP) and trichromatic (composed of the default RFP, along with both CFP and YFP). A dichromatic color can arise from a single recombination event, while a trichromatic signature requires at least two independent recombination events within the same cell. (**d**) The Cre-recombination induced trichromatic cell can remain as a single cell, indicating quiescence, or divide to form a multicellular clone. The cells belonging to the multicellular clone share a similar color signature. Figure adapted from [3]

## 2 Materials

The section outlines the reagents required for all the steps of protocol, from generating a transgenic line to image acquisition and analysis. To adapt the protocol to different tissues, certain steps would need to be modified, and tips for doing so are presented in the Notes section.

### 2.1 Construction of ins:BB1.0L; cryaa: RFP (Beta-bow)

1. Thy1-BB1.0L plasmid (Addgene Plasmid Number: 18725) (*see* **Note 1** for variants of multicolor reporter system).
2. ins:MCS; cryaa:RFP vector (Addgene Plasmid Number: 110282). This vector designed in author's lab contains the zebrafish insulin promoter with a beta-globin intron upstream of a multiple cloning site (MCS); alongside the zebrafish crystalline, alpha A (cryaa) promoter driving red fluorescent protein (RFP). The cryaa:RFP helps mark the transgenic lines with red fluorescence, which is clearly visible in the eye. The entire plasmid is flanked with I-SceI sites to facilitate transgenesis [12] (*see* **Note 2** for tissue-specific adaptations).
3. Restriction enzymes for cloning: Eco53kI, NsiI, and EcoRV.
4. DNA ligation and cloning setup.
5. PCR primers:

   insulin_Vector_FP: ACACCCTGGTCATCATCCTG

   BB1.0L_RP: CACCTTGTAGATCAGCGTGC

### 2.2 Transgenesis

1. Endotoxin-free DNA Maxiprep kit.
2. I-SceI meganuclease enzyme.
3. Zebrafish Microinjection System, as outlined in [13, 14].

### 2.3 Inducing Cre-Lox Recombination for Generation of Multiple Colors

1. Cre-driver line: Tg(ins:Cre-$ER^{T2}$; cryaa:CFP) [3].
2. 4-Hydroxytamoxifen (4-OHT): 2 mM solution in 100% ethanol. Prepare 20 μL aliquots of the 2 mM stock solution in 0.5-mL Eppendorf tubes and store in −20 °C. Thaw aliquots at room temperature before use. Avoid multiple freeze-thaw of single aliquot. Carefully observe any precipitation in the stock solution. Precipitated 4-OHT can be re-dissolved by heating the solution to 65 °C for 5–10 min [15].
3. Zebrafish embryo medium (E3 medium) [16].
4. 6-well clear plastic plates.

### 2.4 Beta-bow Sample Preparation

1. Tricaine methane sulfonate (MS222): 200 mg/l solution in fish water.
2. Phosphate-buffered saline (PBS).

3. Paraformaldehyde: 4% in PBS. Can be prepared in bulk, aliquoted in 15-mL plastic tubes and frozen for long-term storage. For use, thaw one aliquot and use the appropriate amount. The remaining solution can be stored at 4 °C for 1 week.
4. Triton X-100.
5. Fine dissecting forceps.
6. 90-mm petri dish.
7. Glass microscope slides: 76 × 26 mm (length × width).
8. Precision glass coverslips: 24 × 60 mm (length × width), 170 μm ± 5 μm thickness.
9. Mounting medium without DAPI (purchase "Fluoromount-G" or "Vectashield Antifade Mounting Medium for Fluorescence (H-100)" or similar).
10. Stereo microscope equipped with fluorescence lamp and red filter cube (TRITC: excitation, 532–554 nm; emission, 570–613 nm, or Texas Red: excitation, 540–580 nm; emission: 592–667 nm).

### 2.5 Confocal Imaging

1. Confocal microscope (ZEISS LSM780 or similar).
2. Lasers for exciting CFP (450 nm), YFP (514 nm), and RFP (581 nm).
3. Objective lens: 40 × 1.2 NA, water immersion (or similar).

### 2.6 Beta-bow Image Analysis

1. Fiji, downloaded from https://fiji.sc/
2. Beta-bow analysis scripts, downloaded from the author's github page: https://github.com/sumeetpalsingh/Beta-bow
3. ImageJ plugin for 3D object counting [17], downloaded from http://imagej.net/3D_Objects_Counter
4. R, downloaded from https://www.r-project.org/
5. RStudio, downloaded from https://www.rstudio.com/

## 3 Methods

Here, we outline the procedure for multicolor lineage tracing of beta-cells derived from the dorsal bud until 30 days post-fertilization (dpf). Maintain the zebrafish under standard conditions at 28 °C, and carry out animal experiments in accordance to the animal welfare laws.

### 3.1 Generation of the Beta-bow Transgenic Zebrafish

1. Digest 5 μg of Thy1-BB1.0L plasmid with Eco53kI and NsiI at 37 °C for 2 h. The digestion will yield six DNA fragments of approximate length 4.5 kb, 3.3 kb, 2.7 kb, 1 kb, 322 bp, and 273 bp. The 3.3-kb fragment contains the BB1.0L construct (*see* **Note 3**).

2. Digest 5 μg of ins:MCS; cryaa:RFP vector with EcoRV and NsiI at 37 °C for 2 h. The digestion will linearize the vector and yield a DNA fragment of approximately 9 kb.
3. Run the digested products on a 0.7% agarose DNA electrophoresis gel. To efficiently resolve the five digestion products from Thy1-BB1.0L digestion, run at a moderate voltage (90 V).
4. Cut the 3.3-kb band from the Thy1-BB1.0L digestion and the 9-kb band from the ins:MCS; cryaa:RFP digestion. Collect the cut gel pieces in separate 1.5-mL Eppendorf tubes, and purify the fragments using a DNA gel extraction kit.
5. Ligate the two fragments using T4 DNA ligase. For this, prepare a 20 μl reaction that contains 1× T4 DNA Ligase Buffer, 2 μL T4 DNA Ligase, and volume of the two fragments that provide the following amount of DNA: 50 ng of Vector (ins:MCS; cryaa:RFP, 9 kb), 50 ng of Insert (Thy1-BB1.0L, 3.3 kb). This amount of DNA fragments used for ligation provides a 1:3 molar ratio for Vector:Insert. Carry out the ligation reaction for 2 hours to overnight at room temperature.
6. Transform 5 μL of the ligation reaction into chemically competent bacteria using the protocol provided by the manufacturer (*see* **Note 4**). Recover the transformed bacteria for 1 h in 1 mL SOB media. Plate 200 μL of recovered cells on ampicillin containing LB agar plates, and let the bacteria grow overnight at 37 °C.
7. Identify the correctly cloned colonies by PCR using the insulin_Vector_FP and BB1.0L_RP. Colonies with correctly cloned plasmid will yield a band of approximately 650 bp (Fig. 3a). The positive colonies now contain the ins:BB1.0L; cyaaa:RFP (abbreviated as Beta-bow) plasmid (Fig. 3b).
8. Inoculate two correct colonies into 100 mL of LB media containing ampicillin and maxiprep the plasmid using the manufacturer's instructions. Verify the success of the cloning by sequencing the plasmid using insulin_Vector_FP primer.
9. The Beta-bow plasmid is flanked with I-SceI meganuclease sites for efficient transgenesis (*see* **Note 5**). Inject the Beta-bow plasmid into the one-cell stage zebrafish along with I-SceI meganuclease for transgenesis as outlined in [13]. The injected embryos will give rise to mosaic animals with random integration of the Beta-bow plasmid, termed F0 animals.
10. Screen the F0 mosaic embryos for red fluorescence in the eye (red eye) at 5 days post-fertilization (dpf). Keep the red eye embryos for growing to adulthood.
11. Once the F0 animals reach sexual maturity (around 3 months post-fertilization (mpf)), screen them for germline transmission of the transgene. For this, cross individual F0 animals to a

wild-type animal, and screen the embryos (F1) for red eye marker at 5 dpf. The F1 embryos with red eye represent Beta-bow transgenic line: Tg(ins:BB1.0L; cryaa:RFP), abbreviated as Beta-bow. The animals are grown to adulthood and regularly crossed to wild-type animals to maintain the line (*see* **Note 6**).

### 3.2 Treatment with 4-OHT to Induce Stochastic Recombination

1. Cross Beta-bow animals to the Cre-driver line, Tg(ins:Cre-ER$^{T2}$; cryaa:CFP), to obtain embryos. To ensure accurate staging of embryos, separate male and female fish in the evening prior to the mating. In the morning of the cross, set the males and females together in one mating tank. After 1 hour, collect the embryos that were deposited from the cross, and note the time of collection. Store the embryos in clean E3 medium.
2. Approximately 25% of the embryos from the cross will be positive for both transgenes. However, the presence of the transgenes can only be observed after 3 days post-fertilization (dpf), when the red and cyan fluorescent protein signals (from Beta-bow and Cre-driver line, respectively) become clearly expressed in the retina. Since the induction of Cre-based recombination needs to be done prior to 3 dpf, use all the embryos from the cross for 4-OHT treatment.
3. At 24 h post-fertilization (hpf), manually dechorionate the embryos by gently pulling the egg shell apart with forceps.
4. Transfer 25 dechorionated embryos to a single well of a six-well plastic dish. Fill multiple wells if needed (*see* **Note 7**).
5. For each well, prepare 10 mL of 4-OHT working solution at 1 μM concentration by dissolving 5 μL of 2 mM stock solution in E3 medium (*see* **Note 8**). For multiple wells, a larger volume of working solution can be prepared by appropriately scaling up the volume of 4-OHT.
6. Replace the solution in the six-well plate with the 4-OHT working solution. Protect the plate from light.
7. After 6 h, transfer the embryos to a new dish containing E3 medium (*see* **Note 9**).
8. Rinse the embryos three times for 30 min each and again transfer to a new dish. At this step, embryos from multiple wells can be combined and placed in a 90-mm plastic dish.

### 3.3 Beta-bow Image Acquisition

For multicolor analysis of DBC proliferation, the islets are imaged using a confocal microscope. This requires dissection, fixation, and mounting on an imaging slide. We perform imaging at three time points: 3.5 dpf, 15 dpf, and 30 dpf (Fig. 2a). The 3.5 dpf stage provides a control stage for labeling efficiency, i.e., the samples

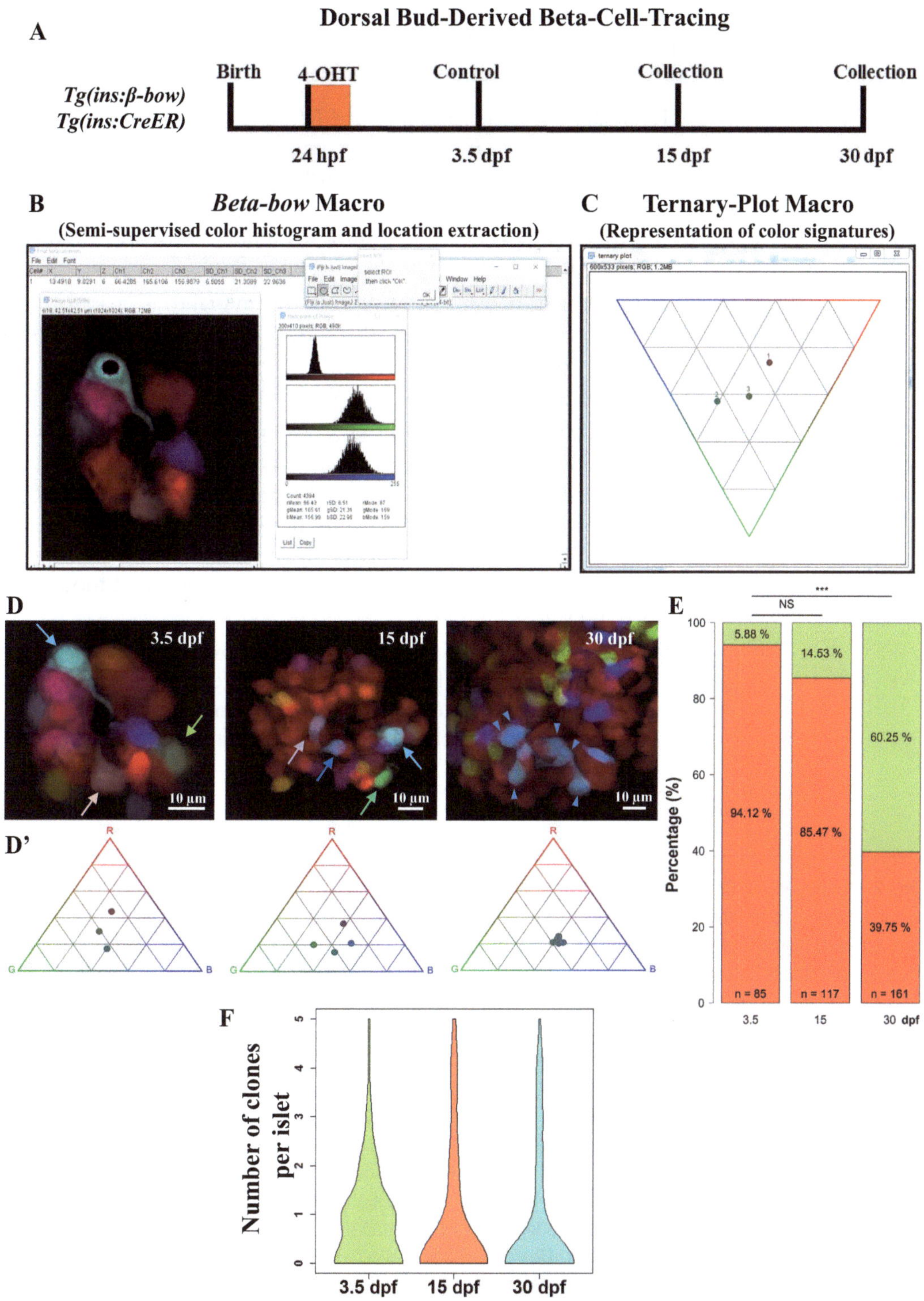


**Fig. 2** Clonal analysis of dorsal bud-derived beta-cells using the *Beta-bow* system. (**a**) A schematic depicting the timeline for lineage tracing of DBCs. The DBCs are labeled by 4-OHT treatment of *Tg(ins:beta-bow); Tg(ins: CreERT2)* embryos from 24 to 30 hpf. Samples are collected at 3.5, 15, and 30 dpf. The primary islets from the

mostly contain individual trichromatic cells. At 15 and 30 dpf, we quantify the clones obtained from individual cells to understand the diversity in proliferation of individual DBCs.

1. Isolate the islets from *Tg(ins:BB1.0L; cryaa:RFP)*; *Tg(ins:Cre-ERT2; cryaa:CFP)* double transgenic fish at the appropriate stage. For this, euthanize the fish using prolonged incubation in MS222. Gently transfer the fish in a petri dish containing the MS222 solution and wait until the animal shows no opercular (gills) movement. Typically, this takes 5 min.
2. Fix the entire fish by transferring the animal to 4% paraformaldehyde + 1% Triton X-100 solution. Fix the fish for 2 days at 4 °C (*see* **Note 10**).
3. Replace the 4% paraformaldehyde + 1% Triton X-100 solution with PBS, and perform three to five quick PBS washes.
4. Manually dissect the pancreas and isolate the primary islet. The islet will be bright red due to the fluorescence from the RFP expressed in the beta-cells. Thus, perform the dissections under a stereo microscope equipped with fluorescence lamp and a filter cube for RFP (*see* **Note 11** for dissection and mounting of 3.5 dpf islet).
5. Clean the primary islet by carefully removing the surrounding tissue. Use the red fluorescence as a guide and take precaution not to injure or poke the islet. After clearing, the individual cells on the islet surface become discernable.

**Fig. 2** (continued) DBCs labeled animals were imaged using a confocal microscope and analyzed using ImageJ macros (**b**, **c**). (**b**) The *Beta-bow* macro determines the color signatures of user-selected beta-cells. The macro requires a manual selection of the nucleus of each multicolor beta-cell. After the selection, the color information is automatically extracted and saved for future processing. (**c**) The color signatures of beta-cells can be plotted on a ternary plot using the *TernaryPlot* macro, which provides a rapid visual representation. (**d**) The labeled islets were analyzed at 3.5, 15, and 30 dpf. Analysis of the trichromatic cells within each islet revealed two kinds of labeled DBCs: ones that remained as single cells (marked with arrows) and others that proliferated to form multicellular clones (marked with arrowheads). A visual representation of the color signature for trichromatic cells is provided as ternary plots (**d′**). (**e**) Quantification showing the clonal distribution of the traced DBCs as a percentage of cells that remain as single cells or form multicellular clones. The total number ($n$) of traced DBCs is depicted at the bottom of the barplots. The clonal distribution at 3.5 dpf consisted primarily of single cells (94.1%). The distribution did not significantly change till 15 dpf. However, at 30 dpf, the distribution diverged significantly as compared to the 3.5 dpf stage and consisted of only 40% single cells (Fisher's exact test, NS: $p > 0.05$; $***p \leq 0.001$). (**f**) Frequency of trichromatic events for different stages. Quantification of the presence of trichromatic events (single cells or multicellular clones) within the primary islets from animals labeled to trace the DBCs. The quantification depicts violin plots, which shows the probability distribution for islets from 3.5, 15, and 30 dpf stages ($n > 50$ for each stage). The frequency of trichromatic events per islet remains similar between the three stages (mean equals 0.8 (3.5 dpf), 0.87 (15 dpf), and 0.77 (30 dpf)). Figure adapted from [3]

6. Prepare a glass slide with a drop (5–10 μL) of mounting media, and transfer the islets. Carefully cover the islet preparation with a glass cover slip while avoiding any air bubbles (*see* **Note 12**).
7. In a confocal microscope, use 40 × 1.2 NA water immersion objective for imaging the Beta-bow islet. Locate the islet using the red fluorescence to guide you, and start the acquisition phase.
8. Set up a sequential acquisition for RFP, YFP, and CFP fluorescence. Excite the RFP, YFP, and CFP using the 561-, 514-, and 450-nm laser lines, respectively (*see* **Note 13**). Capture their signal by setting the detectors approximately as follows: for RFP, 570–650 nm, false color "red"; for YFP, 524–555 nm, false color "green"; for CFP, 460–505 nm, false color "blue." With this setting, the red channel will record the default RFP fluorescence, while the green and blue channels will record fluorescence signal generated by 4-OHT-based recombination.
9. Set up a Z-stack by moving through the islet in red channel. Image the entire thickness of the primary islet with a Z-thickness of 0.51 μm. Set the image resolution at 1024 × 1024 pixels, averaging at 1.
10. Acquire the channels sequentially and save the images for post-processing in Fiji.

### 3.4 Clonal Analysis

1. Open the image file in FIJI using "LSM Toolbox." For this, select "Plugin" → "LSM Toolbox" → "Show LSM toolbox." In the "LSM toolbox," click "Open LSM" and select the image file (*see* **Note 14**).
2. Obtain color signatures of individual beta-cells that underwent recombination. Such cells would express the YFP or CFP and be marked by signal in the green or blue channels. Color signatures are obtained by extracting the normalized fluorescent intensities corresponding to the three fluorescent channels from within the nuclear region of the labeled cells. For this, the nuclei of beta-cells are selected manually, and the color profile (intensity for red, green, and blue (RGB) channels) is obtained using a custom macro "betabow.ijm." betabow.ijm provides a semi-supervised interface for recording the color profiles of multiple selected regions. To this end, run the betabow.ijm macro after opening the image file. Select the nucleus of individual beta-cells that display recombination. Move through the confocal stack to mark every beta-cell one by one and extract the color profiles. Be careful to select each cell only once. The macro will keep a record of the mean and the standard deviation for RGB channels, along with the X, Y, and Z coordinates of the selected region (Fig. 2b). If a wrong region is unintentionally selected, the macro has the option for ignoring the selection.

3. The fluorescence signal from the red channel (the default color) can bleed-through into the green and blue channels. To account for such bleed-through, background subtraction is performed using non-recombined beta-cells. For background subtraction, select three unrecombined cells at the end of the analysis. The unrecombined cells will be used to estimate the background signal in the green and blue channels. The mean green and blue intensities from unrecombined cells represent the background intensity, which is subtracted from the color profile of recombined cells.
4. Background removal and normalization is done automatically by the analysis macros. For normalization, background subtracted RGB values for each cell are transforming into percentages as follows: Cnormalized = Cmean/(Rmean + Gmean + Bmean) × 100, where C is R, G, or B. (RGB)normalized defines the color signature of a particular cell.
5. To visualize the color signatures, plot the normalized values onto a ternary plot [11] (Fig. 2c). A ternary plot consists of an equilateral triangle, in which the vertices represent one particular dimension, a single color in this case, while the sides travel between the two adjacent dimensions. A point on the side of the triangle would only have two colors, while points toward the middle would contain all three colors. For this, the output of betabow.ijm is passed onto the "TernaryPlot.ijm" macro. The macro will normalize the color profile and plot a ternary plot with the color signature. Each point on the plot will correspond to one beta-cell selected during the color extraction step (*see* **Note 15**).
6. To statistically test clonality within the islet, run the "Equivalence.ijm" macro on the output from betabow.ijm. The Equivalence.ijm macro utilizes "Equivalence Acceptance Criteria" by performing "two one-sided test (TOST) procedure" on the normalized RGB values. TOST, with the null hypothesis that the two observations are not equivalent, utilizes confidence intervals based on mean and standard deviation of RGB values for calculating similarity between two observations. TOST allows to test whether trichromatic beta-cells in an islet share a similar color signature. Specifically, two beta-cells that pass TOST with $p < 0.05$ are paired together. Pairs of beta-cells that share a common beta-cell are also considered to be within one clone (*see* **Note 15**). Trichromatic cells that do not belong to any pair in the islet are considered to be single cells.

### 3.5 Analysis of Clonal Dynamics with Age

1. To understand the clonal dynamics of DBCs during zebrafish development, label the beta-cells at 24 hpf and analyze the islets at 3.5, 15, and 30 dpf (Fig. 2a).

2. For each stage of analysis, obtain multiple (at least 30) trichromatic events. Each trichromatic event would consist of either a single cell or a multicellular clone (Fig. 2d). A single trichromatic cell would have a color signature that was never repeated in the entire islet that contained the cell. On the other hand, the cells within a multicellular clone would share a similar trichromatic color signature. Typically, such cells occur adjacent to each other due to limited mixing of cells.
3. To analyze proliferative histories, compare the percentage of multicellular clones at 3.5 dpf with 15 and 30 dpf by performing Fisher's exact test. For this, use the function in R: fisher.test(x = matrix(c($Single_{3.5dpf}$, $Multi_{3.5dpf}$, $Single_{Stage}$, $Multi_{Stage}$), nrow = 2, ncol = 2), alternative = "two.sided"). Here, $Single_{3.5dpf}$ and $Multi_{3.5dpf}$ are the number of single cells and multicellular clones obtained at 3.5 dpf, respectively, and $Single_{Stage}$ and $Multi_{Stage}$ are the number of single cells and multicellular clones obtained at the test stage, respectively. The function will test if the clonal distribution (single cells versus multicellular clones) at 15 or 30 dpf varies significantly from 3.5 dpf.
4. For instance, at 3.5 dpf, 94.1% of the labeled DBCs remained as single cells (Fig. 2e). Moreover, at 15 dpf, the traced DBCs showed a similar clonal distribution, with 85.5% of traced beta-cells remaining as single cells. This indicates that a majority of DBCs remain quiescent from 3.5 to 15 dpf. In contrast, at 30 dpf, only 40% of the traced DBCs remain as single cells, while the rest (60%) form multicellular clones. These data provide two key conclusions: first, a subset of DBCs proliferate between 15 and 30 dpf, and second, a subset of DBCs do not contribute to the increase in beta-cell numbers. These observations together point to the proliferative heterogeneity among the nominally homogenous DBCs.
5. Additionally, keep a record of the number of trichromatic events (single cells or multicellular clones) within an islet. The frequency of trichromatic events recorded per islet should remain similar between stages (Fig. 2f). A stable frequency suggests lack of de novo recombination. This increases the confidence in the analysis of uniquely labeled cells (*see* **Note 16**).

#### *3.6 Estimating the Volume of Individual Beta-Cells*

1. Open the Beta-bow in Fiji and separate the color channels using "Image" → "Color" → "Split Channels."
2. Close the Red channel, and merge the Green and Blue channels using "Image" →"Merge Channels." The use of only the Green and Blue channels allows analysis of individual cells that have distinct labeling from their neighbors.

3. Convert the image to 8-bit using "Image" → "Type" → "8-bit."
4. Open the 3D Objects Counter Menu using "Plugin" → "3D Object Counter" → "3D-OC Set Measurements," and only select "Volume" in "Parameters to calculate."
5. Run the 3D Objects Counter plugin using "Plugin" → "3D Object Counter" → "3D Object Counter." Use the "Threshold" slider to define the intensity value which would separate the image pixels into two populations: background (intensities below the selected value) and object. Change the "Slice" slides to move through the Z-stack. Pick a threshold value where the cells are clearly defined with the background removed (Fig. 3a).

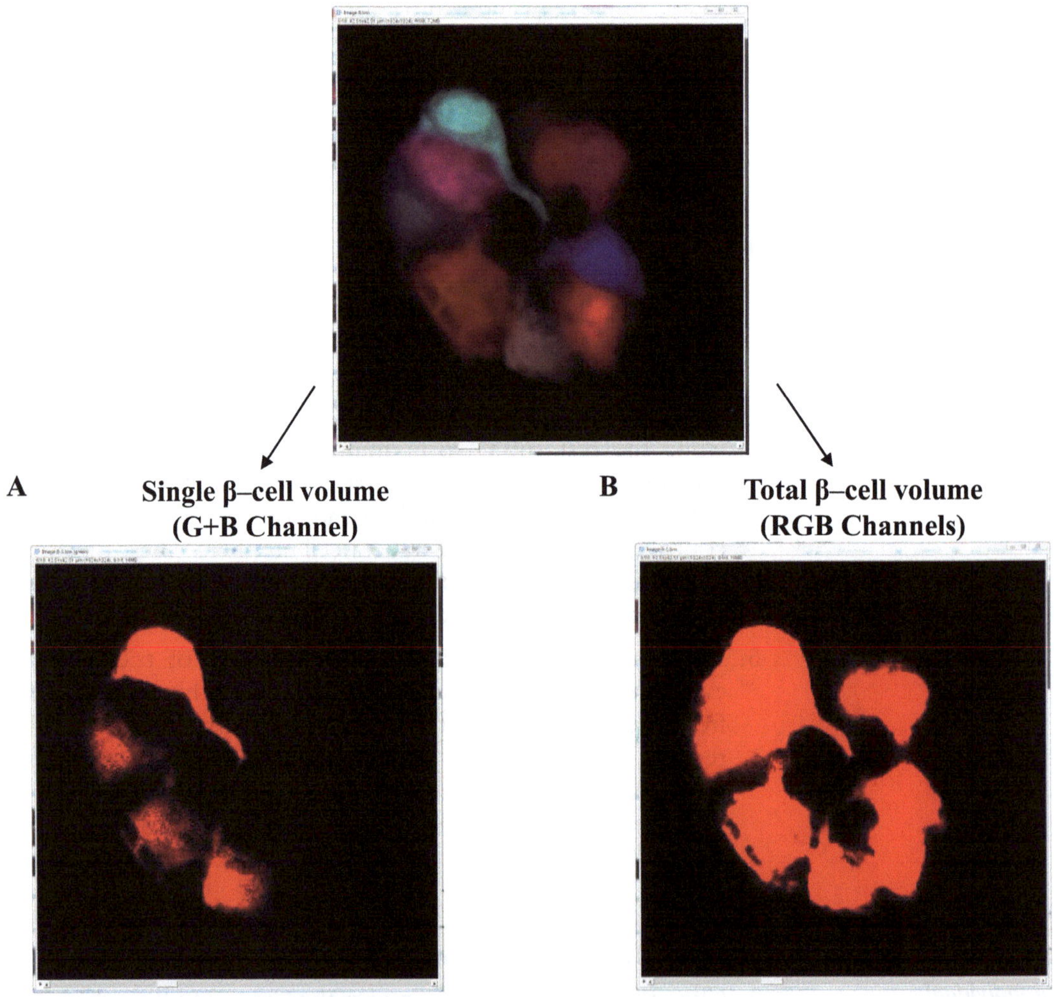


**Fig. 3** Estimation of cell volume from *Beta-bow* images. The confocal images of labeled DBCs were used for measuring the volume of single beta-cells (**a**), or the total volume occupied by beta-cells within the primary islet. (**b**) The analysis of cell volume utilizes the recombined colors (blue and green) to distinguish individual cells, while the measurement of total volume utilizes the red color for estimating the total volume occupied by all beta-cells. Figure adapted from [3]

6. Run the 3D Object Counter by pressing “OK.” The plugin generates a list of object and its corresponding volume.
7. Manually check the generated results by clicking on the object’s number in the result window, and remove results that comprise two cells. Adjacent cells with similar color profiles are not successfully separated, and thus should be excluded from the analysis.
8. The volume of individual cells from 3.5, 15, and 30 dpf islets will provide a distribution of individual beta-cell volumes for each stage. The distribution can be used to extract parameters about single beta-cell volume, such as the mean volume at each stage ($V_{\text{mean, Stage}}$).
9. Compare the distribution of beta-cell volumes between each stage using ANOVA. A lack of significant increase in the volumes of beta-cells would imply that the beta-cell expansion is not driven by cellular hypertrophy.

### 3.7 Estimating the Total Volume of Beta-Cells in the Primary Islet

1. Open the Beta-bow in Fiji and convert the image to 8-bit using “Image” → “Type” → “8-bit.” For estimating the total volume of the beta-cells in the primary islet, the fluorescence signal from all three (RFP, CFP, YFP) channels is used to label the cells.
2. Follow **steps 4–6** from Subheading 3.6. Pick a threshold value such that the entire area covered by beta-cells is selected (Fig. 3b).The plugin would now generate multiple objects corresponding to connected regions between beta-cells in the primary islet.
3. Add the volume of all the objects to obtain the total volume of beta-cells in the primary islet. Compile the data from each stage to obtain the change in the total volume of beta-cells belonging to the primary islet during zebrafish development.
4. **Steps 1–3** are automated using the macro “Process_Folder_-vol.ijm” available in the github folder. The macro is capable of processing a folder with images from Beta-bow primary islets and returns a single results table with total volume for each image.

### 3.8 Controls for 4-OHT-Based Recombination

Recombination from the Cre-Lox system can occur without 4-OHT treatment. To characterize the background recombination in the system, we carry out control treatments in two ways: Firstly, we treat a separate set of embryos with ethanol, which is the 4-OHT vehicle; and secondly, we treat the embryos before the onset of insulin gene expression.

1. For vehicle treatment, treat 25 dechorionated embryos with 0.05% ethanol in a single well of a six-well dish for 6 h (*see* **Note 7**).

2. For labeling the embryos before the onset of insulin, treat the embryos with 1 μM 4-OHT from 6 hpf to 12 hpf. Insulin expression starts around 14 hpf in zebrafish [5]. Thus, treatment from 6 to 12 hpf should not activate the Cre-ER$^{T2}$, which is expressed from the insulin promoter in the Cre-driver line.
3. For both controls, transfer the embryos to a new dish after the treatment and rinse with E3 medium three times for 30 min each. After the final rinse, transfer the embryos for raising till 30 dpf, and obtain images for control samples as in Subheading 3.3 (Fig. 4).

### 3.9 Estimation of Background Recombination Frequency

To estimate the frequency of background recombination in the Beta-bow animals, calculate the probability for appearance of recombined cells in the primary islet from 30 dpf animals used as controls for 4-OHT-based recombination (Subheading 3.8).

1. Mount and image primary islets from 30 dpf control animals. Manually count the number of cells with CFP or YFP expression ($n_{CFP,YFP}$) (*see* **Note 17**).
2. Calculate the volume occupied by the beta-cells in the primary islets using RFP fluorescence ($V_{total}$).
3. Estimate the number of beta-cells in the primary islet by dividing the total volume of beta-cells to the average volume of single beta-cells at 30 dpf ($n_{total} = V_{total}/V_{mean,\ 30\ dpf}$) (*see* **Note 18**). The average volume of single beta-cells is obtained from Subheading 3.6.
4. Calculate the probability for single-recombination events as the ratio of the number of cells with CFP or YFP expression to the number of beta-cells ($p_{single} = n_{CFP,YFP}/n_{total}$).The probability for double-recombination event within the same cell, which is required for generation of trichromatic cells, equals the square of the probability for single-recombination event ($p_{double} = p_{single}{}^2$). $p_{double}$ represents the background recombination frequency for Beta-bow.

## 4 Notes

1. Multiple variants of the multicolor reporter systems have been developed (Table 1). These variants allow labeling with different combinations of fluorescent proteins as compared to BB1.0L system. A distinct advantage of recently developed Brainbow-3 versions is the possibility of signal amplification using antibody staining [18]. BB1.0L contains variants of GFP, and hence all three fluorescent proteins (CFP, YFP, RFP) are recognized by the same antibody. The inability to amplify the native fluorescence signal in histological samples can limit the ability to trace cells deep within a tissue or couple the

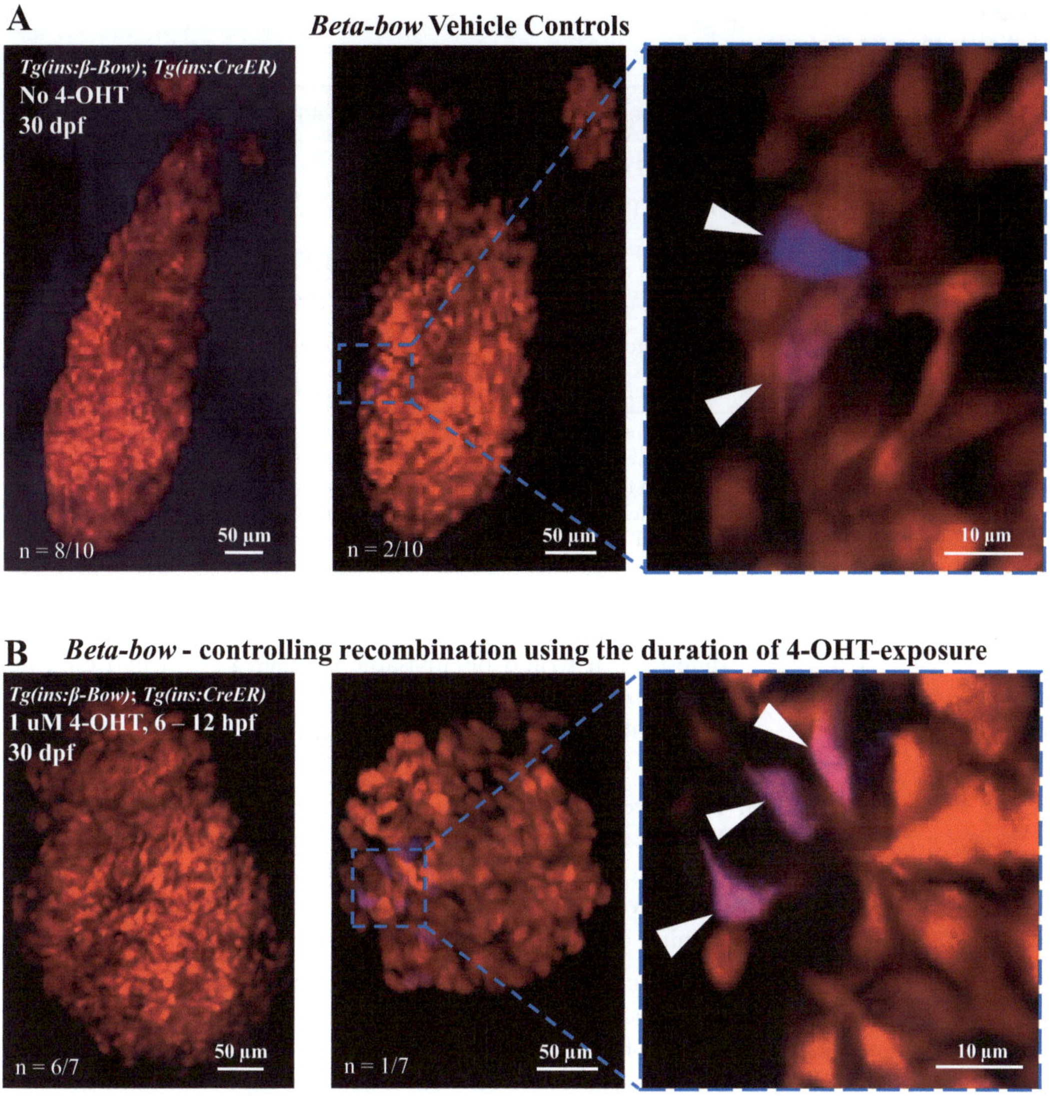


**Fig. 4** Background recombination in *Beta-bow* islets. Maximum intensity projection of the islets from 30 dpf *Tg (ins:beta-bow); Tg(ins:CreERT2)* animals that were treated during the embryonic stage for assessment of background recombination. Animals were either treated with vehicle (0.05% ethanol) from 24 to 30 hpf (**a**) or with 1 μM 4-OHT from 6 to 12 hpf (**b**). 8/10 and 6/7 islets from vehicle and 4-OHT treatment, respectively, showed no leaky recombination. Islets with background recombination displayed beta-cells with dichromatic labeling (arrowheads). No trichromatic cells were observed under both treatments. Figure adapted from [3]

multicolor lineage tracing with a staining protocol that destroys GFP fluorescence. However, protocols that improve the optical transparency of fixed tissue using tissue clearing could be used to help in analyzing cells deep inside the sample [19].

2. We have described the generation of ins:BB1.0L construct that is useful for lineage tracing of zebrafish beta-cells. To generate constructs for other tissues, the BB1.0L construct would need to be cloned downstream of a tissue-specific promoter. Since

**Table 1**
**Multicolor reporter systems**

| Multicolor reporter systems | Fluorescent proteins | Antibody-based amplification |
|---|---|---|
| Brainbow-1.0 [10] | RFP, CFP, YFP | Not possible |
| Brainbow-1.1 [10] | OFP, RFP, YFP, CFP | Not possible |
| Brainbow-2.1 [10] | EGFP, YFP, RFP, CFP | Not possible |
| dBrainbow [21] | EGFP-V5, EBFP-HA, mKO2-myc | Possible |
| Cytbow [11] | EBFP2, tdTomato, mCerulean, EYFP | Not possible |
| Brainbow-3 [18] | mOrange2, EGFP, mKate2 | Possible |

The table lists various multicolor reporter systems available for lineage tracing. The fluorescent proteins contained within each system and the possibility to amplify fluorescent signal with antibody staining are reported. V5, HA, and myc in dBrainbow are epitope tags which can be used for antibody staining. Abbreviations for fluorescent proteins: RFP (red fluorescent protein), CFP (cyan fluorescent protein), YFP (yellow fluorescent protein), OFP (orange fluorescent protein), EGFP (enhanced green fluorescent protein), EBFP (enhanced blue fluorescent protein), mKO2 (monomeric Kusabira-Orange2)

BB1.0L fragment is prepared using EcoRV/NsiI restriction digestion, it would be advisable to have a blunt-end enzyme site followed by a site against NsiI, PstI, or SbfI downstream of the tissue-specific promoter. PstI and SbfI generate sticky ends that are compatible with NsiI.

3. We have found it extremely difficult to PCR amplify the BB1.0L construct. This is possibly due to the repetitive nature of the construct, which contains three copies of GFP variants. Thus, we suggest a restriction digestion-based approach for cloning of the BB1.0L construct.
4. We have had better success using chemically competent cells over electro-competent cells for this particular ligation. Using 5 μL of ligation solution directly for transformation with electro-competent cells leads to sparking, while using lower amount of ligation solution decreases the efficiency of transformation. One way to circumvent the problem is to remove the salts from the ligation solution using DNA desalting columns.
5. I-SceI meganuclease-based transgenesis leads to random integration of the injected plasmid as a concatemer (repeated copies of the DNA fragment at a single genomic locus) [12]. This is beneficial for multicolor lineage tracing as increasing the numbers of BB1.0L copies generates greater color diversity and helps with cellular barcoding.
6. Generate five to ten independent Beta-bow transgenic lines for testing. Each independent Beta-bow transgenic line will contain different number of ins:BB1.0L copies integrated into the genome and a different locus of integration. The number of copies and integration location can influence the efficiency of recombination. It is better to generate multiple (five to ten)

independent lines and characterize the recombination efficiency for each one separately. Choose the line that yields multiple colors upon induction of recombination according to the steps outlined in Subheadings 3.2–3.4. Alternatively, the transgenic lines can be prioritized by the number of BB1.0L copies inserted in the genome. The number of insertions can be characterized by quantitative PCR (qPCR) as outlined in [20]. The line with the highest number of insertions can be utilized for experiments.

7. Since the embryos are not sorted for the transgene expression, only 25% of the treated embryos will be double positive. Thus, each batch of 25 embryos will give approximately six animals for analysis. Set up multiple wells to increase the sample number.
8. To obtain the appropriate concentration of 4-OHT for induction of trichromatic cells, we suggest performing a pilot experiment with different doses (0.1, 1, 10 μM) of 4-OHT for 6 hrs. Analyze the presence of trichromatic cells at 48 h after labeling. A low dose of 4-OHT will lead to limited color diversity and possible absence of trichromatic cells. Importantly, a strong induction of recombination with high dose of 4-OHT will also reduce the color diversity as cells lose the red fluorescence (default color from Brain-bow construct) and only retain green or blue fluorescence [20]. Once an appropriate concentration of 4-OHT is determined using the pilot experiment, the time period of labeling can be reduced below 6 h by appropriately increasing the concentration of 4-OHT (e.g., incubation in 1 μM 4-OHT for 6 h was equivalent to an incubation in 2 μM 4-OHT for 3 h (data not shown)). Reducing the time period of incubation can help restrict the recombination to a short time window of development.
9. It is preferable to remove the embryos from the dish in which the 4-OHT treatment took place to avoid any recombination from leftover 4-OHT.
10. Permeabilization with 1% Triton X-100 during fixation is optional. However, it makes dissection of the tissue easier.
11. For 3.5 dpf samples, the dissection of the primary islet from the rest of the pancreas may be difficult. For such samples, dissection and mounting of the entire pancreas are sufficient to image the islet. In the larvae, the pancreas lies on the right side of the zebrafish body. Carefully remove the skin and tissues surrounding the pancreas and transfer the entire organ to a slide containing the appropriate mounting media.
12. We do not use any spacers between the glass slide and cover slip to mount an islet. This flattens the tissue and reduces the time needed to image the entire islet.

13. It is possible to excite the CFP using the DAPI laser line (405 nm), but the signal is very weak and diminishes rapidly with the thickness of the tissue.
14. For formats not supported by the "LSM Toolbox," convert them first to TIFF for analysis.
15. The ternary plot macro ("TernaryPlot.ijm") and clonal analysis macro ("Equivalence.ijm") restrict their analysis to trichromatic cells. To ascertain the presence of all three colors within a cell, the mean of fluorescent intensity is tested to be statistically different from zero. To do so with $p < 0.01$, which corresponds to a distance of 2.58 standard deviation (sd) from the mean, the following formulation was used: Cmean - 2:58 * Csd >0, where C is R, G, and B. A cell which satisfies the criteria for all three channels is considered as triple positive.
16. Restricting the lineage analysis to trichromatic cells only comes at a cost. Some samples do not contain clonal information. However, the analysis of trichromatic cells provides sparse labeling of beta-cells.
17. We did not observe any trichromatic cells in the control samples. The leaky recombination yields cells with either RFP + YFP expression or RFP + CFP expression.
18. For our samples, the average volume of an individual beta-cell at 30 dpf ($V_{\text{mean, 30 dpf}}$) equaled 184.70 μm$^3$ [3]. This number was obtained from the analysis of primary islets that originated from animals which were recombined at 24 hpf with 1 μM 4-OHT treatment for 6 h.

**Zebrafish Husbandry**: *The protocol mentioned in this chapter was conducted in accordance with the Animal Welfare Act and with permission of the Landesdirektion Sachsen, Germany (AZ 24–9168, TV38/2015, A12/2016, A13/2016, and all corresponding amendments). Ensure compliance with the local animal welfare laws before conducting any experiments.*

## Acknowledgment

We apologize to our colleagues in the field for omitted citations due to restrictions on space and number of references. This work was supported by funding from the DFG–Center for Regenerative Therapies Dresden, Cluster of Excellence at TU Dresden, and the German Center for Diabetes Research (DZD), as well as research grants from the German Research Foundation (DFG), the European Foundation for the Study of Diabetes (EFSD), and the DZD to N.N.

## References


1. Kinkel MD, Prince VE (2009) On the diabetic menu: zebrafish as a model for pancreas development and function. BioEssays 31:139–152
2. Pisharath H, Rhee JM, Swanson MA et al (2007) Targeted ablation of beta cells in the embryonic zebrafish pancreas using E. coli nitroreductase. Mech Dev 124:218–229
3. Singh SP, Janjuha S, Hartmann T et al (2017) Different developmental histories of beta-cells generate functional and proliferative heterogeneity during islet growth. Nat Commun 8:664
4. Tehrani Z, Lin S (2011) Endocrine pancreas development in zebrafish. Cell Cycle 10:3466–3472
5. Biemar F, Argenton F, Schmidtke R et al (2001) Pancreas development in zebrafish: early dispersed appearance of endocrine hormone expressing cells and their convergence to form the definitive islet. Dev Biol 230:189–203
6. Ninov N, Borius M, Stainier DYR (2012) Different levels of Notch signaling regulate quiescence, renewal and differentiation in pancreatic endocrine progenitors. Development 139:1557–1567
7. Wang Y, Rovira M, Yusuff S, Parsons MJ (2011) Genetic inducible fate mapping in larval zebrafish reveals origins of adult insulin-producing β-cells. Development 138:609–617
8. Ninov N, Hesselson D, Gut P et al (2013) Metabolic regulation of cellular plasticity in the pancreas. Curr Biol 23:1242–1250
9. Hesselson D, Anderson RM, Beinat M, Stainier DYR (2009) Distinct populations of quiescent and proliferative pancreatic beta-cells identified by HOTcre mediated labeling. Proc Natl Acad Sci U S A 106:14896–14901
10. Livet J, T a W, Kang H et al (2007) Transgenic strategies for combinatorial expression of fluorescent proteins in the nervous system. Nature 450:56–62
11. Loulier K, Barry R, Mahou P et al (2014) Multiplex cell and lineage tracking with combinatorial labels. Neuron 81:505–520
12. Thermes V, Grabher C, Ristoratore F et al (2002) I-SceI meganuclease mediates highly efficient transgenesis in fish. Mech Dev 118:91–98
13. Grabher C, Joly J-S, Wittbrodt J (2004) Highly efficient zebrafish transgenesis mediated by the meganuclease I-SceI. Methods Cell Biol 77:381–401
14. Rosen JN, Sweeney MF, Mably JD (2009) Microinjection of zebrafish embryos to analyze gene function. J Vis Exp 25:1115
15. Felker A, Nieuwenhuize S, Dolbois A et al (2016) In vivo performance and properties of Tamoxifen metabolites for CreERT2 control. PLoS One 11:e0152989
16. Westerfield M (2007) The Zebrafish book. A guide for the laboratory use of zebrafish (Danio rerio), 5th edn., Univ. of Oregon Press, Eugene
17. Bolte S, Cordelieres FP (2006) A guided tour into subcellular colocalisation analysis in light microscopy. J Microsc 224:13–232
18. Cai D, Cohen KB, Luo T et al (2013) Improved tools for the Brainbow toolbox. Nat Methods 10:540–547
19. Richardson DSS, Lichtman JWW (2015) Clarifying tissue clearing. Cell 162:246–257
20. Pan YA, Freundlich T, Weissman TA et al (2013) Zebrabow: multispectral cell labeling for cell tracing and lineage analysis in zebrafish. Development 140:2835–2846
21. Hampel S, Chung P, McKellar CE et al (2011) Drosophila Brainbow: a recombinase-based fluorescence labeling technique to subdivide neural expression patterns. Nat Methods 8:253–259

# Chapter 13

# Generating Beta-Cell-Specific Transgenic Mice Using the Cre-Lox System

**Lorna I. F. Smith, Thomas G. Hill, and James E. Bowe**

## Abstract

Beta-cell-specific transgenic mice provide an invaluable model for dissecting the direct signaling mechanisms involved in regulating beta-cell structure and function. Furthermore, generating novel transgenic models is now easier and more cost-effective than ever, thanks to exciting novel approaches such as CRISPR.

Here, we describe the commonly used approaches for generating and maintaining beta-cell-specific transgenic models and some of the considerations involved in their use. This includes the use of different beta-cell-specific promoters (e.g., pancreatic and duodenal homeobox factor 1 (Pdx1), rat insulin 2 promoter (RIP), and mouse insulin 1 promoter (MIP)) to drive site-specific recombinase technology. Important considerations during selection include level and uniformity of expression in the beta-cell population, ectopic transgene expression, and the use of inducible models.

This chapter provides a guide to the procurement, generation, and maintenance of a beta-cell-specific transgene colony from preexisting Cre and *loxP* mouse strains, providing methods for crossbreeding and genotyping, as well as subsequent maintenance and, in the case of inducible models, transgenic induction.

**Key words** Beta-cell, Transgenic mouse, Cre-Lox, FLP-*FRT*, Tissue specific

## 1 Introduction

Beta-cell-specific transgenic mice provide an invaluable tool for dissecting the direct signaling mechanisms that regulate islet function in vivo. Global knockout or knock-in mouse models often exhibit altered metabolism, fertility, and, in extreme cases, increased morbidity and mortality, due to phenotypic effects in other tissues, confounding interpretation of beta-cell physiology. Furthermore, beta-cell-specific knockouts remove the possibility of indirect effects on beta-cell function via other tissues, either locally to create structural changes to the pancreas, or via endocrine effects by organs such as the hypothalamus.

Aileen J. F. King (ed.), *Animal Models of Diabetes: Methods and Protocols*, Methods in Molecular Biology, vol. 2128, https://doi.org/10.1007/978-1-0716-0385-7_13, 

### 1.1 Beta-Cell-Specific Driver Lines

Recent years have seen a rapid development in the accessibility, efficiency, and power of genetic tools to generate novel transgenic mouse lines, and advances such as CRISPR will increasingly influence the development of tissue-specific mouse models going forward. However, for the purposes of this guide, we focus on the generation and use of the most commonly used and well-established approaches for generating beta-cell transgenics.

Cre-*loxP* and similar site-specific recombinase (SSR) technologies such as FLP-*FRT*, originating from bacteriophage P1 and *Saccharomyces cerevisiae*, respectively [1, 2], provide a crucial tool for both beta-cell-specific knock-in and knockout of genes. These systems work via a two-gene mechanism. A tissue-specific promoter-driven Cre or FLP recombinase enzyme is inserted into the genome. Separately, *loxP* or *FRT* target sequences are inserted flanking a gene of interest or section of coding DNA. Tissue-specific expression of Cre or FLP catalyzes the recombination of DNA between the flanking *loxP* or *FRT* target sequences, deleting or inserting the gene of interest specifically in cells with recombinase activity (Fig. 1). A more detailed description of the mechanisms by which Cre-*loxP* and other systems cause site-specific transgenic recombination can be found elsewhere [3, 4]. Generation of these mouse strains provides a versatile tool which can be reused for many other research purposes, looking at the same gene in other tissues of interest, by combining with another promoter-driven recombinase mouse strain, or for studying a range of knock-in/knockout recombination for the same or different genes within the same tissue. Though both Cre-*loxP* and FLP-*FRT* are highly similar systems, Cre-*loxP* is currently accepted as the better tool for use in

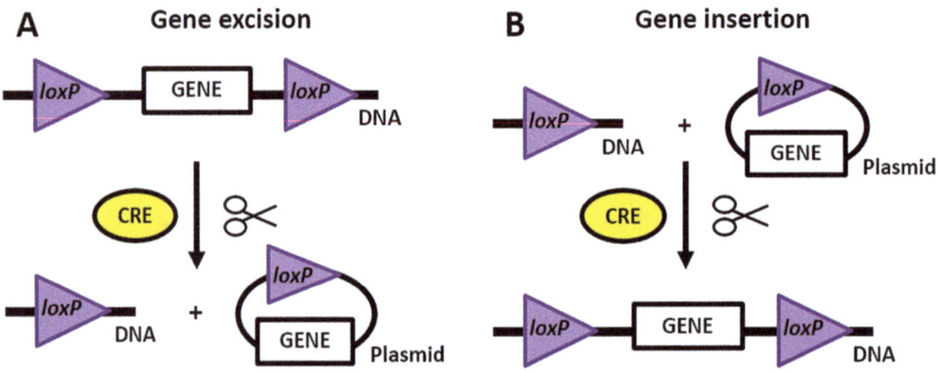


**Fig. 1** Common methods of DNA recombination by Cre recombinase and *loxP* sites. During gene excision (**a**), where two *loxP* sites are in the same direction, Cre recombinase expression will cause the DNA region and gene located between the two sites to be excised from the DNA strand. During gene insertion (**b**), the presence of a *loxP* site in the DNA strand will become the site of insertion for a circular DNA plasmid also containing a *loxP* sequence and the gene of interest. The action of Cre has a greater tendency toward gene excision but can be shifted toward gene insertion by increasing the concentration of *loxP* gene plasmids. Figure adapted from [3, 4]

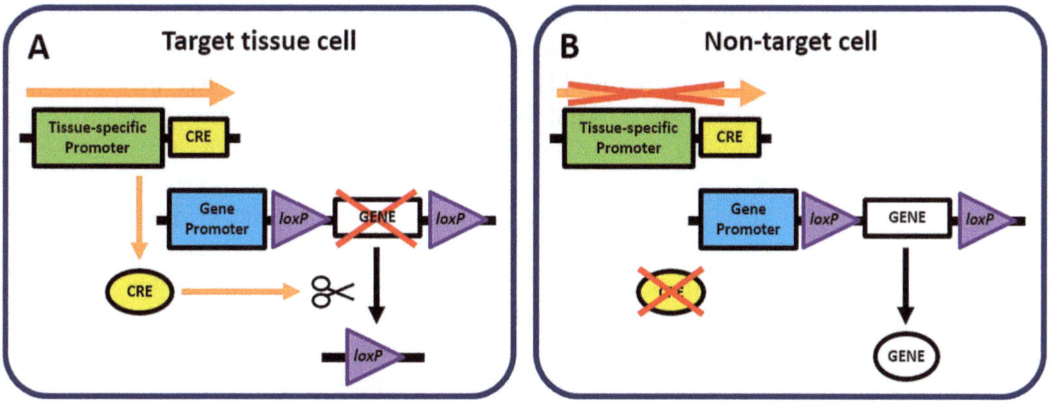


**Fig. 2** Mechanisms of a tissue-specific Cre-*loxP* mouse knockout model. In the cells of the target tissue (**a**), the tissue-specific promoter promotes the transcription and expression of Cre recombinase, which then excises the gene of interest from between the *loxP* sites to cause gene deletion and knockout. In the rest of the body (**b**), the tissue-specific promoter is not active, and therefore Cre recombinase is not expressed. Therefore, the gene of interest remains present in the DNA and is expressed as normal in these nontarget cells. Figure adapted from [85]

mammalian genomes due to its higher affinity for the *loxP* site than FLP for *FRT* [3, 5]. The flexibility of this system has led to the rapid creation of expansive databases of tissue-specific Cre and floxed gene mouse lines available to researchers [6, 7], with the database for FLP-*FRT* mice available also growing.

Generation of a successful beta-cell-specific transgenic mouse relies upon the promoter gene which drives Cre or FLP recombinase activity being expressed exclusively in the beta-cells (Fig. 2). Furthermore, the selected promoter must regulate a gene which is highly and consistently expressed in the beta-cell, in order to ensure sufficient expression of the transgene. Currently, fragments from three gene promoters are most commonly used, that of pancreatic and duodenal homeobox factor 1 (Pdx1), rat insulin 2 promoter (RIP), and the more recently developed mouse insulin 1 promoter (MIP). Details of these and other beta-cell-specific transgene drivers (in addition to those for use in other pancreatic cell types) have been discussed in more detail elsewhere [8].

The majority of beta-cell-specific transgenic lines employ RIP fragments, most likely due to their high levels of expression in beta-cells and short fragment size of approximately 0.6 kb [8, 9]. However, high levels of ectopic expression have been observed for many strains of transgenic mice which use this driver in brain areas related to glucose homeostasis, such as the cortex, the striatum, and throughout the hypothalamus, with particularly high concentrations in the arcuate nucleus [10, 11]. Previous attempts to increase cell specificity by increasing the promoter length and regulatory units significantly decreased the population of beta-cells in which transgenic expression was maintained [8]. Furthermore, in

RIP-Cre-*loxP* models, beta-cell specificity was increased by mutating Cre to decrease activity, though this reduced recombination efficiency by 35% [12]. Despite these issues, the RIP promoter can still provide a useful tool for the study of beta-cell-specific function; different RIP-driven transgene strains have different patterns of ectopic expression [11] and may therefore have ectopic expression in neurons only where the manipulated gene in question is not expressed or may not be relevant to their role. Regardless, the ways in which ectopic transgenic activity may influence results should be investigated, accounted, and, if possible, controlled for. Furthermore, there is some evidence that the presence of the RIP-Cre transgene has negative effects on beta-cell mass and function [13–15].

Expression of Pdx1 in the adult mouse is restricted to the beta-cell and approximately 20% of pancreatic delta cells. However, Pdx1 is essential for fetal development of the entire endocrine pancreas, with expression in the pancreatic endoderm domain from E8.5 [16], and is expressed in both alpha and beta progenitor cells in mice even at P21 [17]. Therefore, the use of a typically 4.5-kb Pdx1 fragment in the generation of beta-cell-specific transgenic mice is most useful when using an inducible transgenic model, in order to delay recombination to when Pdx1 expression is restricted mainly to beta-cells. Inclusion of a 1-kb fragment of Pdx1 promoter containing regulatory areas I and II, such as in the Tg(Pdx1-cre)$^{PBMga}$ and *pdx1*$^{PB}$CreER™, can also help to improve beta-cell specificity [18, 19]. Leakage of expression has also been observed in the hypothalamus of some *Pdx1*-driven transgenic strains, although seen in fewer hypothalamic regions than the RIP-driven strains [10, 11].

The beta-cell-specific driver most recently developed, likely due to its much larger promoter fragment size required of 8.3–8.5 kb, is that of MIP. This driver was first used in the development of a MIP-GFP mouse [20] and was subsequently used to produce the inducible MIP-CreERT$^{1Lphi}$ mouse strain. The MIP-CreERT$^{1Lphi}$ was superior to established RIP-Cre and PDX1-Cre lines in terms of tissue specificity, due to a complete lack of off-target Cre activity, including a lack of ectopic brain expression [11, 21]. However, some ectopic GFP expression in the MIP-GFP mouse was seen under stress conditions [22], which implies that not necessarily all developed versions of a MIP-driven transgene will be immune from the problem of ectopic expression. The insulin 1 gene appears to have low expression in approximately 10–35% of mature mouse beta-cells [23] and appears weaker/more variable than that of insulin 2 expression [9, 24, 25]. Unfortunately, while the MIP--CreERT$^{1Lphi}$ mouse does have excellent tissue specificity, it also exhibits inappropriate human growth hormone (hGH) expression in β-cells, due to the use of a hGH minigene for the insertion of the Cre transgene [26], an issue also affecting some strains of RIP-Cre and Pdx1-Cre mice [27]. These low levels of hGH expression lead to activation of the prolactin receptor and subsequent increased islet serotonin levels [26, 28, 29]. Thus, the presence of the Cre transgene in the MIP-CreERT$^{1Lphi}$ mice has beta-cell effects in itself, and this must be accounted for in any transgenic mice generated using this line. In order to circumvent these intrinsic hGH-mediated effects, additional beta-cell-specific transgenic lines have since been developed, the Ins1$^{tm1.1(cre)Thor}$ mouse (known as Ins1$^{Cre}$) and the inducible Ins1$^{tm2.1(cre/ERT2)Thor}$ mouse (known as *Ins1*$^{CreERT2}$) [30]. These mouse lines also have Cre activity driven by the Ins1 promoter and conditionally avoid ectopic Cre expression within the brain but importantly lack the hGH minigene fragment [30].

### 1.2 Inducible Transgenic Models

The development of inducible Cre-*loxP* and Flp-*FRT* transgenics has allowed researchers to study the importance of a gene during a specific stage of life, avoiding potential effects of tissue-specific knockout during development. The first inducible system to be developed was the Tet-Off system [31], rapidly followed by Tet-On [32]. This complementary pair of systems depends on the administration of tetracycline, or more commonly the tetracycline derivative, doxycycline (dox), which binds to a transcriptional activator protein, expressed in the target tissue, in order to either induce (Tet-On) or suppress (Tet-Off) transcription of Cre (Fig. 3). Tissue-specific expression is regulated instead by using the beta-cell-specific promoter to drive expression of the tetracycline-controlled transcriptional activator protein (tTA), which in turn regulates the transcription of Cre [33]. In the Tet-Off system, activity of Cre is suppressed during administration of dox, which inhibits tTA from binding the Cre promoter. Cre is therefore only able to alter target gene expression once administration of dox is discontinued [33]. In the Tet-On system, reverse tetracycline-controlled transactivator protein (rtTA) is non-functional without dox, which binds rtTA and allows it to transcribe Cre. Cre-mediated gene knock-in or knockout can therefore only occur once dox is administered [4, 33].

The tamoxifen-inducible system is the more commonly utilized inducible system [34] and instead targets Cre function at the protein level (Fig. 4). Cre recombinase is fused to a mutated estrogen receptor (ER) ligand-binding domain, which is unable to bind endogenously expressed 17β-estradiol, and instead can only be activated by administration of synthetic ligand tamoxifen [35]. Without tamoxifen, the Cre-ER$^{T}$ fusion protein is sequestered in the cytoplasm bound to heat-shock protein (Hsp) 90, until the active metabolite of the administrated tamoxifen, 4-hydroxytamoxifen, displaces Hsp90 and allows Cre-ER$^{T}$ to enter the nucleus, where it can then cause recombination at the *loxP* sites [4, 36, 37].

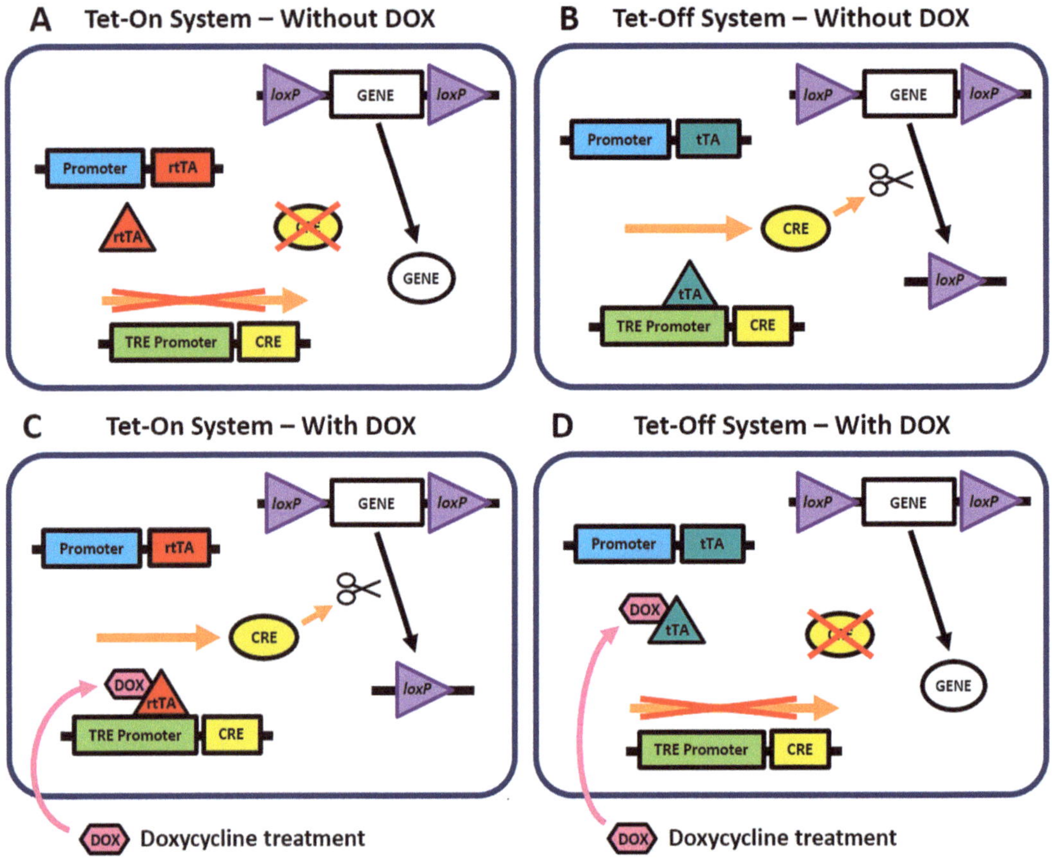


**Fig. 3** Tetracycline -On and -Off knockout models. In these systems, Cre recombinase expression is dependent on the tetracycline response element (TRE) in its promoter. In the Tet-On system (**a**, **c**), reverse transcription-activating (rtTA) protein is unable to bind to the Cre promoter, and normal target gene expression occurs (**a**). Treatment with doxycycline allows rtTA to bind the Cre promoter and elicit target gene deletion and knockout (**c**). In the Tet-Off system (**b**, **d**), Cre is expressed by default, and target gene knockout occurs in the absence of doxycycline (**b**). Doxycycline treatment prevents the binding of tetracycline transcription-activating (tTA) protein to the Cre promoter (**d**), suppressing Cre recombinase expression, with the target gene maintaining normal expression until doxycycline is removed. Figure adapted from [4, 33]

While highly advantageous, inducible models do come with some disadvantages. Firstly, target gene expression is altered through the exogenous administration of a chemical, of which both the chemical and the method through which it is administered can lead to side effects. Furthermore, knockout within the target tissue is often not as effective, with a gene knockdown efficiency of 50–90% common [19, 37–39], as well as basal leakage of expression [39–41]. It is important to note that the larger the dose or the longer the duration of administration, the more effective the knockdown is likely to be, but also the greater the potential side effects. Tamoxifen is a selective estrogen receptor modulator (SERM), and therefore it has a vast array of side effects, including

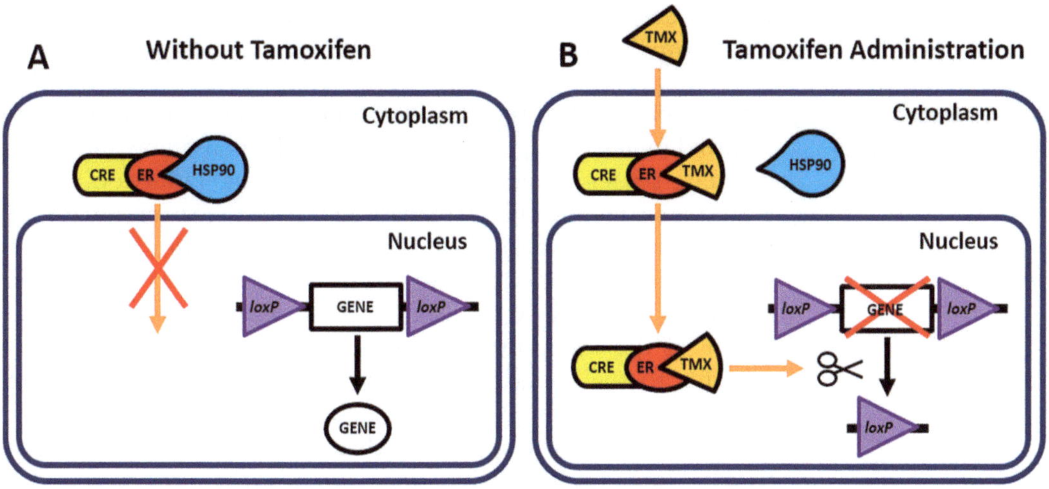


**Fig. 4** Tamoxifen-inducible Cre-*loxP* mouse knockout model. Cre recombinase is fused to the estrogen receptor (Cre-ER) and remains in the cytoplasm, bound to heat-shock protein 90 (HSP90) (**a**). Cre-ER is therefore unable to access the DNA, and gene expression continues as normal. Upon administration of the SERM tamoxifen (TMX) (**b**), its metabolite 4-hydroxytamoxifen displaces HSP90 to bind Cre-ER. This allows the Cre-ER fusion protein to translocate into the nucleus and excise the target gene from between the two *loxP* sites, causing gene deletion and knockout. Figure adapted from [36]

on fertility [19, 42], cardiac tissue [43, 44], bone turnover [45], and more importantly on adipogenesis, obesity, and beta-cell proliferation and function [28, 46, 47]. As an antibiotic, tetracycline has cytotoxic effects on mitochondrial function, lipid metabolism, and the liver [48–50], in addition to effects on the gut flora [51], which can alter food absorption. It is therefore essential the use of these exogenous chemical inducers be controlled for.

### *1.3 CRISPR and Other Approaches*

Although the process of creating transgenic knock-in or knockout mice has been established for over 30 years [52], this process has recently been simplified due to the development of clustered regularly interspaced short palindromic repeats/CRISPR associated protein 9 (CRISPR/Cas9) genome editing technology. Originating from bacterial genomes, where it functions as a prokaryotic antiviral system [53], the ability of CRISPR/Cas9 to target any region of the genome occurs through non-coding guiding RNA (gRNA), which uses complementary base pairing to bind the target DNA, guiding the Cas9 nuclease to introduce a controlled break in the DNA, through which gene insertion or deletion can then occur [54]. The methodology through which CRISPR is utilized for the generation of transgenic mouse strains is described in detail elsewhere [55–57]. Importantly, it can be used as a tool to more rapidly expand the library of "floxed" genes, of which only 25% of the mouse genome has so far been targeted [55]. Thus, if researchers intend to target a beta-cell gene for which no established LoxP or

FLT model exists, CRISPR enables the relatively easy generation of these novel lines allowing subsequent beta-cell manipulation.

### 1.4 Validation of Gene Manipulation

This chapter covers the considerations, options, and techniques involved in generating a beta-cell transgenic mouse line, including breeding strategies, genotyping, and transgenic induction. However, it is also important to validate any novel tissue-specific model to confirm effective and specific gene manipulation. There are several different options for validating successful beta-cell-specific manipulation of the target allele, with the choice depending on the nature of the end protein being knocked down, the techniques available to the researcher, and the specific mouse lines used in the generation of the new model. Describing these approaches in detail is beyond the scope of this chapter; however, it is important to note the options available.

Once the experimental mice have been generated, islet isolation allows for direct analysis of the genomic DNA, mRNA expression, and protein content of the islets. Detailed technical guides to islet isolation in mice have been previously published describing this methodology [58]. While the isolated islets will contain other non-beta-cells, the predominance of beta-cells within the isolated islets allows the simplest approach to validating the knockout model. However, keep in mind that many islet preparations will contain heterogeneous mixtures of cell types. Therefore, some level of unrecombined product may be observed in successful transgenic beta-cells.

Analysis of the genomic DNA in isolated islets will confirm the presence of gene recombination in transgenic mouse islets. This requires that you have a PCR assay that will distinguish the transgenic allele from the wild-type allele. Most genotyping protocols for floxed mice have not been designed for this purpose; therefore, you may need to check the literature for such an assay or design your own. Other tissues from transgenic mice and islets from control mice should also be included in this analysis to confirm specificity. In addition to genomic DNA analysis, islet mRNA can also be extracted and analyzed to assess transcription of the target allele. Quantitative PCR requires specific primers for the target gene of interest and will validate the extent of gene knockdown or upregulation. More detailed methodology covering the analysis of genomic DNA or mRNA can be found in previous dedicated guides [59].

Analysis of islet protein expression is also useful for validation of transgenic models and is commonly done either through western blotting of protein samples from isolated islets or through histological approaches. Both approaches rely on the availability of specific antibodies against the target peptide. Western blotting will offer an approximately quantitative measure of islet protein based on changes in band density, though it has the disadvantage that it is not possible to discriminate between beta-cells and other cell types. If islet isolation is not available as a technique, then histological approaches are a suitable option for validating a novel beta-cell transgenic model and have the advantage that protein expression specifically within the beta-cells can be assessed. The use of fixed pancreas sections from transgenic mice to stain for both insulin and the target peptide can be used to effectively validate the model. Detailed methodology covering both western blotting and histological approaches can again be found in separate specific guides [60–62]. Finally, if reporter genes, such as LacZ, have been incorporated into the transgenic model, then these can be an invaluable tool for the histological validation of both the efficacy and tissue specificity of the model [63].

## 2 Materials

### 2.1 Colony Maintenance and Breeding

#### 2.1.1 Initial Generation of Your Cre-loxP Crossed Mouse Colony

Having selected and located your chosen mouse strain or strains, options for their procurement are varied. Some strains may be maintained in repositories as active colonies, and live mice may be purchased directly. However, frequently, strains can only be bought as embryos or sperm. In this case, the mice will need to be rederived in house, for which 3–6 months must be allowed. Many institutes will have their own system for the process of rederivation.

1. Chosen Cre-expressing mouse of breeding age (*see* **Notes 1** and **2**).
2. Gene of interest *loxP* mouse of breeding age.

#### 2.1.2 Experimental Colony Maintenance

1. Cre-*loxP* mice.
2. *loxP* mice (*see* **Note 3**).
3. Cre mice (*see* **Note 4**).

### 2.2 Genotyping

All solutions are to be prepared using ultrapure water (prepared by purifying deionized water, to a sensitivity of 15 MΩ-cm at room temperature) and analytical grade reagents (unless otherwise stated). All reagents are prepared and stored at room temperature (unless otherwise stated). Disposal of waste materials should adhere to all waste disposal regulations.

#### 2.2.1 DNA Extraction from Ear Biopsy

1. Ear punch.
2. Forceps.
3. 1.5-mL microcentrifuge tubes.
4. 0.1 M sodium hydroxide (NaOH).
5. 1 M Tris, pH 7.4.
6. Absolute ethanol (*see* **Note 5**).

7. 3 M Sodium acetate ($C_2H_3NaO_2$).
8. 70% Ethanol solution in water.

#### 2.2.2 Genotyping: PCR

1. DNA polymerase master mix; store at −20 °C (*see* **Note 6**).
2. Cre Forward and Reverse Primers; store at −20 °C (*see* **Note 7**).
3. *loxP* Forward and Reverse Primers; store at −20 °C.
4. Molecular Grade Water.

#### 2.2.3 Gel Electrophoresis

1. 10× TAE: 400 mM Tris, 200 mM acetic acid, and 10 mM EDTA (pH 8.0) (*see* **Note 8**).
2. Agarose.
3. SYBR Safe (*see* **Note 9**).
4. DNA electrophoresis gel mold and comb set.
5. 6× Loading buffer: 0.25% bromophenol blue, 0.25% xylene cyanol FF, 30% glycerol in Molecular Grade Water; store at 4 °C (*see* **Note 10**).
6. DNA ladder; store at 4 °C (*see* **Note 11**).

### 2.3 Induction of Knockout in Inducible Models

#### 2.3.1 Tamoxifen

1. Tamoxifen free base (*see* **Note 12**).
2. Peanut oil (*see* **Note 13**).
3. 1-mL Sterile syringe.
4. 27G × 20 mm/0.75″ sterile needle.

#### 2.3.2 Tetracycline

1. 5% sucrose in drinking water solution.
2. Doxycycline.

## 3 Methods

### 3.1 Colony Maintenance and Breeding

#### 3.1.1 Initial Generation and Maintenance of Your Cre-loxP Crossed Mouse Colony

1. Following rederivation, establishing a colony of your Cre-*loxP* mouse requires the generation of mice which have homozygous expression of the *loxP*-flanked or "floxed" site (fl/fl) and heterozygous expression of the beta-cell-specific Cre (Cre/+) (*see* **Note 14**). This is achieved by performing two generations of crossbreeding (Fig. 5).
2. The initial founder mice will be in two distinct colonies; the Cre-expressing colony which is heterozygous for Cre expression driven by the beta-cell-specific driver of choice (Cre/+) and the *loxP* colony which is homozygous for the floxed gene of interest (fl/fl) (*see* **Note 15**).
3. Breeding between these colonies (*see* **Note 2**) will produce a first generation of offspring which are 100% heterozygous for

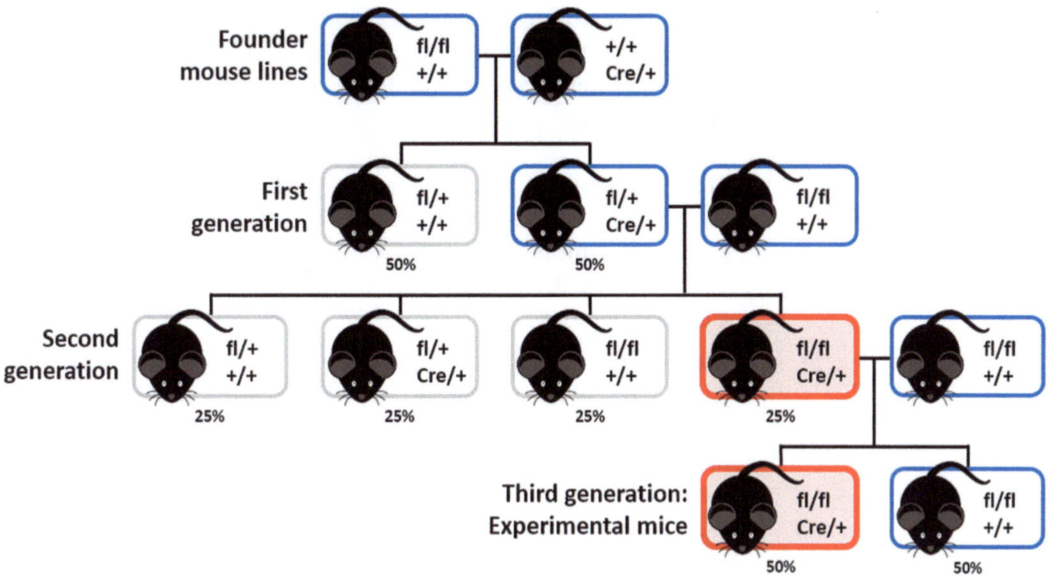


**Fig. 5** Breeding strategy for obtaining and maintaining Cre-*loxP* mice from two separate colonies of the Cre strain and the *loxP* strain. Founder mice from the chosen Cre colony (genotype: heterozygous Cre (Cre/+)) are crossed with mice from the relevant *loxP* colony (genotype: homozygous floxed for the gene of interest (fl/fl)). This produces a first-generation litter where all mice will be heterozygous for the floxed gene (fl/+), but only 50% of the offspring will also be heterozygous for Cre (Cre/+). The first-generation offspring with the fl/+, Cre/+ genotype can then be crossed with a homozygous floxed mouse (fl/fl) from the original *loxP* colony. This produces the second generation of offspring which may have one of four different genotypes: 25% fl/fl, Cre/+; 25% fl/fl, +/+; 25% fl/+, Cre/+; 25% fl/+, +/+. The fl/fl, Cre/+ genotype is the desired genotype for experimental mice, highlighted here in red. These second-generation mice can either be directly used experimentally or mated with further homozygous floxed mice to produce third-generation offspring with a 50% experimental genotype

expression of the floxed gene (fl/+), while 50% will have heterozygous expression of the Cre transgene (Cre/+).

4. Once adult, the first generation of offspring is then used for a second round of breeding. A heterozygous-Cre, heterozygous-floxed mouse can be mated with a mouse (fl/fl) from the original *loxP* colony. This will generate a second generation of offspring, of which 50% will be homozygous floxed (fl/fl) and 50% heterozygous floxed (fl/+), while 50% will have heterozygous expression of the Cre transgene (Cre/+). Thus, 25% of the second-generation offspring will be experimental mice containing both homozygous floxed (fl/fl) and heterozygous Cre (Cre/+) expression.

5. The *Cre-loxP* colony can now be maintained by breeding experimental mice (fl/fl: Cre/+) with a further homozygous floxed (fl/fl) mouse, which will produce 50% heterozygous Cre, homozygous floxed (fl/fl: Cre/+) and 50% Cre negative, homozygous floxed (fl/fl: Cre/+) mice (*see* **Note 3**).

*3.1.2 Maintaining Your Cre Colony (See* ***Note 4****)*

1. Cross heterozygous Cre mice with wild-type mice from the appropriate background strain. This produces 50% offspring which are heterozygous for the Cre transgene and 50% wild-type offspring.

*3.1.3 Reverse Generating Your Cre Colony*

1. If the Cre colony is not maintained and experiments to test the Cre control are later required, or if the generation of a new Cre-*loxP* combination is desired, the Cre mouse strain can be reverse generated from the Cre-loxP colony by the steps detailed below.
2. Cross a wild-type background strain mouse with a heterozygous Cre, homozygous floxed mouse (+/+: +/+ X Cre/+: fl/fl) to produce 50% offspring heterozygous Cre, heterozygous floxed (Cre/+: fl/+) and 50% offspring Cre negative, heterozygous floxed (+/+: fl/+).
3. Cross a wild-type background strain mouse with a heterozygous Cre, heterozygous floxed offspring from **step 1** (+/+: +/+ X Cre/+: fl/+). This will produce 25% target mice of heterozygous Cre, floxed gene negative (Cre/+: +/+) for restoring your Cre colony.

## *3.2 Genotyping*

*3.2.1 DNA Extraction from Ear Biopsy (See* ***Note 16****)*

1. Ear punch biopsies may be taken for genotyping from as early as postnatal day 14, with only a 2-mm-diameter sample required for obtaining an appropriate genomic DNA (gDNA) yield [64]. Biopsies may be processed immediately as described below, stored at 4 °C for up to a week, or stored at −20 °C for longer-term storage.
2. To the tissue, add 30 μL 0.1 M NaOH and incubate at 95 °C for approximately 10 min (*see* **Note 17**).
3. Allow to cool to room temperature for a minimum of 5 min.
4. Add 5 μL 1 M Tris (pH 8.0) and vortex the sample for approximately 5 s.
5. Briefly spin down samples at maximum speed for 10 s.
6. For most genotyping protocols, these steps will be sufficient to generate a gDNA sample for effective genotyping. In this case, the gDNA may be stored at −20 °C until ready for PCR; 3 μL of sample can be added to the PCR reaction. However, if a more purified gDNA sample is required for more efficient genotyping, then the following optional **steps 7–13** may also be performed.
7. Add 70 μL ethanol and 12 μL 3 M sodium acetate.
8. Spin 20 min at 4 °C, maximum speed (approximately 18,000 × *g*).
9. Remove supernatant.
10. Wash pellet with 250 μL 70% ethanol (*see* **Note 18**).

11. Spin 10 min at 4 °C, maximum speed (approximately 12,000 × *g*).
12. Remove supernatant.
13. Allow pellet to air dry before resuspending in 20 μL Molecular Grade Water (*see* **Note 19**).
14. Store gDNA at −20 °C until ready for PCR; 2–3 μL of sample can be added to the PCR (*see* **Note 20**).

#### 3.2.2 Genotyping: PCR

1. A typical PCR reaction mix is given below (*see* **Note 21**):

| Polymerase chain reaction | | |
|---|---|---|
| For 1 reaction | | 30 μL |
| 2× DNA polymerase master mix (*see* **Note 6**) | | 15 μL |
| Forward primer | 1 μM | 1.5 μL |
| Reverse primer | 1 μM | 1.5 μL |
| Molecular Grade Water | | 9 μL |
| Sample gDNA/positive or negative control DNA | | 3 μL |

2. The reaction mix for each gDNA sample can then be briefly spun down, placed in a thermal cycler, and any product DNA amplified through the following steps:
   (a) Initial denaturation step at 94–98 °C of 1–3 min (*see* **Note 22**).
   (b) Denaturation step at 94–98 °C of 15 s–1 min.
   (c) Annealing step at approximately 5 °C below the primer $T_m$ for approximately 30 s (*see* **Note 23**).
   (d) Elongation step, the temperature will be dictated by the DNA polymerase used and the length of time determined by the DNA polymerase used and primer product length (*see* **Note 24**).
   (e) Repeat **steps 2–4** (25–30 cycles are generally sufficient).
   (f) A final elongation step of approximately 5 min.
   (g) Cool samples to 4 °C and hold until the samples can be removed and stored at −20 °C or used immediately for gel electrophoresis.

#### 3.2.3 Gel Electrophoresis

1. Make the working 1× concentration for TAE or TBE buffer by diluting 100 mL 10× TAE/TBE in 900 mL distilled water.
2. Make a 2% agarose gel by adding 2 g agarose to 100 mL 1× TAE in a conical flask (*see* **Note 25**). Dissolve agarose by heating; this is typically performed in a microwave for

approximately 2 min under supervision to prevent boiling over, with regular intervals for mixing by swirling the flask.

3. Allow to cool slightly, and then add 5 μL SYBR Safe (*see* **Notes 9** and **26**), before pouring gel into the mold with an appropriate comb to set.
4. Place the gel inside the electrophoresis tank and fill the tank with appropriate buffer (TAE or TBE) so the gel is completely submerged.
5. Prepare appropriate volumes of gDNA sample (approximately 10 μL per well) and DNA ladder and add loading buffer, if not already present (*see* **Note 10**), at the ratio indicated.
6. Load samples into the gel wells with a DNA ladder (*see* **Note 11**) loaded into DNA wells adjacent to sample wells (*see* **Note 27**).
7. Run gel at 100–150 V for 40 min–1 h, depending on the band size and whether you want to detect separation of different sized bands (*see* **Note 28**).
8. Visualize your gel using a UV transilluminator or gel imaging system, being sure to use appropriate eye protection, and capture gel images if possible. Individual mice can be genotyped based on the presence or location of bands (*see* **Note 29**).

### 3.3 Induction of Knockout in Inducible Models

#### 3.3.1 Tamoxifen (See Note 30)

1. Tamoxifen is hazardous and should be handled with care (*see* **Note 31**).
2. Dissolve tamoxifen in peanut oil at a concentration of 20 mg/mL by agitation or shaking or rotation overnight at 37 °C. The solution is light sensitive and should be kept in an amber or foil wrapped container. Once in solution, store at 4 °C for the duration of injections.
3. 100 μL of the 20 mg/mL tamoxifen/peanut oil solution should be administered via intraperitoneal (i.p.) injection to the adult mouse (2 mg/mouse) in order to induce recombination (*see* **Note 32**).
4. This i.p. administration of 100 μL of the 20 mg/mL tamoxifen/peanut oil solution should be repeated for 4 consecutive days, adjusting slightly the location of injection in the lower left quadrant of the mouse peritoneal cavity, in order to minimize irritation at the puncture site (*see* **Note 33**).
5. Mice should be housed in filter top cages and isolated in a ventilation cabinet, marked as biohazards for the duration of the injection period and for a minimum of 72 h after administration of the final tamoxifen injection. They should be observed for any adverse reactions during this time, with any animals which die or are culled placed in biosafety bags for infectious waste incineration. All contaminated bedding should

be removed within a ventilated area while wearing appropriate PPE and disposed of as hazardous materials. Cages should also be placed within sealed bags labeled "Cytotoxic – Tamoxifen" and transported to the washroom where they can be placed into the cage wash while wearing PPE. These precautions should be observed until 72 h after final injection, at which time cages and bedding should be changed. You should communicate your intention to use tamoxifen and discover any further potential precautions or procedures enforced by your institute (*see* **Note 31**).

6. Following final administration, mice should be left for a minimum of 7 days prior to initiation of experimentation (*see* **Note 34**).

### 3.3.2 Tetracycline

1. Make up drinking water with 5% sucrose (*see* **Note 35**).
2. Dissolve 2 mg/mL of the tetracycline analogue doxycycline in drinking water with sucrose [65, 66]. The solution is light sensitive and should be kept in an amber or foil wrapped water bottle.
3. The water bottle should then be provided to experimental mice ad libitum for 3 days for Tet-On mice to induce recombination (*see* **Note 36**) or be provided continuously to Tet-Off mice, exchanging with a fresh bottle every 3 days, until the time at which you wish Cre-*loxP* recombination to occur.

## 4 Notes

1. This chapter will describe the techniques required when working with an inducible Cre-*loxP* mouse model for generating beta-cell-specific knockout. These methods are fully applicable to the use of the alternative FLP-*FRT* SSR, while also adaptable for the use with other tissue-specific knock-in and knockout mouse models. In the analogous FLP-*FRT* system, the Cre-expressing colony is your FLP mice and the *loxP* colony is your *FRT* mice.
2. Sexual maturity in mice is typically from 6 weeks of age, though this varies between strains. Earlier pairings can result in smaller litter sizes, and mice are typically mated between 6 and 8 weeks. Reproductive lifespan also varies between individual mouse strains, ranging between 22 and 38 weeks.
3. Due to the heterozygous expression of Cre, the *loxP* colony without the presence of the Cre gene is automatically maintained in the process of maintaining the Cre-*loxP* colony. These littermates make important controls for comparing against the effects of the Cre knockout and the effects of inducible

knockout agents such as tamoxifen and tetracycline. Additionally, the process of insertion of the *loxP* sites can inadvertently alter expression of the target gene globally; using these Cre-negative *loxP* littermates as controls will ensure any effect measured is due to knockout of the gene in your selected tissue and not the result of changes in expression elsewhere due to global disruption by the presence of the *loxP* sites.

4. It is advisable to maintain the colony of Cre mice by cross-breeding with wild-type individuals from the background strain. This can provide an important control for examining any effects of the Cre transgene itself on beta-cell function [13]. Additionally, it is important to prevent excessive inbreeding over time, which can lead to genetic abnormalities. Excessive inbreeding and overuse of individual breeders can lead to reduced litter sizes, smaller offspring, and greater mortality and morbidity in neonates. New animals can be introduced from the Cre colony to the Cre-*loxP* colony every 6–12 months to prevent inbreeding.
5. Subheading 2.2.1, **items 6–8** are optional if requiring a more purified gDNA sample.
6. DNA polymerase often needs to be accompanied by a specific buffer providing optimal conditions for its activity, as well as deoxynucleotides (dNTPs). Reagents such as $Mg^{2+}$, which are required by thermostable DNA polymerases such as *Taq*, $K^+$, which is often reduced from its normal final concentration of 50 mM when longer PCR products are required, and DMSO which can lower melting temperature ($T_m$) of base pairs, may also be present [67]. Many commercially available DNA polymerases can be purchased as a "master mix," combining the enzyme with its appropriate buffer and containing the optimal concentrations of these additional reagents.
7. Primer sequences for genotyping the appropriate Cre strain and floxed gene can usually be obtained from either the laboratory or the commercial supplier that supplied the mouse lines. A separate PCR will be required for each transgene; therefore, each mouse sample will undergo one reaction for the Cre gene and another for the floxed gene.
8. There are two commonly used buffers for gel electrophoresis: TAE: Tris (2-amino-2-[hydroxymethyl]-1,3-propanediol), acetic acid, and EDTA (ethylenediaminetetraacetic acid), or TBE buffer: Tris, boric acid, and EDTA, which both work well for separation of DNA fragments ranging from several hundred to several thousand bp, a range encompassing the typical product size of genotyping primers [68]. Both can be made at a 10× stock for storage and can then be diluted in distilled water for use in electrophoresis. Opinions over which

buffer gives better band definition differ, but it is generally agreed 10× TAE stock solution lasts longer, while 10× TBE precipitates over time. 10× TBE: 890 mM Tris, 890 mM boric acid, 20 mM EDTA (pH 8.0).

9. Or equivalent noncarcinogenic alternative to ethidium bromide.

10. Some DNA polymerase master mixes and DNA ladders already contain a loading buffer, so check first before adding. There are many recipes for loading buffers also containing loading dyes such as Orange G and Cresol Red, alternative density agents such as sucrose or Ficoll, as well as DNA protective stabilizers such as 10 mM Tris–HCl (pH 7.5) and 50 mM EDTA (pH 8.0). The dyes each have different migration through different percentage agarose gels which can be an important consideration when looking at a particular PCR product size.

11. Ensure this ladder is appropriate for the size of the PCR product, typically a 100-bp or 1-kb DNA ladder is appropriate.

12. This is metabolized to the active 4-hydroxytamoxifen in vivo.

13. This is a delivery vehicle for highly lipophilic compounds such as tamoxifen and can be exchanged for other non-polar solvents such as corn oil.

14. The abbreviation "fl" is used to refer to the "floxed" gene, while the wild-type or unmodified version of a gene is denoted as +.

15. In the analogous FLP-*FRT* system, the Cre-expressing colony is your FLP mice, heterozygous for FLP expression and homozygous negative for an FRT-flanked site. The *loxP* colony is your *FRT* mice, homozygous negative for FLP expression and homozygous positive for an *FRT*-flanked site.

16. The preferred method for obtaining a tissue sample for genotyping, used in 85% of labs, is to perform an ear biopsy, using either an ear punch or notch [69]. This method also provides a means of identification, based on the number and location of the punches or notches on each ear. Tail biopsies have been used historically; however, their use should be considered as a last resort for obtaining tissue, due to complications associated with damage to the tail vertebrae and associated nervous tissue. Tail biopsies should not be taken from mice significantly younger than 3 weeks or older than 4 weeks of age without good scientific justification, and similarly, toe amputation should not be considered as a method for obtaining gDNA at any age. Collection of a 20–50-μL blood sample after 2 weeks of age is an alternative in mice where identification is not required, with the additional advantage of being easily repeated [64], though there are some added technical problems when preparing the

DNA due to the low proportion of nucleated white blood cells [70].

Noninvasive alternatives can provide a method of genotyping in individuals younger than 2 weeks of age; however, they can often present with their own problems. Collection of gDNA from hair follicles [71] can be subject to cross contamination between litter mates, and care not to damage the rectum is needed when obtaining rectal epithelial samples. Collection of oral epithelial cells via an oral swab [72] or intestinal epithelial cells from stools [73, 74] can provide adequate material for genotyping and is strongly encouraged by bodies such as the British Veterinary Association Animal Welfare Foundation (BVA AWF). The presence of nucleases and other enzymes in stool samples means a commercial kit is required for their removal, in order to protect the gDNA from degradation and to ensure the downstream PCR are not inhibited [75]. Furthermore, the PCR amplification products from stool samples can still be weaker than from those seen from tissue biopsy, with the added problem that where a form of identification is required, noninvasive identifiers such as use of marker pens or fur shaving are only visible for 1–2 weeks [75].

17. The following visual signs of tissue degradation indicate the tissue has been incubated long enough; edges of the tissue should appear paler or translucent, and in some cases, the solution will begin to appear brown after mixing by vortex. If unsure, vortex the sample, briefly spin down, and return the sample to 95 °C for another 10 min.

18. If you have trouble visualizing the pellet, replace **step 10** with the following: add 1 μL 20 μg/μL glycogen and 250 μL ethanol to sample; store at −20 °C for 30 min.

19. If water is added before all ethanol has evaporated, you may have trouble getting the DNA back into solution. You can try mixing samples by vortex or incubating for 1 h at 55 °C in order to assist dissolution.

20. For a more highly purified DNA preparation, a phenol/chloroform extraction following overnight DNA digestion with Proteinase K can be performed [76].

21. False-positive bands can often be as a result of contaminated primers; therefore, it is advisable to aliquot multiple primer stocks in order to minimize risk of contamination and easily replace contaminated stocks. It is also advisable to create a "master mix" combining first water and the DNA polymerase master mix and then adding the forward and reverse primers, in order to prevent primer contamination of the DNA polymerase, particularly when you intend to use it to detect both Cre and your floxed gene. This master mix can then be distributed

to your tubes before addition of the sample DNA, in order to prevent cross contamination between samples being measured.

22. Any longer and the enzymatic activity of the DNA polymerase may be destroyed. "Hot start" PCR can help to reduce non-specific bands and increase amplicon yield [67]. Under these conditions, this denaturation step lasts up to 9 min, and the DNA polymerase is added to the reaction after this cycle is completed or alternatively is specifically a hot start-specific polymerase which is activated by the denaturation step.
23. This may need to be optimized by raising the temperature by 1 °C until reaching the official primer annealing temperature and can last between 15 s and 1 min.
24. For example, Taq DNA polymerase has an optimal temperature of 72–74 °C and requires 1 min per 1 kb of DNA to be amplified.
25. Smaller (75–2000 bp) DNA products seen when genotyping work best using a 2–4% agarose gel. The volume should be adjusted depending on the size of the agarose gel mold.
26. If fluorescence across agarose gel is uneven, affecting clarity of bands, try post-staining gel with 20 μL SYBR Safe in 100 mL TBE or TAE for 30 min under gentle agitation.
27. Check for leakage of sample out from well by looking for dye entering neighboring wells or floating into buffer as this could cause contamination of neighboring samples. This can be avoided by dispensing loading volume slowly and avoiding expulsion of bubbles from the pipette tip, or by decreasing the loading volume. If necessary, leave empty wells between wells loaded with samples.
28. Voltage can be adjusted and lowered to reduce heating of the gel and provide higher resolution and prevent asymmetrical running of samples, helping to produce a more distinct separation of bands. If upon imaging of the gel, further separation of DNA bands is required, the gel can be returned to the tank and voltage reapplied. A more detailed description of optimization techniques for gel electrophoresis is discussed elsewhere [68].
29. While the presence of the Cre transgene in a mouse gDNA sample appears as a binary presence or absence of a product band in an agarose gel, detection of a floxed gene is most often detected by a shift in the position of the DNA band, unless one primer spans part of the *loxP* site itself. This is due to insertion of the *loxP* site sequence into the target gene, making the product slightly larger than the unmodified, wild-type version of the gene. This slightly larger DNA product does not travel as quickly through the gel during electrophoresis and can therefore be detected higher up on the gel. A heterozygous gDNA

sample will therefore be seen as two DNA bands appearing closely together as both the wild-type and floxed gene variants will be present. It is therefore important to include a homozygous floxed, a homozygous wild-type, and a heterozygous control sample on the gel, in order to ensure the gel has been run long enough for adequate separation of the bands to occur and to correctly interpret the results.

30. We have selected tamoxifen administration by i.p. injection, due to its relative ease of administration and reliability of dosing. However, other suitable methods of tamoxifen administration include oral gavage [77], subcutaneous injections [78], and tamoxifen-supplemented feed [79].
31. PPE, including lab coat, gloves, and face mask, should be worn while handling tamoxifen, which is a hazardous substance. Preparation of syringes and injections should be performed within a minimum Class II Biosafety cabinet, with appropriate PPE. Used needles and syringes can be placed in a sharps container, while any materials coming into contact with tamoxifen should be disposed of as cytotoxic waste or decontaminated with disinfectant and water. You should review the MSDS for further information and your institution's own Standard Operating Procedures or Environmental Health and Safety policies.
32. As a selective estrogen receptor modulator, we recommend administration of tamoxifen after onset of sexual maturity at 6–8 weeks in mice to avoid confounding factors. However, if necessary, mice younger than this should be administered with tamoxifen at 75 mg/kg body weight.
33. Duration of injections may vary; research has indicated 90% knockdown from as little as two consecutive 1 mg/mouse/day i.p. injections in some tamoxifen-inducible Cre mouse models, which has beneficial implications for improving animal welfare and reducing workload. However, a small but significant increase in knockdown was seen between two and five consecutive injections [79].
34. It is advisable to factor in a longer period of washout, taking into account the effects and tissues being measured. Evidence for tamoxifen-induced recombination in beta-cells at 3 months after a series of three subcutaneous tamoxifen injections [78] and in white adipose tissue at 2 months after 5 consecutive days of oral gavage or i.p. injection [47] emphasizes the need for tamoxifen-administered controls and for the effects of tamoxifen to be taken into account when analyzing results (*see* Subheading 1.3).
35. This most commonly used method of administration of doxycycline is through the drinking water, which requires a

sweetener in order to improve palatability [80]. This will impact glucose homeostasis studies, as well as behavioral studies such as the Sucrose Preference Test and 5% sucrose should be provided to any control mice for the same duration. Alternatively, sweeteners such as Splenda, Aspartame, and Xylitol can be used at 4 mg/mL. Alternative methods of administration include doxycycline-containing food pellets [33, 81], oral gavage [82], or i.p. injection [83]. Comparisons of doxycycline delivery methods can be seen elsewhere [80].

36. Duration and dosing of the doxycycline administration in Tet-On systems varies from 2 to 3 days [65, 66, 80] to 2 weeks [84] and may need to be altered depending on knockdown efficiency of the strain being used.

## References


1. Hoess RH, Ziese M, Sternberg N (1982) P1 site-specific recombination: nucleotide sequence of the recombining sites. Proc Natl Acad Sci U S A 79(11):3398–3402
2. McLeod M, Craft S, Broach JR (1986) Identification of the crossover site during FLP-mediated recombination in the Saccharomyces cerevisiae plasmid 2 microns circle. Mol Cell Biol 6(10):3357–3367
3. Tahimic CGTS, Sakurai K, Aiba K, Nakatsuji N (2013) Cre/loxP, Flp/FRT systems and pluripotent stem cell lines. In: Renault SD, Duchateau P (eds) Site-directed insertion of transgenes, Topics in current genetics, vol 23, pp 189–209. https://doi.org/10.1007/978-94-007-4531-5_7
4. Zhang J, Zhao J, Jiang WJ, Shan XW, Yang XM, Gao JG (2012) Conditional gene manipulation: Cre-ating a new biological era. J Zhejiang Univ Sci B 13(7):511–524. https://doi.org/10.1631/jzus.B1200042
5. Ringrose L, Lounnas V, Ehrlich L, Buchholz F, Wade R, Stewart AF (1998) Comparative kinetic analysis of FLP and cre recombinases: mathematical models for DNA binding and recombination. J Mol Biol 284(2):363–384. https://doi.org/10.1006/jmbi.1998.2149
6. BCBC (2015) Beta cell biology consortium resource catalog. http://www.betacell.org. Accessed 08 Aug 2019
7. IMSR (2019) International mouse strain resource. http://www.findmice.org. Accessed 08 Aug 2019
8. Magnuson MA, Osipovich AB (2013) Pancreas-specific Cre driver lines and considerations for their prudent use. Cell Metab 18 (1):9–20. https://doi.org/10.1016/j.cmet.2013.06.011
9. Johnson JD (2014) A practical guide to genetic engineering of pancreatic beta-cells in vivo: getting a grip on RIP and MIP. Islets 6(3): e944439. https://doi.org/10.4161/19382014.2014.944439
10. Song J, Xu Y, Hu X, Choi B, Tong Q (2010) Brain expression of Cre recombinase driven by pancreas-specific promoters. Genesis 48 (11):628–634. https://doi.org/10.1002/dvg.20672
11. Wicksteed B, Brissova M, Yan W, Opland DM, Plank JL, Reinert RB, Dickson LM, Tamarina NA, Philipson LH, Shostak A, Bernal-Mizrachi E, Elghazi L, Roe MW, Labosky PA, Myers MG Jr, Gannon M, Powers AC, Dempsey PJ (2010) Conditional gene targeting in mouse pancreatic ss-Cells: analysis of ectopic Cre transgene expression in the brain. Diabetes 59(12):3090–3098. https://doi.org/10.2337/db10-0624
12. Gong GC, Fan WZ, Li DZ, Tian X, Chen SJ, Fu YC, Xu WC, Wei CJ (2015) Increased specific labeling of INS-1 pancreatic beta-cell by using RIP-driven Cre mutants with reduced activity. PLoS One 10(6):e0129092. https://doi.org/10.1371/journal.pone.0129092
13. Lee JY, Ristow M, Lin X, White MF, Magnuson MA, Hennighausen L (2006) RIP-Cre revisited, evidence for impairments of pancreatic beta-cell function. J Biol Chem 281 (5):2649–2653. https://doi.org/10.1074/jbc.M512373200
14. Pomplun D, Florian S, Schulz T, Pfeiffer AF, Ristow M (2007) Alterations of pancreatic beta-cell mass and islet number due to Ins2-controlled expression of Cre recombinase: RIP-Cre revisited; part 2. Horm Metab Res

39(5):336–340. https://doi.org/10.1055/s-2007-976538

15. Teitelman G, Kedees M (2015) Mouse insulin cells expressing an inducible RIPCre transgene are functionally impaired. J Biol Chem 290 (6):3647–3653. https://doi.org/10.1074/jbc.M114.615484
16. Bernardo AS, Hay CW, Docherty K (2008) Pancreatic transcription factors and their role in the birth, life and survival of the pancreatic beta cell. Mol Cell Endocrinol 294(1–2):1–9. https://doi.org/10.1016/j.mce.2008.07.006
17. Herrera PL (2000) Adult insulin- and glucagon-producing cells differentiate from two independent cell lineages. Development 127(11):2317–2322
18. Wiebe PO, Kormish JD, Roper VT, Fujitani Y, Alston NI, Zaret KS, Wright CV, Stein RW, Gannon M (2007) Ptf1a binds to and activates area III, a highly conserved region of the Pdx1 promoter that mediates early pancreas-wide Pdx1 expression. Mol Cell Biol 27 (11):4093–4104. https://doi.org/10.1128/MCB.01978-06
19. Zhang H, Fujitani Y, Wright CV, Gannon M (2005) Efficient recombination in pancreatic islets by a tamoxifen-inducible Cre-recombinase. Genesis 42(3):210–217. https://doi.org/10.1002/gene.20137
20. Hara M, Wang X, Kawamura T, Bindokas VP, Dizon RF, Alcoser SY, Magnuson MA, Bell GI (2003) Transgenic mice with green fluorescent protein-labeled pancreatic beta -cells. Am J Physiol Endocrinol Metab 284(1): E177–E183. https://doi.org/10.1152/ajpendo.00321.2002
21. Tamarina NA, Roe MW, Philipson L (2014) Characterization of mice expressing Ins1 gene promoter driven CreERT recombinase for conditional gene deletion in pancreatic beta-cells. Islets 6(1):e27685. https://doi.org/10.4161/isl.27685
22. Kojima H, Fujimiya M, Matsumura K, Nakahara T, Hara M, Chan L (2004) Extrapancreatic insulin-producing cells in multiple organs in diabetes. Proc Natl Acad Sci U S A 101(8):2458–2463
23. Szabat M, Pourghaderi P, Soukhatcheva G, Verchere CB, Warnock GL, Piret JM, Johnson JD (2011) Kinetics and genomic profiling of adult human and mouse beta-cell maturation. Islets 3(4):175–187
24. Hay CW, Docherty K (2006) Comparative analysis of insulin gene promoters: implications for diabetes research. Diabetes 55 (12):3201–3213. https://doi.org/10.2337/db06-0788
25. Roderigo-Milne H, Hauge-Evans AC, Persaud SJ, Jones PM (2002) Differential expression of insulin genes 1 and 2 in MIN6 cells and pseudoislets. Biochem Biophys Res Commun 296 (3):589–595
26. Oropeza D, Jouvet N, Budry L, Campbell JE, Bouyakdan K, Lacombe J, Perron G, Bergeron V, Neuman JC, Brar HK, Fenske RJ, Meunier C, Sczelecki S, Kimple ME, Drucker DJ, Screaton RA, Poitout V, Ferron M, Alquier T, Estall JL (2015) Phenotypic characterization of MIP-CreERT1Lphi mice with transgene-driven islet expression of human growth hormone. Diabetes 64 (11):3798–3807. https://doi.org/10.2337/db15-0272
27. Brouwers B, de Faudeur G, Osipovich AB, Goyvaerts L, Lemaire K, Boesmans L, Cauwelier EJ, Granvik M, Pruniau VP, Van Lommel L, Van Schoors J, Stancill JS, Smolders I, Goffin V, Binart N, in't Veld P, Declercq J, Magnuson MA, Creemers JW, Schuit F, Schraenen A (2014) Impaired islet function in commonly used transgenic mouse lines due to human growth hormone minigene expression. Cell Metab 20(6):979–990. https://doi.org/10.1016/j.cmet.2014.11.004
28. Carboneau BA, Le TD, Dunn JC, Gannon M (2016) Unexpected effects of the MIP-CreER transgene and tamoxifen on beta-cell growth in C57Bl6/J male mice. Physiol Rep 4(18). https://doi.org/10.14814/phy2.12863
29. Kim H, Kim H, Kim K, German MS, Kim H (2018) Ectopic serotonin production in beta-cell specific transgenic mice. Biochem Biophys Res Commun 495(2):1986–1991. https://doi.org/10.1016/j.bbrc.2017.12.005
30. Thorens B, Tarussio D, Maestro MA, Rovira M, Heikkila E, Ferrer J (2015) Ins1 (Cre) knock-in mice for beta cell-specific gene recombination. Diabetologia 58(3):558–565. https://doi.org/10.1007/s00125-014-3468-5
31. Gossen M, Bujard H (1992) Tight control of gene expression in mammalian cells by tetracycline-responsive promoters. Proc Natl Acad Sci U S A 89(12):5547–5551
32. Gossen M, Freundlieb S, Bender G, Muller G, Hillen W, Bujard H (1995) Transcriptional activation by tetracyclines in mammalian cells. Science 268(5218):1766–1769
33. Toselli C, Hyslop CM, Hughes M, Natale DR, Santamaria P, Huang CT (2014) Contribution of a non-beta-cell source to beta-cell mass during pregnancy. PLoS One 9(6):e100398. https://doi.org/10.1371/journal.pone.0100398

34. Metzger D, Clifford J, Chiba H, Chambon P (1995) Conditional site-specific recombination in mammalian cells using a ligand-dependent chimeric Cre recombinase. Proc Natl Acad Sci U S A 92(15):6991–6995
35. Feil R, Brocard J, Mascrez B, LeMeur M, Metzger D, Chambon P (1996) Ligand-activated site-specific recombination in mice. Proc Natl Acad Sci U S A 93 (20):10887–10890
36. Tian Y, James S, Zuo J, Fritzsch B, Beisel KW (2006) Conditional and inducible gene recombineering in the mouse inner ear. Brain Res 1091(1):243–254. https://doi.org/10.1016/j.brainres.2006.01.040
37. Chow LM, Tian Y, Weber T, Corbett M, Zuo J, Baker SJ (2006) Inducible Cre recombinase activity in mouse cerebellar granule cell precursors and inner ear hair cells. Dev Dyn 235(11):2991–2998. https://doi.org/10.1002/dvdy.20948
38. Casanova E, Fehsenfeld S, Lemberger T, Shimshek DR, Sprengel R, Mantamadiotis T (2002) ER-based double iCre fusion protein allows partial recombination in forebrain. Genesis 34 (3):208–214. https://doi.org/10.1002/gene.10153
39. Kistner A, Gossen M, Zimmermann F, Jerecic J, Ullmer C, Lubbert H, Bujard H (1996) Doxycycline-mediated quantitative and tissue-specific control of gene expression in transgenic mice. Proc Natl Acad Sci U S A 93 (20):10933–10938
40. Agha-Mohammadi S, O'Malley M, Etemad A, Wang Z, Xiao X, Lotze MT (2004) Second-generation tetracycline-regulatable promoter: repositioned tet operator elements optimize transactivator synergy while shorter minimal promoter offers tight basal leakiness. J Gene Med 6(7):817–828. https://doi.org/10.1002/jgm.566
41. Liu Y, Suckale J, Masjkur J, Magro MG, Steffen A, Anastassiadis K, Solimena M (2010) Tamoxifen-independent recombination in the RIP-CreER mouse. PLoS One 5 (10):e13533. https://doi.org/10.1371/journal.pone.0013533
42. Patel SH, O'Hara L, Atanassova N, Smith SE, Curley MK, Rebourcet D, Darbey AL, Gannon AL, Sharpe RM, Smith LB (2017) Low-dose tamoxifen treatment in juvenile males has long-term adverse effects on the reproductive system: implications for inducible transgenics. Sci Rep 7(1):8991. https://doi.org/10.1038/s41598-017-09016-4
43. Bersell K, Choudhury S, Mollova M, Polizzotti BD, Ganapathy B, Walsh S, Wadugu B, Arab S, Kuhn B (2013) Moderate and high amounts of tamoxifen in alphaMHC-MerCreMer mice induce a DNA damage response, leading to heart failure and death. Dis Model Mech 6 (6):1459–1469. https://doi.org/10.1242/dmm.010447
44. Koitabashi N, Bedja D, Zaiman AL, Pinto YM, Zhang M, Gabrielson KL, Takimoto E, Kass DA (2009) Avoidance of transient cardiomyopathy in cardiomyocyte-targeted tamoxifen-induced MerCreMer gene deletion models. Circ Res 105(1):12–15. https://doi.org/10.1161/CIRCRESAHA.109.198416
45. Zhong ZA, Sun W, Chen H, Zhang H, Lay YE, Lane NE, Yao W (2015) Optimizing tamoxifen-inducible Cre/loxp system to reduce tamoxifen effect on bone turnover in long bones of young mice. Bone 81:614–619. https://doi.org/10.1016/j.bone.2015.07.034
46. Guillaume M, Handgraaf S, Fabre A, Raymond-Letron I, Riant E, Montagner A, Vinel A, Buscato M, Smirnova N, Fontaine C, Guillou H, Arnal JF, Gourdy P (2017) Selective activation of estrogen receptor alpha activation function-1 is sufficient to prevent obesity, steatosis, and insulin resistance in mouse. Am J Pathol 187(6):1273–1287. https://doi.org/10.1016/j.ajpath.2017.02.013
47. Ye R, Wang QA, Tao C, Vishvanath L, Shao M, McDonald JG, Gupta RK, Scherer PE (2015) Impact of tamoxifen on adipocyte lineage tracing: inducer of adipogenesis and prolonged nuclear translocation of Cre recombinase. Mol Metab 4(11):771–778. https://doi.org/10.1016/j.molmet.2015.08.004
48. Morimoto M, Kopan R (2009) rtTA toxicity limits the usefulness of the SP-C-rtTA transgenic mouse. Dev Biol 325(1):171–178. https://doi.org/10.1016/j.ydbio.2008.10.013
49. Begriche K, Massart J, Robin MA, Borgne-Sanchez A, Fromenty B (2011) Drug-induced toxicity on mitochondria and lipid metabolism: mechanistic diversity and deleterious consequences for the liver. J Hepatol 54 (4):773–794. https://doi.org/10.1016/j.jhep.2010.11.006
50. Moullan N, Mouchiroud L, Wang X, Ryu D, Williams EG, Mottis A, Jovaisaite V, Frochaux MV, Quiros PM, Deplancke B, Houtkooper RH, Auwerx J (2015) Tetracyclines disturb mitochondrial function across eukaryotic models: a call for caution in biomedical research. Cell Rep. https://doi.org/10.1016/j.celrep.2015.02.034
51. Yin J, Zhang XX, Wu B, Xian Q (2015) Metagenomic insights into tetracycline effects on

microbial community and antibiotic resistance of mouse gut. Ecotoxicology 24 (10):2125–2132. https://doi.org/10.1007/s10646-015-1540-7

52. Thomas KR, Capecchi MR (1987) Site-directed mutagenesis by gene targeting in mouse embryo-derived stem cells. Cell 51 (3):503–512
53. Szczepankowska A (2012) Role of CRISPR/cas system in the development of bacteriophage resistance. Adv Virus Res 82:289–338. https://doi.org/10.1016/B978-0-12-394621-8.00011-X
54. Singh P, Schimenti JC, Bolcun-Filas E (2015) A mouse geneticist's practical guide to CRISPR applications. Genetics 199(1):1–15. https://doi.org/10.1534/genetics.114.169771
55. Quadros RM, Miura H, Harms DW, Akatsuka H, Sato T, Aida T, Redder R, Richardson GP, Inagaki Y, Sakai D, Buckley SM, Seshacharyulu P, Batra SK, Behlke MA, Zeiner SA, Jacobi AM, Izu Y, Thoreson WB, Urness LD, Mansour SL, Ohtsuka M, Gurumurthy CB (2017) Easi-CRISPR: a robust method for one-step generation of mice carrying conditional and insertion alleles using long ssDNA donors and CRISPR ribonucleoproteins. Genome Biol 18(1):92. https://doi.org/10.1186/s13059-017-1220-4
56. Ohtsuka M, Sato M, Miura H, Takabayashi S, Matsuyama M, Koyano T, Arifin N, Nakamura S, Wada K, Gurumurthy CB (2018) i-GONAD: a robust method for in situ germline genome engineering using CRISPR nucleases. Genome Biol 19(1):25. https://doi.org/10.1186/s13059-018-1400-x
57. Burgio G (2018) Redefining mouse transgenesis with CRISPR/Cas9 genome editing technology. Genome Biol 19(1):27. https://doi.org/10.1186/s13059-018-1409-1
58. Ramirez-Dominguez M (2016) Isolation of mouse pancreatic islets of langerhans. Adv Exp Med Biol 938:25–34. https://doi.org/10.1007/978-3-319-39824-2_3
59. Biassoni R, Raso A (2016) Quantitative real-time PCR: methods and protocols. Springer, New York
60. Kalyuzhny AE (2017) Signal transduction immunohistochemistry : methods and protocols. Methods in molecular biology, 2nd edn. Humana Press, New York, NY, p 1554. 9781493967575
61. Kurien BT, Scofield RH (2006) Western blotting. Methods 38(4):283–293. https://doi.org/10.1016/j.ymeth.2005.11.007
62. Lin F, Prichard J (2015) Handbook of practical immunohistochemistry: frequently asked questions, 2nd edn. Springer, New York
63. Loughna S, Henderson D (2007) Methodologies for staining and visualisation of beta-galactosidase in mouse embryos and tissues. Methods Mol Biol 411:1–11
64. Bonaparte D, Cinelli P, Douni E, Herault Y, Maas M, Pakarinen P, Poutanen M, Lafuente MS, Scavizzi F, Federation of European Laboratory Animal Science Associations Working Group (2013) FELASA guidelines for the refinement of methods for genotyping genetically-modified rodents: a report of the Federation of European Laboratory Animal Science Associations Working Group. Lab Anim 47(3):134–145. https://doi.org/10.1177/0023677212473918
65. Cheng Y, Su Y, Shan A, Jiang X, Ma Q, Wang W, Ning G, Cao Y (2015) Generation and characterization of transgenic mice expressing mouse Ins1 promoter for pancreatic beta-cell-specific gene overexpression and knockout. Endocrinology 156(7):2724–2731. https://doi.org/10.1210/en.2015-1104
66. Rao P, Monks DA (2009) A tetracycline-inducible and skeletal muscle-specific Cre recombinase transgenic mouse. Dev Neurobiol 69(6):401–406. https://doi.org/10.1002/dneu.20714
67. Lorenz TC (2012) Polymerase chain reaction: basic protocol plus troubleshooting and optimization strategies. J Vis Exp 63:e3998. https://doi.org/10.3791/3998
68. Sanderson BA, Araki N, Lilley JL, Guerrero G, Lewis LK (2014) Modification of gel architecture and TBE/TAE buffer composition to minimize heating during agarose gel electrophoresis. Anal Biochem 454:44–52. https://doi.org/10.1016/j.ab.2014.03.003
69. Mazlan NH, Lopez-Salesansky N, Burn CC, Wells DJ (2014) Mouse identification methods and potential welfare issues: a survey of current practice in the UK. Animal Technology and Welfare 13(1): 1–10
70. Winberg G (1991) A rapid method for preparing DNA from blood, suited for PCR screening of transgenes in mice. PCR Methods Appl 1 (1):72–74
71. Schmitteckert EM, Prokop CM, Hedrich HJ (1999) DNA detection in hair of transgenic mice – a simple technique minimizing the distress on the animals. Lab Anim 33 (4):385–389. https://doi.org/10.1258/002367799780487922
72. Meldgaard M, Bollen PJ, Finsen B (2004) Non-invasive method for sampling and

extraction of mouse DNA for PCR. Lab Anim 38(4):413–417. https://doi.org/10.1258/0023677041958981

73. Broome RL, Feng L, Zhou Q, Smith A, Hahn N, Matsui SM, Omary MB (1999) Non-invasive transgenic mouse genotyping using stool analysis. FEBS Lett 462 (1–2):159–160

74. Chen Z, Mantha RR, Chen JS, Slivano OJ, Takahashi H (2012) Non-invasive genotyping of transgenic animals using fecal DNA. Lab Anim (NY) 41(4):102–107. https://doi.org/10.1038/laban0412-102

75. Hamann M, Lange N, Kuschka J, Richter A (2010) Non-invasive genotyping of transgenic mice: comparison of different commercial kits and required amounts. ALTEX 27(3):185–190

76. Jackson DP, Lewis FA, Taylor GR, Boylston AW, Quirke P (1990) Tissue extraction of DNA and RNA and analysis by the polymerase chain reaction. J Clin Pathol 43(6):499–504

77. Seibler J, Zevnik B, Kuter-Luks B, Andreas S, Kern H, Hennek T, Rode A, Heimann C, Faust N, Kauselmann G, Schoor M, Jaenisch R, Rajewsky K, Kuhn R, Schwenk F (2003) Rapid generation of inducible mouse mutants. Nucleic Acids Res 31(4):e12

78. Reinert RB, Kantz J, Misfeldt AA, Poffenberger G, Gannon M, Brissova M, Powers AC (2012) Tamoxifen-induced Cre--loxP recombination is prolonged in pancreatic islets of adult mice. PLoS One 7(3):e33529. https://doi.org/10.1371/journal.pone.0033529

79. Andersson KB, Winer LH, Mork HK, Molkentin JD, Jaisser F (2010) Tamoxifen administration routes and dosage for inducible Cre-mediated gene disruption in mouse hearts. Transgenic Res 19(4):715–725. https://doi.org/10.1007/s11248-009-9342-4

80. Cawthorne C, Swindell R, Stratford IJ, Dive C, Welman A (2007) Comparison of doxycycline delivery methods for Tet-inducible gene expression in a subcutaneous xenograft model. J Biomol Tech 18(2):120–123

81. Redelsperger IM, Taldone T, Riedel ER, Lepherd ML, Lipman NS, Wolf FR (2016) Stability of doxycycline in feed and water and minimal effective doses in tetracycline-inducible systems. J Am Assoc Lab Anim Sci 55(4):467–474

82. Le YZ, Zheng W, Rao PC, Zheng L, Anderson RE, Esumi N, Zack DJ, Zhu M (2008) Inducible expression of cre recombinase in the retinal pigmented epithelium. Invest Ophthalmol Vis Sci 49(3):1248–1253. https://doi.org/10.1167/iovs.07-1105

83. Schonig K, Schwenk F, Rajewsky K, Bujard H (2002) Stringent doxycycline dependent control of CRE recombinase in vivo. Nucleic Acids Res 30(23):e134

84. Sinha M, Lowell CA (2017) Efficiency and specificity of gene deletion in lung epithelial doxycycline-inducible Cre mice. Am J Respir Cell Mol Biol 57(2):248–257. https://doi.org/10.1165/rcmb.2016-0208OC

85. Chai OH, Song CH, Park SK, Kim W, Cho ES (2013) Molecular regulation of kidney development. Anat Cell Biol 46(1):19–31. https://doi.org/10.5115/acb.2013.46.1.19

# Chapter 14

# The Glucose Tolerance Test in Mice


## Patricia Fonseca Pedro, Anastasia Tsakmaki, and Gavin A. Bewick



## Abstract

Type 2 diabetes is characterized by glucose intolerance, caused by insulin resistance in peripheral metabolic tissues and by impaired glucose-stimulated insulin secretion, the hallmark of beta-cell dysfunction. The glucose tolerance test is used in clinic and research to identify individuals with impaired glucose tolerance and overt type 2 diabetes. It is the most routinely used physiological test for first pass assessment of glucose homeostasis in rodents because of its simplicity. The GTT measures changes in blood glucose concentration over a 2-h period following the administration of a bolus of glucose. However, this simplicity belies several important considerations which need to be addressed, to aid reproducibility and produce interpretable data. Here, we describe in detail how to perform a GTT using four different routes of glucose administration: intraperitoneal, oral, voluntary oral, and intravenous.



**Key words** Diabetes, Glucose tolerance, Insulin, Beta-cell


## 1 Introduction

The glucose tolerance test (GTT) measures the clearance of a glucose load from the body. It is used to detect pathological changes in glucose metabolism which are associated with diabetes and metabolic disease. Animals are fasted and blood glucose levels determined before a solution of glucose is administered. Subsequently, blood glucose concentrations are measured across a 2-h time frame. It is the most common physiological test carried out in metabolic research particularly for phenotyping transgenic mice and/or in response to metabolic challenges such as high-fat feeding or other stimuli. The procedure is simple to carry out and does not require specific technical training beyond confidence in basic animal handling and substance administration. However, there are several important considerations to address which affect the design of a GTT protocol.

The route of administration is a key consideration when designing a GTT. Choice of route will be dictated by the research question to be answered. Oral administration of a glucose bolus can be


Aileen J. F. King (ed.), *Animal Models of Diabetes: Methods and Protocols*, Methods in Molecular Biology, vol. 2128, https://doi.org/10.1007/978-1-0716-0385-7_14,

achieved by intragastric delivery via feeding needle or catheter or by training mice to self-administer glucose in a short time frame either in liquid or gel form. Oral delivery has the advantage of mimicking physiology and engaging the incretin response, but the rate of gastric emptying can have a profound effect on glucose absorption and therefore on glucose tolerance. These parameters can be removed from the interpretation by delivering the glucose bolus either intraperitoneally or intravenously. It should be noted that the peak plasma glucose will be lower for OGTT compared to IP or IVGTT. There are other considerations to account for before embarking on experimentation including age, diet, strain, sex, glucose dose, protocol timings, etc. Each of these is discussed in the notes section accompanying the detailed methods below.

The GTT is a simple test which can be used to generate fast accurate metabolic data regarding glucose tolerance. Its interpretation is most powerful when performed in conjunction with the measurement of plasma insulin during the GTT and with an understanding of insulin sensitivity derived either from an insulin tolerance test or clamp studies.

## 2 Materials

Prepare all solutions using ultrapure water (prepared by purifying deionized water, to attain a sensitivity of 18 MΩ-cm at 25 °C) and analytical grade reagents. Prepare and store all reagents as indicated.

### 2.1 Intraperitoneal and Intravenous Glucose Tolerance Test

1. 45% D-(+)-Glucose solution in $H_2O$, sterile-filtered.
2. Dulbecco's Phosphate Buffered Saline Modified, without calcium chloride and magnesium chloride, liquid, sterile-filtered.
3. 1-mL syringe without needle.
4. 27G or 30G needle.
5. Glucose monitor.
6. Glucose strips.
7. Restraint device (optional).
8. Timer.

### 2.2 Oral Glucose Tolerance Test

1. D-(+)-Glucose solution, 45% in $H_2O$, sterile-filtered.
2. Dulbecco's Phosphate Buffered Saline Modified, without calcium chloride and magnesium chloride, liquid, sterile-filtered, suitable for cell culture.
3. 1-mL syringe without needle.
4. Feeding needle for mouse.
5. 27G or 30G needle.

6. Glucose monitor.
7. Glucose strips.
8. Restraint device (optional).
9. Timer.

***2.3 Voluntary Oral Glucose Tolerance Test***

1. Gelatine, any commercially available for baking.
2. D-(+)-Glucose.
3. 27G or 30G needle.
4. Glucose monitor, Roche.
5. Glucose strips, Roche.
6. Parafilm.
7. 30-mm petri dish or small weigh boat.
8. Restraint device (optional).
9. Timer.

## 3 Methods

All experimental procedures that involve animals (rodents) should receive approval from the local ethical committee. Humane treatment of animals should be practiced at all times; best practices can be found at www.Lasa.co.uk, and we encourage the use of the ARRIVE guidelines for transparency and reproducibility; these can be found at https://www.nc3rs.org.uk/arrive-guidelines. Animals can be obtained from either a commercial supplier or a local breeding program. For non-genetic studies, we use the inbred C57Bl/6 or the outbred CD1 strain (*see* **Note 1**). To determine the effect of exogenous substances on glucose tolerance, we routinely use young adult male mice between 8 and 12 weeks of age housed under temperature controlled (22 ± 2 °C) and light conditions (12-h light:12-h dark cycle) with ad libitum access to drinking water and standard rodent chow prior to the experiment and ad libitum access to drinking water during the experiment. However, mice of any age can be used; typically, high-fat-fed mice with obesity and insulin resistance will be in the region of 20–30 weeks of age depending on the feeding protocol used (*see* **Note 2** for more information on age, sex, and diet).

***3.1 Administered Glucose Tolerance Tests (IP, IV, OGTT)***

1. We recommend acclimatizing mice to all handling procedures (e.g., scruffing, or appropriate restraint method) and administration protocols where possible to minimize stress (*see* **Note 3**).
2. Weigh the mice. For mice of differing fat mass, we recommend dosing the bolus of glucose on lean mass **if** body composition data is available (*see* **Note 4**).

3. Fast the mice for 6 h (e.g., from 8 am to 2 pm) (*see* **Note 5** for duration of fast discussion).
4. Prepare an experiment data sheet, glucose strips, syringes, and needles. Calculate the volume of glucose solution required for each individual mouse. To inject 2 g of glucose/kg body weight, the injection volume required is 30% glucose (μL) = 6.667 × body weight (g). Calculate the total volume of glucose required for all mice receiving glucose plus 20%, to account for dead space in the syringes. Make up this volume of 30% glucose solution by diluting the 45% solution 1 in 3 with DPBS, filter sterilize.
5. For IP and IV glucose tolerance testing, the syringes can be preloaded in the lab before the study begins and transported to the animal facility. For OGTT, the volume required for each mouse should be recorded on the data sheet; syringe loading is carried out for each mouse during the procedure.
6. Blood sampling can be achieved from the tip of the tail using a scalpel, scissors, or needle. We recommend using a 27 or 30G needle to make a small puncture wound at the very tip of the tail as this method is least traumatic. A small drop of blood, <5 μL, is placed on the glucose strip in the glucose monitor. Record the glucose concentration in the data sheet. This represents baseline $t = 0$ and should be measured for each mouse.
7. Administer the glucose load:
   (a) IPGTT: Restrain the animal appropriately, scruffing is simplest and quickest. Inject the animal in the lower right quadrant of the ventral flank.
   (b) IVGTT: Restrain the mouse appropriately (we use a plastic restrain tube). Inject the glucose bolus into the tail vein. Warming the tail to achieve vasodilation can aid the intravenous injection.
   (c) OGTT: Restrain the animal appropriately; scruffing is simplest and quickest. Deliver the glucose bolus intragastrically using a feeding needle.
8. Routinely blood glucose concentrations are measured at 15, 30, 60, and 120 min after glucose delivery (*see* **Note 6**). For each time point, bleeding can be restarted by removing the clot by gentle brushing with a tissue. The tail can be lightly massaged if the blood flow is insufficient. Results are recorded in the data sheet. We do not recommend restraining the mouse for glucose measurement, but if required, a restraint device may be used as long as the animals are acclimatized to it.
9. If plasma insulin measurements are required, collect 40 μL of whole blood in a suitable blood collection tube. Insulin should only be measured at relevant time points, often $t = 0$ and a

combination of either $t = 5$, 15, or 30. Spin the blood for 1 min at 16 kg and collect plasma in a clean tube (approx. 20 μL). Store plasma on ice and long term at −20 °C.

10. At the end of the study, ensure that the incision is clotting by briefly applying pressure to the tail tip.
11. Place the mouse in a clean cage with water and food available ad libitum.
12. Monitor the animals carefully to observe any abnormal behavior(s).
13. The results are presented as average blood glucose concentration against time. Additionally, the data can be presented as area under the curve above baseline; this is particularly useful if fasting baseline glucose concentrations differ between groups as is often the case in diabetic/obese insulin-resistant mice (*see* Fig. 1 for examples).

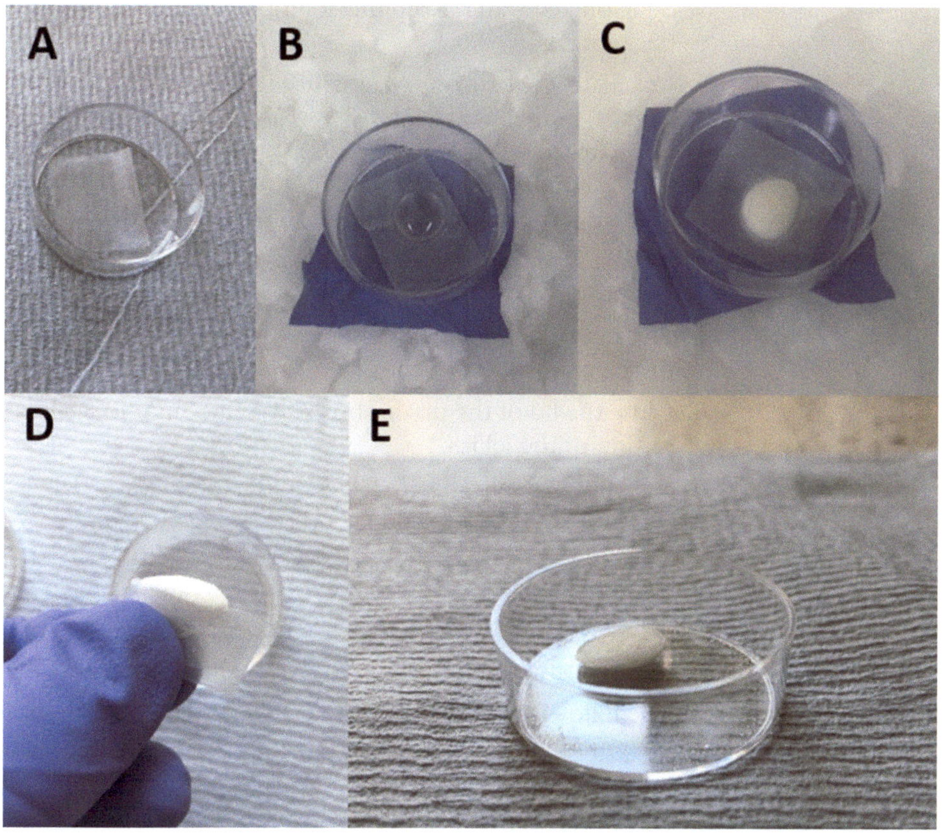


**Fig. 1** Casting a glucose jelly: A small square of parafilm is placed at the bottom of a 3-mm petri dish and placed on top of dry ice (**a**). The appropriate volume of jelly is dispensed by pipette onto the parafilm (**b**) and a lid placed on the dry ice. Jellies should set in 5 min (**c**, **d**) and can be stored at −20 °C until the day of the GTT when it is presented to the mice within a petri dish (**e**)

### 3.2 Voluntary Glucose Tolerance Test (Self-Administered Glucose)

#### 3.2.1 Preparation of Solutions for Glucose Jelly

1. Make up a 20% gelatine solution.
   (a) Weigh 10.0 g of gelatine and transfer to a 250-mL beaker.
   (b) Add 50 mL of deionized water.
   (c) Use a microwave to warm up the solution until gelatine is completely dissolved and clear (*see* **Note 7**). Aliquot the solution while warm (*see* **Note 8**) and store at −20 °C. For thawing, the solution should be carefully warmed in a heat block until clear and runny.
2. Make up a 60% glucose solution.
   (a) Weigh 60.0 g of D-glucose and transfer to a glass beaker.
   (b) Make up to 100 mL of deionized water. Use a stirring hot plate to stir the solution with gentle heating to fully dissolve the glucose powder (*see* **Note 9**).

#### 3.2.2 Preparation of Glucose Jelly (30% Glucose, 30 mg in 100 μL)

To make approximately 10 jellies (200 μL each)

1. Mix 1 mL of 60% glucose, 500 μL of 20% gelatine, and 500 μL of deionized water. Keep this solution warm on a heating block (*see* **Note 10**).
2. Calculate the glucose weight needed for each mouse on the basis of individual weight (2 g glucose/kg) and subsequently calculate the jelly volume needed for each animal (e.g., a 30.0-g mouse requires 60.0 mg of glucose which can be delivered in a 200-μL jelly; jelly volume in μL = body weight in g × 6.667)
3. Place 15 mm approx. squares of parafilm at the bottom of 30-mm petri dishes and transfer these to dry ice (Fig. 2).
4. Pipette each individual volume of glucose jelly solution onto the parafilm in each petri dish.
5. Close the lid of the dry ice box and wait 5 min, to facilitate fast setting of the jellies.
6. Separate parafilm from jellies and store clearly labelled at −20 °C until ready to use.

#### 3.2.3 Conditioning Mice for the Glucose Tolerance Test

The voluntary administration protocol requires mice to be conditioned to eat the glucose jellies in a defined timeframe.

1. Singly house the mice and fast overnight (from 6 pm).
2. The following morning, between 9 and 9:30 am, each mouse is presented with a petri dish containing a 200-μL glucose jelly.
3. Leave the jelly in each cage for 30 min. All mice may not eat the jelly in this time.
4. For the following two mornings (without fasting), present mice with a 200-μL glucose jelly between 9 and 9:30 am. On the third morning, all mice should eat the jelly in less than

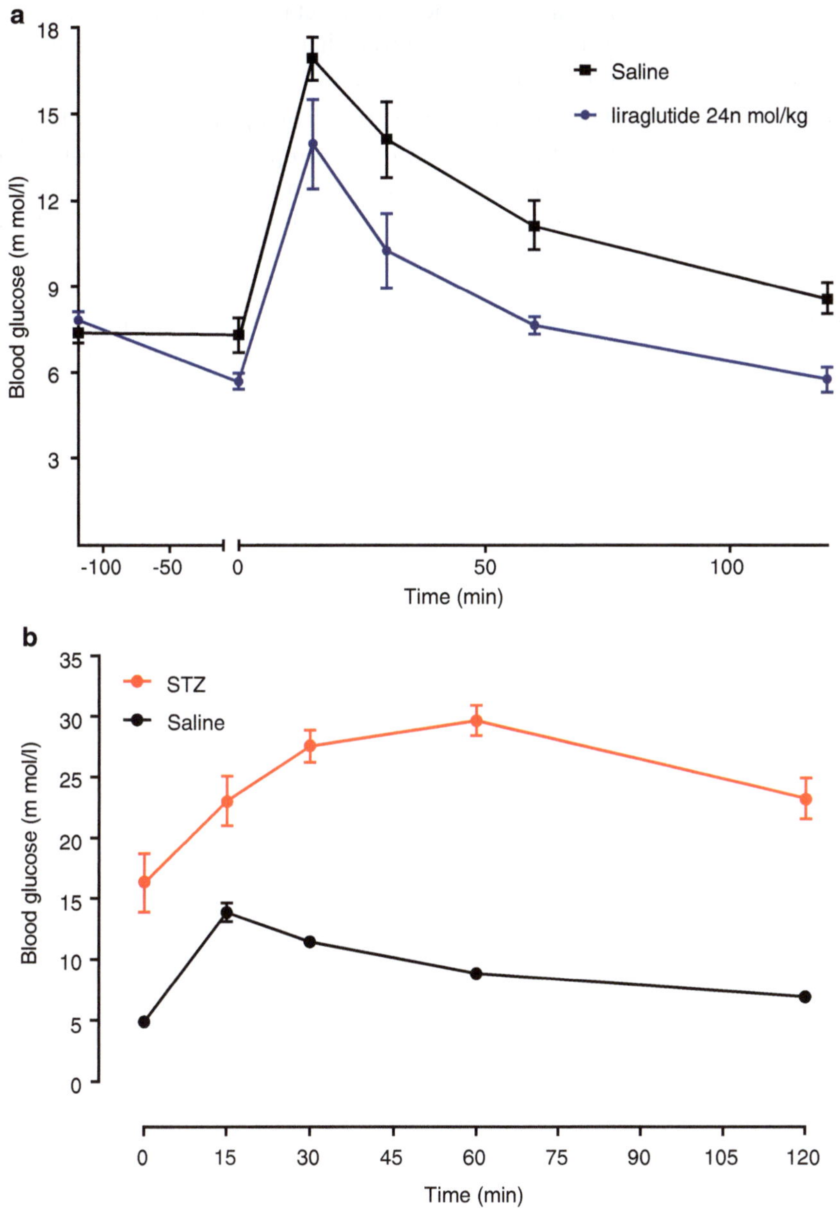


**Fig. 2** Typical glucose tolerance curves: (**a**) Glucose excursion following liraglutide, a long-acting GLP-1 agonist which stimulates insulin secretion $n = 8$. Liraglutide was delivered by intraperitoneal injection 2 h prior to IP glucose bolus (2 g/kg). Glucose was measured at $-120$, 0, 15, 30, 90, and 120 min post IP glucose bolus. (**b**) Glucose excursion comparing control mice ($n = 10$) with mice treated with low-dose streptozotocin (STZ), a beta-cell toxin. A GTT was conducted 25 days after STZ administration when overt diabetes in the STZ group had been established. Glucose was measured at 0, 15, 30, 90, and 120 min following glucose (1.5 g/kg) bolus administration

5 min, and most will finish the jelly in a much shorter time frame. Mice which do not finish the jelly within 5 min should be excluded from the cohort. GTTs can be conducted on the conditioned cohort 1 week later (*see* **Note 11**).

5. In our experience, mice retain their conditioned memory for glucose jellies for more than a month without having to repeat the conditioning process.
6. The day before the study (or up to a few days before), weigh the mice. For mice of differing fat mass, we recommend dosing the bolus of glucose on lean mass if body composition data is available. Cast jellies for each mouse according to body weight or lean mass as described above.

*3.2.4 Voluntary Glucose Tolerance Test*

1. Fast the mice for 6 h (e.g., from 8 am to 2 pm) transferring the mice to individual cages, if not already housed in this way (*see* **Note 5**).
2. Prepare an experiment data sheet and all apparatus required.
3. Blood sampling can be achieved from the tip of the tail using a scalpel, scissors, or 27G needle. We recommend using a needle to make a small puncture wound at the very tip of the tail as this method is least traumatic. A small drop of blood, <5 μL, is placed on the glucose strip in the glucose monitor. Record the glucose concentration in the data sheet. This represents baseline $t = 0$ and should be measured for each mouse.
4. Present each mouse with the correct volume of jelly. Any mouse which does not eat the jelly within 5 min should be excluded. The petri dish can be removed at the first sampling time point.
5. Routinely, blood glucose concentrations are measured at 15, 30, 60, and 120 min after glucose delivery. For each time point, bleeding can be restarted by removing the clot with gentle brushing with a tissue. The tail can be lightly massaged if the blood flow is insufficient.
6. If plasma insulin measurements are required, collect 40 μL of whole blood in a suitable collection tube with heparin or EDTA to prevent clotting. Insulin should only be measured at relevant time points, often $t = 0$ and a combination of either $t = 5$, 15, or 30. Spin the blood for 1 min at 16 kg and collect plasma in a clean tube (approx. 20 μL). Store plasma on ice and long term at −20 °C.
7. At the end of the study, ensure that the incision is clotting by briefly applying pressure to the tail tip.
8. Place the mouse in a clean cage with water and food available ad libitum.
9. Monitor the animals carefully to observe any abnormal behavior(s).
10. The results are presented as average blood glucose concentration against time. Additionally, the data can be presented as

area under the curve above baseline; this is particularly useful if fasting baseline glucose concentrations differ between groups as is often the case in diabetic/obese insulin-resistant mice.

## 4 Notes

1. The choice of mouse strain is a key consideration. Outbred versus inbred mice exhibit differences in sensitivity, while comprehensive investigation of glucose metabolism in commonly used inbred strains has identified strain-dependent differences in insulin secretion and sensitivity. Furthermore, the background of a genetically modified animal is an important determining factor for its glucose tolerance. Not to mention the well-documented effect of Cre recombinase expression on whole-body glucose homeostasis particularly when expressed in metabolically relevant tissues and cells. It is therefore of critical importance to run the appropriate controls and to clearly document the strains used. Additionally, efforts should be made to reduce genetic drift which can occur in long-term isolated colonies.
2. Insulin sensitivity coupled with glucose tolerance is markedly reduced with advancing age, and there are clear differences in glucose tolerance between sexes. As such, it is recommended that age- and sex-matched mice are used when evaluating glucose tolerance. Furthermore, dietary composition is an important modifier of glucose tolerance as is body composition, particularly degree of adiposity, and body weight. It is therefore important from both an intra- and inter-lab reproducibility stand point to include manufacture details of the diet composition, to use weight-matched mice where possible, and to provide a measure of body composition.
3. Stress has a large impact on delicate physiological parameters; for example, the stress hormone cortisol acts as a functional antagonist to insulin while stress readily impacts food intake. It is essential to acclimatize mice to the handling procedures and wherever possible to the administration route to reduce variability between mice during a GTT.
4. Routinely, the glucose dose is calculated by weight, often between 1 and 2 g/kg. This has the advantage of reducing variability but may suffer in the context of obesity where differences in body weight are driven by fat mass and not lean mass, the major site of glucose utilization. Correcting for lean mass, if data is available, may be more relevant than correcting dose for body weight. Andrikopoulos et al. concluded 2 g/kg dose according to body weight was an appropriate dose for discriminating differences in glucose tolerance between high-fat- and chow-fed mice [1].

5. Mice are nocturnal and exhibit circadian entrainment of metabolically active tissues which influences insulin sensitivity and beta-cell function. A large percentage of their night time feed is taken just after lights-out and at a time relatively close to lights-on with grazing in between. Day time feeding is characterized by grazing periods which can constitute up to 30% of total daily intake. For practical reasons, unless reverse lighting is available, most labs study glucose tolerance during the daytime when mice eat the least and are unused to handling meal challenges. To reduce variations in baseline glucose, mostly dictated by recency of food intake, it is common practice to fast mice before administering a glucose tolerance test. The most routinely used fasting paradigm is a 14–18-h overnight fast with the GTT carried out in the morning after lights-on (8–10 am). This establishes a very reproducible plasma glucose baseline but at the expense of most of the liver glycogen stores and is perhaps more representative of starvation than physiological glucose regulation. To account for this, 5–6-h fasting is becoming more common place. Indeed, a 6-h fast provides a more physiological context, preserving liver glycogen and preventing the increase in insulin-stimulated glucose utilization following prolonged fasting in mice. A 6-h fast generated clearer data with more statistical significance, when comparing high-fat- and chow-fed C57Blk/6 mice [1].
6. Blood glucose can be additionally sampled at $t = 5, 10, 45$, and 90 min if required.
7. Use a low-power setting on the microwave; if the microwave is used on a high-power setting, the gelatine solution is prone to spill.
8. Gelatine solution solidifies at room temperature; aliquot before it reaches room temperature.
9. Solution should be aliquoted while warm and stored at −20 °C. For thawing, solution should be carefully warmed in a heat block.
10. The gelatine and glucose solutions should be clear and runny to ensure proper mixing and ease of pipetting.
11. Not all mice are amenable to conditioning although these mice are rare. It is important to start the conditioning with two or three extra mice to account for any untrainable mice within a cohort.

## Reference

1. Andrikopoulos S, Blair AR, Deluca N, Fam BC, Proietto J (2008) Evaluating the glucose tolerance test in mice. Am J Physiol Endocrinol Metab 295:E1323–E1332

# Chapter 15

# Assessment of Insulin Tolerance In Vivo in Mice

Irene Cózar-Castellano and Germán Perdomo

## Abstract

Insulin resistance in humans and mice is an important hallmark of metabolic diseases. Therefore, assessment of insulin sensitivity/resistance in animal models provides valuable information in the pathophysiology of diabetes and obesity. Depending on the nature of the information required, we can choose between direct and indirect techniques available for the determination of insulin sensitivity. Thus, the complex hyperinsulinemic-euglycemic glucose clamps and the insulin suppression test assess insulin-mediated glucose utilization under steady-state conditions, whereas less complex methods, such as the insulin tolerance test (ITT), rely on measurements of blood glucose levels in animals subjected to intraperitoneal insulin loading. Finally, surrogated indexes derived from blood glucose and plasma insulin levels are also available for assessment of insulin sensitivity/resistance in vivo. In this chapter, we focus on the intraperitoneal insulin tolerance test (IPITT) protocol for measuring insulin resistance in mice.

**Key words** Insulin sensitivity, Insulin resistance, Glucose homeostasis, Metabolism, High-fat diet

## 1 Introduction

Type 2 diabetes mellitus (T2DM) has reached an epidemic level worldwide [1]. The global prevalence of T2DM in the last three decades increased from 4.3% to 7.9–9% in women and men, respectively [2]. It is predicted that by 2040, there will be 642 million people living with diabetes in the world [3]. A hallmark of T2DM is the presence of insulin resistance, a condition characterized by reduced response to circulating levels of insulin in the peripheral tissues, such as skeletal muscle, liver, and adipose tissue [4]. In fact, in the years preceding the onset of T2DM, progressive amelioration of insulin sensitivity in the liver and peripheral tissues arises. In parallel, pancreatic β-cell insulin secretion is increased in the first stage of insulin resistance, which is followed by a failure of β-cell function leading to reduced insulin production and secretion [5].

The epidemics of diabetes and obesity have reinforced the motivation to investigate whole-body glucose metabolism and insulin action in mouse models of obesity and diabetes. However, despite considerable efforts made in the last decades, the

Aileen J. F. King (ed.), *Animal Models of Diabetes: Methods and Protocols*, Methods in Molecular Biology, vol. 2128,
https://doi.org/10.1007/978-1-0716-0385-7_15, 

identification of the molecular mechanisms underlying insulin resistance remains to be fully elucidated.

The "gold standard" for accurately determining insulin resistance is the hyperinsulinemic-euglycemic clamp [6]. However, this technique cannot be implemented on a routine basis in most laboratories. Thus, insulin tolerance test (ITT) was developed and implemented as a simple way to evaluate insulin action in vivo in mice. In brief, ITT consists in intraperitoneal injection of an insulin bolus, and the rate of fall of glucose is assessed every 15 min over a 90–120-min period. The decay in blood glucose levels mirrors the insulin action on peripheral tissues such as liver, skeletal muscle, and adipose tissue [7]. The magnitude in the fall of glucose depends on insulin dose, basal glucose level, and hormone-mediated counter-regulatory mechanisms to restore blood glucose homeostasis [8]. Other considerations include age and sexual dimorphism. Thus, it is recommended that ITT must be assessed in age- and sex-matched mice [7, 8]. Finally, reproducible results rely on experienced personnel to avoid causing stress and anxiety to the mice before and during the ITT.

## 2 Materials

1. Mice (we use C57Bl/6).
2. Insulin solution (we use Humulin® R [100 U/mL]).
3. Blood glucose monitor and test strips for glucose measurement (we use Breeze 2 glucometer).
4. Sterile saline buffer.
5. Scalpel blade.
6. Sterile alcohol pads.
7. Scales suitable for weighing animals.
8. 1-mL syringe and 30-G × 1/2-in. needles.
9. Timer.
10. Clean mouse cages.

## 3 Methods

1. Begin the experiment between 8:00 and 9:00 a.m. by placing mice into a new cage with clean bedding (without feces and food pellets) and free access to water (*see* **Notes 1–4**). A period of acclimatization between placing mice in the new cage and the beginning of the experiment is recommended (e.g., 15 min).

**Table 1**
**Example of an experimental table record for an IPITT**

| Mouse ID# | Body weight (g) | Insulin solution to inject (μL) | Blood glucose levels (mg/dL) | | | | |
|---|---|---|---|---|---|---|---|
| | | | 0′ | 15′ | 30′ | 60′ | 90′ |
| 1 | | | | | | | |
| 2 | | | | | | | |
| 3 | | | | | | | |
| 4 | | | | | | | |
| 5 | | | | | | | |
| 6 | | | | | | | |
| 7 | | | | | | | |
| 8 | | | | | | | |

2. Prepare an experimental record table to register body weight, blood glucose levels at basal ($t = 0$) and at every time points ($t = 0, 15, 30, 90, 120$ min), and insulin volume to be injected in mice (*see* Table 1).
3. Pick up the mouse with gloves and weigh it using the scale. Record the value in the table.
4. Measure baseline glucose levels drawing blood from the saphenous vein or from the tail vein by placing one drop of blood onto the glucometer strip (*see* **Note 5**). Record the value in the table ($t = 0$) (*see* **Notes 5** and **6**).
5. Using mouse weight data, calculate and record in table the volume of insulin that is necessary to inject 0.75 U/kg of body weight for each mouse (*see* **Note** 7).
6. Dilute the stock solution of insulin (typically, for a 25-g mouse, insulin is diluted 1:1000) and prepare a sterile solution of insulin in saline buffer (0.9% NaCl) (*see* **Note 8**).
7. Preload 1-mL syringes with the insulin solution (usually the volume to inject per mouse is ~100–200 μL) being careful to remove air from the syringes.
8. Pick up mouse with gloves and restrain it with the other hand. Alternatively, you can use a mouse restrainer for this purpose (*see* **Note 9**).
9. Proceed with the intraperitoneal insulin injection in each mouse at 60-s interval between mice (depending on your experience, you may increase the time by 1–2 min.) Start the timer when the first mouse is injected. It is important to maintain the time course from one injection to other.

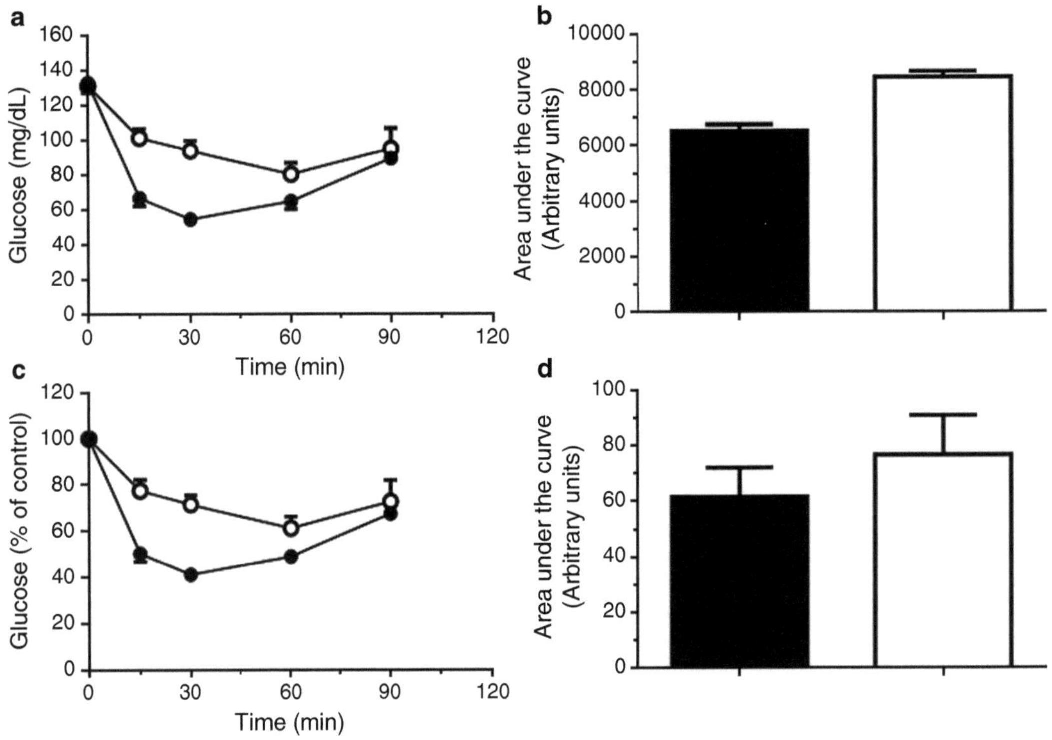


**Fig. 1** Illustration of a representative insulin tolerance test. (**a**) Graph depicts absolute blood glucose levels. (**b**) Area under the curve. (**c**) Graph depicts blood glucose levels relative to initial values in percentage. (**d**) Area under the curve. Data are mean ± SEM for $n = 5$. The group of mice plotted as white symbols is insulin resistant compared to mice plotted with black symbols

10. Starting with the first mouse injected and following the same interval established for injections, repeat blood glucose measurements at 15, 30, 60, 90, and 120 min after insulin injection by gently removing the scab and massaging the tail tip. Record the values of blood glucose levels at each time point (*see* **Notes 10–15**).
11. At the end of the experimental session, clean the mouse tail tip with alcohol pads and return mice into cages with food and water. Monitor mice for the next 2 h checking for normal behavior (i.e., mice get regular access to food and water, avoiding hypoglycemia).
12. Analyze data by plotting glucose values vs. time (*see* **Note 14** and Fig. 1).

## 4 Notes

1. It is recommended to perform the IPITT at the same hour in the morning, particularly when different cohorts of mice are used, because physiological parameters change throughout the

day. In rodents, eating is cyclic and the maximum rate occurs during the night. In fact, feeding mice in the daytime disrupts rhythms in body temperature and activity. Thus, it is very important to perform IPITT at the same hour in the morning. In case you decide to fast for 2 h, start the fasting at 8:00 a.m., and keep consistency with your experiments.

2. Take into consideration the size of the experimental number of animals to minimize technical variations. It is advisable that no more than 15 mice should be tested during the same experimental setting. Ideally, two trained people should participate during the IPITT, each one of them performing specific roles in the experiment (i.e., one person injecting mice and the other measuring blood glucose levels). Insulin should be injected by the same personnel to all mice, even if using different cohorts of animals for the experiment.
3. Whenever possible randomize mice and exclude mice with extreme body weight and/or blood glucose levels, avoiding grouping bias during the experimental setting.
4. Long-term fasting conditions are not recommended for IPITT, due to increased risk of hypoglycemia after insulin injection. Although fasting is not necessary, a 2-h fasting before the IPITT will help to keep blood glucose levels of mice to a similar range of values. Importantly, during the IPITT, remove food and feces, but leave water, placing animals into a clean cage to avoid any intake during the course of the experiment.
5. In case you have two experimental groups (e.g., WT vs. transgenic mouse; regular chow diet vs. high-fat diet), the cohort of mice should be representative of both groups. Avoid assessing IPITT with one of the experimental groups in one day and the next day the second one. It is preferable to have mice from both groups each time an IPITT is carried out. To obtain blood, you may use a scalpel to remove soft tissue from the tip of the tail (~1 mm). The disadvantages of this technique are permanent damage to the tail and pain to the mouse. It is recommended to warm mice using a red light. This technique may be useful if larger volumes of blood are required, for example, to measure hormones. Alternatively, a small (27G or 30G) needle prick to the very end of the tail followed by gently squeezing the tail usually generates a blood droplet large enough for most blood glucose meters. Blood sampling from the lateral saphenous vein may be preferable to avoid tail damage and is relatively quick. Restrain the mouse and puncture the site with a needle (27G or 25G) or lance. Stop the bleeding by gentle finger pressure over the puncture site. Minimize the number of attempts to take blood (no more than three in any one attempt). The scab or blood clot is removed for multiple samples. Anesthesia or sedation is not necessary.

6. In case you are using genetically modified obese mouse models (*db/db*, *ob/ob*) or diet-induced obese mice, you may need to dilute your blood samples to get accurate blood glucose values.
7. The dose of insulin may vary from 0.5 to 7.5 U/kg of body weight depending on the genetic background of mice, body weight, age, nutritional status, and diabetic status. As a rule, older and heavier mice need higher doses of insulin. We typically use 0.75 U/kg insulin for injection in C57BL/6J mice fed with a regular diet, 2–7.5 U/kg in the C57BL/6J diet-induced obesity model, 2–7.5 U/kg in the *db/db* mice, and 0.75 U/kg in the CD-1 mice fed with a regular diet. These suggested insulin doses may be too high (and result in severe or fatal hypoglycemia, below 36 mg/dL) or too low (glucose levels do not fall by ~50%); thus, it is advisable to perform a preliminary test with a small number of mice for your specific mouse strain and conditions of study, and if hypoglycemia is detected, inject glucose (*see* **Note 11**).
8. It is recommended to use a fast-acting insulin because of the short-time duration of the IPITT. Immediately before the experiment, dilute insulin to the desired dose in sterile saline buffer (0.9% NaCl) and keep it on ice. Use always freshly prepared insulin. Avoid using phosphate-buffered saline (PBS) to dilute the insulin because the pH of insulin is low and it becomes neutralized with the PBS and precipitates. In addition, it is important to maintain sterile conditions for both insulin and buffer, because you may need to perform other metabolic tests in the same experimental mice.
9. To perform good intraperitoneal injections, insert the needle with an angle of ~25° to the abdominal wall in the lower right or ~45° in the left quadrant of the abdomen trying to avoid hitting the bladder, the liver, or other internal organs. Likewise, avoid subcutaneous injection. Of note, in obese mice, be aware to avoid fat pad areas. A new needle should be used for each animal, which will reduce the discomfort caused by the procedure and reduce the risk of infections. Furthermore, insulin solution should be injected at body temperature to further reduce mice discomfort.
10. Table 1 shows an example for the experimental record table.
11. Get ready with a glucose solution (20% glucose in water) just in case insulin injections cause hypoglycemia to your mice. If this is the case, rescue your mouse from hypoglycemia by injecting 0.5–1.0 g of glucose/kg.
12. The depth and the rate of fall of glucose depends on insulin dose Thus, in wild-type mice fed with a regular diet, glucose levels typically fall by ~50% from the basal level after 30 min of insulin injection. Afterward, glucose slowly returns toward

basal levels. Other parameters that also affect the glucose levels after insulin injection are the prevailing blood glucose concentration at the beginning of the experiment and the glucose threshold (~80 mg/dL) for hormone counterregulatory processes to maintain glucose homeostasis.

13. You can end the experiment if glucose returns to basal levels at 90 min. Otherwise, prolong the experiment up to 120 min.
14. Whenever possible use the same glucometer and strips, because diverse commercial brands usually give slightly different blood glucose levels or have different upper limits (usually 33–50 mM). This is particularly important if different cohorts of mice are used for the experimental setting. In the case of some glucometers, regular calibration with the standard strip is recommended.
15. The fall in blood glucose levels during the IPITT is often presented as a percentage of basal glucose concentration vs. time. Alternatively, data can be presented by plotting absolute glucose values (mg/dL) in the *y*-axis vs. time (min) in the *x*-axis. In addition, calculations of areas under the curve provide valuable information. If you have values from both sexes, it is recommended to plot the data separately, due to different insulin sensitivity in males vs. females.

## Acknowledgments

This work was supported by grants from the Ministerio de Economía, Industria y Competitividad: SAF2014-58702-C2-1-R and SAF2016-77871-C2-1-R to IC; SAF2014-58702-C2-2-R and SAF2016-77871-C2-2-R to GP. Supported by the EFSD European Research Programme on New Targets for Type 2 Diabetes supported by an educational research grant from MSD to ICC and GP.

### References

1. ACCORD Study Group, Gerstein HC, Miller ME, Genuth S, Ismail-Beigi F, Buse JB, Goff DC Jr, Probstfield JL, Cushman WC, Ginsberg HN, Bigger JT, Grimm RH Jr, Byington RP, Rosenberg YD, Friedewald WT (2011) Long-term effects of intensive glucose lowering on cardiovascular outcomes. N Engl J Med 364 (9):818–828
2. Collaboration NCDRF (2016) Worldwide trends in diabetes since 1980: a pooled analysis of 751 population-based studies with 4.4 million participants. Lancet 387 (10027):1513–1530
3. Ogurtsova K, da Rocha Fernandes JD, Huang Y, Linnenkamp U, Guariguata L, Cho NH, Cavan D, Shaw JE, Makaroff LE (2017) IDF Diabetes Atlas: Global estimates for the prevalence of diabetes for 2015 and 2040. Diabetes Res Clin Pract 128:40–50
4. Samuel VT, Shulman GI (2012) Mechanisms for insulin resistance: common threads and missing links. Cell 148(5):852–871

5. Defronzo RA (2009) Banting Lecture. From the triumvirate to the ominous octet: a new paradigm for the treatment of type 2 diabetes mellitus. Diabetes 58(4):773–795
6. Kim JK (2009) Hyperinsulinemic-euglycemic clamp to assess insulin sensitivity in vivo. Methods Mol Biol 560:221–238
7. Muniyappa R, Lee S, Chen H, Quon MJ (2008) Current approaches for assessing insulin sensitivity and resistance in vivo: advantages, limitations, and appropriate usage. Am J Physiol Endocrinol Metab 294(1):E15–E26
8. McGuinness OP, Ayala JE, Laughlin MR, Wasserman DH (2009) NIH experiment in centralized mouse phenotyping: the Vanderbilt experience and recommendations for evaluating glucose homeostasis in the mouse. Am J Physiol Endocrinol Metab 297(4):E849–E855

# Chapter 16

# Continuous Glucose Monitoring in Conscious Unrestrained Mice

**Aileen J. F. King, Matilda R. Kennard, and Manasi Nandi**

## Abstract

Measurement of blood glucose concentration is a common end point in studies using animal models of diabetes. Usually a blood glucose meter is used to measure non-fasted blood glucose concentrations, typically at frequencies of between 1 and 7 times per week. This process involves pricking the tip of the tail to collect a small blood sample (0.5–5 μL), which could potentially cause a stress response and affect blood glucose concentrations. Moreover, with blood glucose concentrations constantly fluctuating in response to feeding and activity, a single-point measurement can easily misrepresent the actual glycemic control of the animal. In this chapter, we discuss the use of continuous glucose monitoring in mice by radiotelemetry which allows second-by-second changes in blood glucose to be captured without restraining the mouse. Glucose excursions rather than single-point measurements may prove more useful in detecting effects of treatments, and lack of handling may avoid stress responses causing artefacts. We outline what is involved in implanting such devices into mice including some practical tips to maximize success.

**Key words** Continuous glucose monitoring, Mice, Glucose telemetry

## 1 Introduction

Measurements of blood glucose concentrations are often used in animal models of diabetes to monitor the development of the disease or the efficacy of a treatment. Although this has been standard practice for many years, there are limitations of single-point measurements. This includes the requirement to handle the animal in order to obtain a blood sample from the tail, which could cause stress artefacts [1]. In addition, there is a limit to how often blood glucose concentrations can be measured as repetitive sampling can cause damage to the tail, and limited volumes of blood should be extracted from a welfare point of view [2]. Moreover, blood glucose concentrations of all animals fluctuate in response to feeding and activity levels [3, 4], and this variability is not captured by traditional single-point measurements. Monitoring the excursions in blood glucose that occur during active feeding phases and

Aileen J. F. King (ed.), *Animal Models of Diabetes: Methods and Protocols*, Methods in Molecular Biology, vol. 2128,
https://doi.org/10.1007/978-1-0716-0385-7_16, 

inactive resting phases or in response to therapeutic interventions will enhance our mechanistic understanding of diseases such as diabetes and provide more detailed information on the effects of novel therapeutics or interventions.

Recent technological developments now allow us to measure blood glucose in unrestrained rodents, through the use of indwelling devices coupled to radio-telemetry transmitters. Using this method, blood glucose concentrations, activity, and temperature can be measured continuously (e.g., second-by-second data with averages reported at user-defined intervals such as every 10 or 60 s) over several weeks. The benefit of this method is that circadian variability is captured and the impact of therapeutic interventions can be monitored in real time. Furthermore, animals can act as their own controls as baseline readings can be acquired prior to any intervention. Other benefits include the ability to measure blood glucose without restraining the animal, the ability to measure blood glucose during the active night phase, and perhaps most importantly being able to measure glucose excursions. Indeed, minimal glucose fluctuations demonstrate good blood glucose control, and it could be argued that that should be the ultimate end point of studies [5, 6]. However, the price of the devices, acquisition equipment, and the need for specialized surgery may become a limiting factor. Nevertheless, we believe the benefits of the technique may outweigh these disadvantages. In this chapter, we will describe the methodology for mice as this is the most commonly used species for diabetes research.

It is important to note that experiments using radio-telemetry to monitor blood glucose concentrations need to be well planned. Considerations before commencing the study include choosing an appropriate animal model and ensuring you have the correct components of the telemetry system. Prior to any study on research animals, a full harm:benefit analysis should be conducted [7].

The study can be divided into the following stages, and blood glucose monitoring can take place over days or weeks.

Stage 1: Surgical implantation of telemetry device into recipient animal under recovery anesthesia (*see* Subheadings 3.2 and 3.3).

Stage 2: Perioperative recovery and health monitoring (*see* Subheading 3.4).

Stage 3: Calibration of blood glucose measurements (*see* Subheading 3.5).

Stage 4: Continuous blood glucose, temperature, and activity monitoring of animal (e.g., 10 s averages) in "baseline" state (*see* Subheading 3.6).

Stage 5: Experimental manipulation of animal if required, e.g., induction of diabetes, administration of pharmacological or dietary agents, glucose tolerance tests, exercise, etc.

Stage 6: Continuous monitoring of animal post experimental manipulation.

Stage 7: Termination of experiment, acquisition of blood samples, tissues, etc., from animal and careful removal of telemetry device (*see* Subheading 3.7).

Glucose telemetry probes can be implanted into naïve (normoglycemic) male or female animals in which diabetes will later develop or be induced (*see* **Note 1**). Alternatively, diabetic animals can be used at the outset. Prior to surgery, the clinical condition of the animals should be inspected to ensure that they are suitable for recovery surgery particularly with respect to hyperglycemic animals. All surgical procedures for recovery animals should be performed under aseptic conditions. The methods detailed are suitable for individuals with previous experience in small animal surgery with additional relevant training which can be obtained from the device manufacturer.

Animals should typically be >20 g in weight (*see* **Note 2**), and investigators should take into account a potential 10–15% weight loss post-surgery and select suitable recipients accordingly. In experienced labs, typically, 80–95% of implanted animals provide reliable data at 28 days. It is therefore recommended that this is considered when designing the study and the sample size calculation reflects this.

Full physical and behavioral examination of the animal should be conducted to ensure it is fit for recovery surgery, e.g., an alert animal showing no signs of pain or distress and of a suitable weight (*see* **Note 2**). The body weight of diabetic animals can often differ from naïve controls. If implanting mice with existing diabetes, it is recommended that stable diabetes (e.g., stable weight and blood glucose over several days) is established before probe implantation. For paired comparisons and to monitor diabetes development, healthy normoglycemic animals are recommended with induction of diabetes after a surgical recovery period (7 days) followed by baseline recordings (5–7 days).

If using carotid artery implantation, the colony should be assessed to ensure the animals have an intact circle of Willis (*see* **Note 3**). In addition, battery life and/or sensor life may vary from manufacturer to manufacturer but needs to be carefully considered when designing the duration of the study.

The glucose sensor tip needs to be positioned such that blood glucose and oxygen come into contact with the surface of the sensor. As blood glucose monitoring may be performed over several weeks, the recommended placement of the sensor would be into a wide vessel with free-flowing blood to minimize risk of clot formation (e.g., the aortic arch in a mouse).

The transmitter body can be positioned subcutaneously (abdominal flank) or intraperitoneally. Accurate core body temperature data acquisition is achieved via intraperitoneal positioning. However, this site may not be suitable if the animal will subsequently undergo further abdominal surgery, such as islet transplantation.

1. For **subcutaneous positioning**, placement of the glucose sensor tip into the **carotid artery precedes** positioning of the transmitter body in the subcutaneous **abdominal flank**.
2. For **intraperitoneal positioning**, the insertion of the transmitter body into the **peritoneal cavity precedes** glucose sensor tip insertion into the **carotid artery**.

Surgery should always be conducted in accord with the device manufacturer's instructions, and full surgical training for this specific surgery is highly recommended.

## 2 Materials

### *2.1 Telemetry Device*

1. **Telemetry devices** are available from commercial suppliers (*see* **Note 4**; Fig. 1). A glucose telemetry probe will typically consist of a **glucose sensor** portion (which is inserted into a blood vessel). The sensor is coupled to a **radio-transmitter device** (positioned in the peritoneal cavity or subcutaneously under the skin on the flank) which relays the blood glucose signal to a **data acquisition system**.

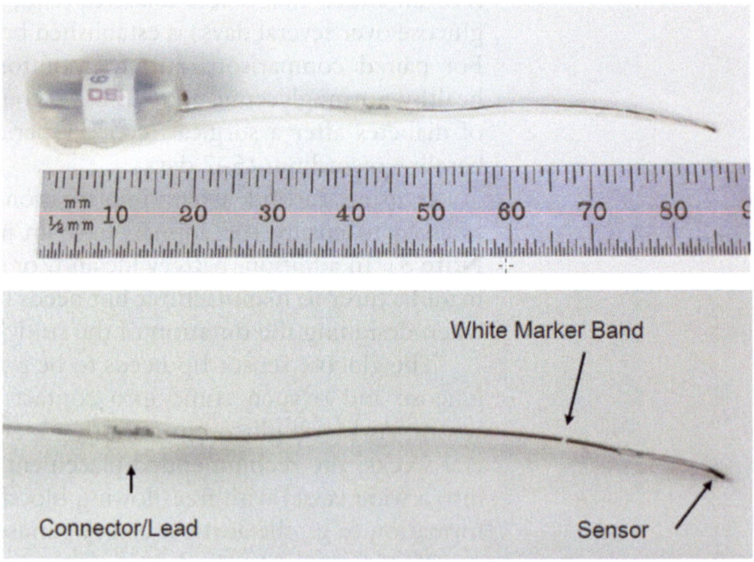


**Fig. 1** HD-XG glucose implant from Data Sciences International. Upper panel: whole implant. Lower panel: magnified view of sensor which is placed in the carotid artery. The white marker band is used to establish which depth the sensor is at within the vessel

2. **Data acquisition system**: a **receiver pad** placed beneath the cage, collects digitized signals from the telemetry device and relays data to a central **data acquisition matrix**. The matrix is connected to a computer with relevant data **analytical software** (*see* **Note 4**).

### 2.2 Surgical Equipment and Supplies

1. Radio-telemetry hardware and software.
2. Binocular surgical magnification and light source (*see* **Note 5**).
3. Supplemental heating, e.g., homeothermic heating blanket or heating pad.
4. Surgical instruments. We typically use:
   (a) Straight scissors (for cutting the skin).
   (b) Elastic stay hooks or other retraction device.
   (c) Two watchmaker forceps.
   (d) Two straight forceps.
   (e) Two blunt curved forceps.
   (f) Two Spencer Wells forceps.
   (g) 25-gauge hypodermic needle.
   (h) Vessel cannulation forceps.
   (i) Needle holder.
5. Disposable surgical supplies. We typically use:
   (a) Sterile gloves.
   (b) Gown.
   (c) Mask.
   (d) Hairnet.
   (e) Skin disinfectant (e.g., iodine).
   (f) Sterile saline.
   (g) 70% Isopropyl alcohol.
   (h) Artificial tears ointment.
   (i) Depilatory cream or shaver.
   (j) Sterile drapes.
   (k) Micropore/surgical tape.
   (l) Sterile gloves.
   (m) Sterile gauge sponges.
   (n) Sterile cotton tipped applicators.
   (o) Surgical sutures for vessel occlusion (nonabsorbable: 5-0 or 6-0).
   (p) Hemostatic agent or device (*see* **Note 6**).

   (q) Surgical suture for wound closure (5-0 with a curved needle) or staples.
   (r) 1, 2, and 5cc syringes.
6. Perioperative care (*see* **Note 7**):
   (a) Antibiotics.
   (b) Analgesics.
   (c) Fluid resuscitation.
   (d) Heated cabinet for recovery.
   (e) Palatable feed.
7. Anesthesia: inhaled anesthetic. Induction: 2.5% isoflurane and 1 L/min oxygen, maintenance: 1.5% isoflurane and 0.5 L/min oxygen. Always seek veterinary advice to ensure appropriate anesthetic regime is used.

## 3 Methods

### *3.1 Switching on the Device*

1. Switch on the device using a magnet.
2. Confirm the device is switched on by listening to an AM radio tuned to its lowest frequency (e.g., 550 kHz) (*see* **Note 8**).

### *3.2 Implantation Surgery: Subcutaneous Device Placement*

The device should be prepared ready for use alongside the surgical field (*see* **Note 9**).

1. Ensure there are no magnets on the surgical table, as this could pull the telemetry device out of the animal during implantation.
2. Anesthesia should be induced according to local guidelines and righting reflex and response to noxious stimulation (e.g., paw pinch) conducted to ensure sufficient depth of anesthesia (*see* **Note 10**).
3. The animal should be laid supine with the head facing the surgeon and the surgical incision site prepared (*see* **Note 11**). The light source and microscope field of view should focus on the neck and thorax region. The anesthetic nose cone should be secured in place (e.g., using micropore tape) and respiration rate (40–60 breaths per minute) and reflexes monitored continuously throughout surgery. The depth of inhaled anesthesia can be controlled rapidly as required (*see* **Note 10**).
4. Preoperative analgesic should be administered (*see* **Note 7**).
5. An ocular lubricant (e.g., gel tears) should be administered to each eye to minimize direct contact with gaseous anesthetic agent and the front paws secured to the mat using micropore tape (if required).

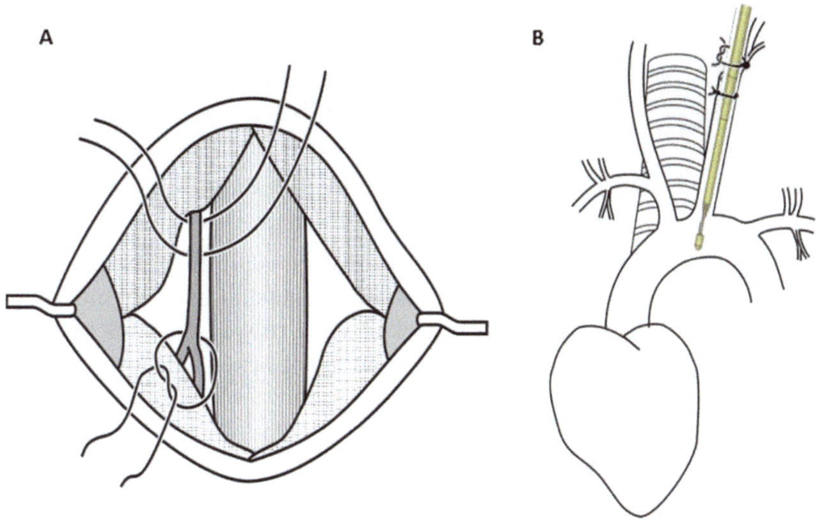


**Fig. 2** Left carotid artery isolated with occlusion sutures in place (**a**). This illustration is orientated with the head at the bottom of the page. Correct placement of the sensor into the aortic arch (**b**)

6. A clean vertical midline incision (~1–2 cm) at the level of the neck should be made using sharp scissors. The salivary glands should be separated using blunt dissection. Stay hooks or retraction devices can be used if required to open up the region.
7. The left carotid artery should be identified and carefully separated from the vagus nerve (*see* **Note 12**). The section of carotid artery should be isolated using blunt forceps and three nonabsorbable sutures placed loosely underneath it (*see* Fig. 2).
8. Saline should be used to moisten tissues at all times and excess solution cleared with a sterile cotton bud.
9. A permanent tie should be made using the suture proximal to the brain (cranial ligation). The middle tie should be kept loose (safety tie), and the suture closest to the heart should temporarily occlude the blood vessel (caudal occlusion). This can be achieved by pulling gently on the sutures using Spencer Wells forceps which are then laid carefully alongside or over the animal.
10. The tip of a 26-gauge needle should be bent to provide a hook to puncture the occluded vessel. A small amount of blood will escape, but a sufficient caudal occlusion will ensure minimal blood loss.
11. The needle tip can be used to widen the hole facilitating introduction of the glucose sensor. Using vessel cannulation forceps, insert the sensor under the needle introducer and into the vessel. Care should be taken to avoid direct contact with the exposed part of the sensor. Once inserted, carefully tighten

the middle safety suture while slowly advancing the sensor toward the aortic arch by loosening the tension on the caudal suture tie. The sensor should advance easily without any significant resistance. Marker bands on the sensor can facilitate correct placement into the aortic arch (Figs. 1 and 2).

12. The caudal tie and middle tie should then be tightened with a double knot around the sensor wire to secure in place.
13. The transmitter portion should now be inserted into the abdominal flank. Loosen the skin using blunt straight forceps from the same neck incision toward the flank on the right-hand side. Once a subcutaneous pocket has been created, flush the pocket with sterile saline and absorb excess fluid using a sterile cotton bud.
14. Carefully insert the transmitter by gently advancing toward the subcutaneous flank pocket by external massage. Care should be taken to avoid excess pressure on the rib cage. The transmitter should sit as low as possible in the flank hindlimb region.
15. The neck incision should be closed using noncontinuous (interrupted) stitches with absorbable sutures. Anesthetic should be lightened at this stage.
16. Finally, administer a local analgesic close to the neck suture site, and we also recommend subcutaneous fluid resuscitation to facilitate rapid recovery (*see* **Note 7**).
17. Remove the animal from the surgical area and place in intended warmed monitoring cage with recommended postoperative care and careful observation (*see* **Notes** 7 and **13**).
18. A quick check that the device is functioning should be performed at this stage, after which the animal should be given a full recovery period (typically 7 days) prior to experimental manipulations, though data acquisition from the implant is recommended throughout.

### 3.3 Implantation Surgery: Intraperitoneal Device Placement

The device should be prepared ready for use alongside the surgical field (*see* **Note 9**).

1. Ensure there are no magnets on the surgical table, as this could pull the telemetry device out of the animal during implantation.
2. Anesthesia should be induced according to local guidelines and righting reflex and response to noxious stimulation (e.g., paw pinch) conducted to ensure sufficient depth of anesthesia (*see* **Note 10**).
3. The animal should be laid supine with the head facing the surgeon (*see* **Note 11**). The light source and microscope field of view should focus on the mid-abdominal region. The anesthetic nose cone should be secured in place (e.g., using

micropore tape) and respiration rate (40–60 breaths/min) and reflexes monitored continuously throughout surgery. The depth of inhaled anesthesia can be controlled rapidly as required (*see* **Note 10**).

4. Preoperative analgesic should be administered (*see* **Note** 7).
5. An ocular lubricant (e.g., gel tears) should be administered to each eye to minimize direct contact with gaseous anesthetic agent and the front paws secured to the mat using micropore tape (if required).
6. A clean vertical midline incision (~2–3 cm) should be made using sharp scissors in the mid-abdominal area. Stay hooks or retraction devices can be used if required to open up the region.
7. The transmitter implant should be placed on top of the intestines and the peritoneal cavity irrigated with sterile saline. The transmitter implant should be parallel to the long axis of the body.
8. Exteriorize the sensor through the abdominal wall using an 18-gauge needle.
9. The abdominal wall should be closed with interrupted sutures with the suture rib of the implant incorporated into the closure and the connector outside the abdominal cavity.
10. A clean vertical midline incision (~1–2 cm) at the level of the neck should be made using sharp scissors.
11. Tunnel the solid trocar subcutaneously from the neck incision to the abdominal incision where the sensor is exteriorized.
12. Slide the hollow trocar sleeve over the trocar and remove the solid trocar.
13. Insert the sensor into the open end of the hollow trocar sleeve and withdraw the sleeve from the neck incision. The sensor should now be exiting the open neck incision.
14. Place a sterile, saline-moistened gauze over the abdominal incision to maintain tissue hydration. The skin incision is left open at this time to allow for adjusting the excess lead into the abdominal cavity.
15. The salivary glands should be separated using blunt dissection. Stay hooks or retraction devices can be used if required to open up the region.
16. The left carotid artery should be identified and carefully separated from the vagus nerve (*see* **Note 12**). The section of carotid should be isolated using blunt forceps and three non-absorbable sutures placed loosely around it (*see* Fig. 2).
17. Saline should be used to moisten tissues at all times and excess solution cleared with a sterile cotton bud.

18. A permanent tie should be made using the suture proximal to the brain (cranial ligation). The middle tie should be kept loose (safety tie), and the suture closest to the heart should temporarily occlude the blood vessel (caudal occlusion). This can be achieved by pulling gently on the sutures using Spencer Wells forceps which are then laid carefully alongside the animal.
19. The tip of a 25- or 26-gauge needle should be bent to provide a hook to puncture the occluded vessel. A small amount of blood will escape, but a sufficient caudal occlusion will ensure minimal blood loss.
20. The needle tip can be used to widen the hole facilitating introduction of the glucose sensor into the vessel. Using vessel cannulation forceps and avoiding handling the tip of the sensor, the sensor can be inserted into the vessel. Once inserted, carefully tighten the middle safety suture while slowly advancing the sensor toward the aortic arch by loosening the tension on the caudal suture tie. The sensor should advance easily without any significant resistance. A white marker band is usually on the sensor to facilitate correct placement into the aortic arch (Figs. 1 and 2).
21. The caudal tie and middle tie should then be tightened around the sensor wire to secure in place.
22. The neck incision should be closed using interrupted stitches with absorbable sutures.
23. A suture should be placed around the lead as it exits the abdominal wall to secure it in place.
24. The abdominal skin incision should be closed using interrupted stitches with absorbable sutures. Anesthetic should be lightened at this stage.
25. Finally, administer a local analgesic close to both suture sites, and we also recommend subcutaneous fluid resuscitation to facilitate rapid recovery (*see* **Note 7**).
26. Remove the animal from the surgical area and place in intended warmed monitoring cage with recommended postoperative care and careful observation (*see* **Notes 7** and **13**).
27. A quick check that the device is functioning should be performed at this stage, after which the animal should be given a full recovery period (typically 7 days) prior to experimental manipulations, though data acquisition from the implant is recommended throughout.

### 3.4 Surgical Recovery

Full postoperative recovery of all experimental animals is necessary prior to collecting any experimental data. This will ensure that post-surgery stress does not confound the results but also will allow accurate calibration of the device.

After immediate postoperative recovery (*see* **Notes 7** and **13**), animals should be carefully monitored. Signs of good health should be apparent such as normal gait and being well-groomed. Animals can initially lose 10–15% weight post-surgery, and a good indication of appropriate recovery from surgery is regaining this weight. In our experience, 7 days is sufficient for animals to return to baseline body weight, and we therefore suggest a 7-day recovery period before device calibration. To minimize confounding factors such as stress, mice should be housed in dedicated telemetry labs with minimal noise and electrical interference (e.g., use of Faraday cages around cage racks), where feasible. Although only one animal per cage can carry the telemetry device, it is suggested that at least one non-implanted or sham-implanted animal is kept in the same cage to avoid the stress of single housing, although if mice are particularly aggressive toward their cage-mate, this may not be advisable [8]. It may be feasible that devices that can transmit at different frequencies will be made available, allowing group housing and monitoring to become standard procedure in the future.

### 3.5 Calibration

The glucose telemetry device requires in vivo calibration in order to have physiologically accurate measurements and ongoing references to adjust for sensitivity loss over time. Device manufacturers recommend that the initial calibration is performed at least 5 days after implantation to allow stabilization of the probe, but we typically conduct this at 7 days (for reasons described above). A blood glucose meter (*see* **Note 14**) is required to measure blood glucose concentrations for the calibration. Blood samples can be acquired by using a 27G or 30G needle to prick the end of the mouse's tail as typically small blood samples (<2 μL) are required for most blood glucose meters. However, it has been suggested that blood collected from the saphenous vein is more representative of core blood glucose concentrations.

Initially, a multipoint calibration should be carried out. Manufacturer's advice should be followed, but blood glucose levels should ideally differ by at least 11 mM between the minimum to maximum points of the multipoint calibration. This should be achieved by either an oral or intraperitoneal glucose tolerance test as intravenous glucose tolerance tests are not recommended for calibrations. If the animal is already overtly hyperglycemic (blood glucose concentrations >20 mM), it may not be advisable to inject more glucose into the animal, and alternatively the blood glucose can be decreased using insulin to achieve a second calibration point (*see* **Note 15**). In both situations, observing the real-time dose response on the data analysis software will ensure the most accurate calibration.

After the initial multipoint calibration, a single-point calibration should be carried out at least twice a week, and a final multipoint calibration should be carried out at the end of the study.

However, this matches the manufacturer's most conservative recommendations, and ideally calibrations should be carried out more frequently. In addition, if the device is switched off for more than a few hours during the study (*see* **Note 16**), a multipoint calibration performed several hours after the device is switched on may be beneficial.

### 3.6 Monitoring

Blood glucose data is recorded every second and typically averaged over 10 s intervals for acute challenges and 1 or 5 min for chronic evaluations. This will require suitable data acquisition software (*see* **Note 3**). The frequency of data acquisition will depend on the study design. Careful notes should be taken of what times any intervention was carried out on the animal (treatments, husbandry, etc.) so that any effect can later be taken into consideration during data analysis. As mentioned previously, the life length of the device is based on the enzymatic reaction of the glucose sensor. This is typically guaranteed to last 4 weeks by the manufacturer, but in our experience, the probes can last two to three times longer than this.

### 3.7 Termination of the Experiment

At the end of the experiment, the animal should be killed according to local ethical guidelines, and any relevant tissues should be removed for analysis. Note that cervical dislocation should not be performed until the probe has been removed. The probe should then be carefully dissected out and cleaned according to manufacturer's specifications. Please note that after implantation, the probe should not be reused in any other animal. However, manufacturers may operate a refurbishment/exchange program.

## 4 Notes

1. Implantation of the probe could potentially alter the onset or progression of the disease. It is therefore recommended that an initial comparison of your model is made with sham-implanted and untreated controls.
2. In our experience using naïve normoglycemic animals, body weight of greater than 23 g increases the chances of successful surgery and recovery.
3. Insertion of the glucose sensor into the left carotid artery means that this vascular supply to the brain will be cut off. Therefore, it is important to check that the colony has an intact circle of Willis (vascular network perfusing brain) such that full perfusion can still be achieved via the right carotid artery. If the circle of Willis is not fully intact, this may result in reduced brain perfusion and associated complications post-recovery.

4. This chapter was based on using the HD-XG implant from Data Sciences International (Fig. 1). It consists of two portions: a transmitter body and a glucose sensor.
   (a) **Transmitter body**, encased in a biocompatible housing, consists of:
      - Electronics module that translates glucose fluctuations into digitized signals and transmits them to a receiver
      - Battery – providing the power supply for the electronics module
      - Reference electrode that acts as an electrical reference for the current being measured by the glucose sensor
      - Magnetically activated switch which is turned off and on by holding a magnet close to the cage without necessitating any restraint of the animal.
   (b) **Glucose sensor section:** This portion extends out of the device body and is encased in silicone tubing and relays blood glucose fluctuations to the electronics in the device body.
      - **Lead** – provides connection to the connector board from the implant housing
      - **Connector board:** provides a connection between the lead and the blood glucose sensor
      - **Glucose sensor:** an enzymatic sensor utilizing glucose oxidase
   (c) Receiver pads and acquisition software are available from supplier. Software examples include Ponemah.
5. We recommend a wall mounted dissecting microscope with cold light source.
6. We sometimes use absorbable 3 $mm^2$ pieces of hemostatic gelatin sponge (Spongostan) to control bleeds if the bleeding exceeded what could be controlled using a cotton bud. However, in most cases, a hemostatic agent or device is not necessary.
7. **Perioperative care**: While we did not use **antibiotics**, they can be used at the discretion of the investigator in conjunction with veterinary advice. Local guidelines should be followed, but to aid recovery, we carried out the following steps:
   (a) We used the following perioperative **analgesic**: subcutaneous NSAID (e.g., Carprieve) prior to surgery and at regular intervals during the recovery phase (approximately 5–7 days) in accordance with veterinary advice. Bupivacaine administered locally at the wound site prior to recovery from anesthesia.

(b) We **fluid resuscitated** animals with 30% glucose in saline (4 × 100 μL subcutaneous injections in four quadrants prior to recovery from anesthesia).

(c) We used DietGel Recovery from Clear$H_2O$, a **palatable feed**, placed on the cage base pre- and post-surgery to allow acclimatization and easy access for 24–48 h following surgery.

(d) Immediately post-surgery, animals were recovered in an incubator set to 27 °C until fully mobile.

8. It is suggested that the battery is switched on immediately prior to surgery while the implant is still within its package. Once the device is implanted, it is recommended that it should not be switched off as this can negatively affect the sensor life.

9. Check device according to manufacturer's instruction. Maintain device in sterile saline until ready for implantation. The device should be switched on using the magnet and handled carefully at all times.

10. **Anesthesia:** We used inhaled isoflurane coupled in a vaporizer coupled to 95% oxygen, 5% carbon dioxide at an appropriate flow rate (typically ~1 L/min). The whole system was coupled to a scavenger to ensure surgeons were protected.

    (a) **Induction:** The isoflurane vaporizer was set to 3–5%. Anesthetic was introduced into the induction chamber prior to the introduction of the animal until righting reflex was no longer apparent.

    (b) **Maintenance**: After removal from the chamber and placement onto the surgical area, anesthesia was maintained via a nose cone, and the isoflurane vaporizer was set to ~1.5–2% and modulated as necessary throughout surgery.

11. The neck region (and for intraperitoneal implantation: the abdomen) can be shaved or fur removed using depilatory cream. Skin should be disinfected using iodine or equivalent solution and sterile saline. A sterile drape should be placed over the animal exposing the surgical field only.

12. The carotid arteries lie parallel and close to the trachea. To isolate, blunt dissection using curved forceps along the left-hand side of the airway, and a large red pulsing carotid vessel should be visible. Care should be applied as carotid artery adheres to the vagus nerve (appearance of dental floss). Damage to the vagus by over manipulation can lead to autonomic complications such as Horner's syndrome which can be diagnosed via the appearance of a drooping eyelid.

13. **Recovery** from anesthetic and return of righting response and full mobility should occur within 30 min following placement of animal in cage. Animals are then transferred to recovery

incubators for between 4 and 12 h. If recovery is not observed in this period or if movements/breathing is labored or restricted, veterinary advice should be sought immediately and animals humanely killed if necessary.

14. We recommend the Nova Biomedical Glucose Meter which is suitable for use in laboratory animals and has an extended monitoring range of up to 50 mM.
15. If a normoglycemic or only slightly hyperglycemic animal is being used, the most appropriate manner in which to create two distinct blood glucose concentrations is to inject glucose (we recommend 2–3 g/kg intraperitoneal administration). If the mouse is already overtly diabetic (blood glucose concentrations >20 mM), it is not appropriate to load more glucose, and thus blood glucose can be decreased using insulin.
16. It is highly recommended the telemetry device is left ON throughout the study as this improves glucose sensor stability and longevity. The operational life length of the implant is dependent on an enzymatic reaction on the sensor rather than battery life, and thus turning the device off does not prolong the length of time that data can be collected. Indeed, the enzymatic reaction continues even if the device is turned off, and product buildup on the sensing electrode may damage it.

## References


1. Balcombe JP, Barnard ND, Sandusky C (2004) Laboratory routines cause animal stress. Contemp Top Lab Anim Sci 43(6):42–51
2. Diehl KH, Hull R, Morton D, Pfister R, Rabemampianina Y, Smith D, Vidal JM, van de Vorstenbosch C (2001) A good practice guide to the administration of substances and removal of blood, including routes and volumes. J Appl Toxicol 21(1):15–23
3. Lo Martire V, Valli A, Bingaman MJ, Zoccoli G, Silvani A, Swoap SJ (2018) Changes in blood glucose as a function of body temperature in laboratory mice: implications for daily torpor. Am J Physiol Endocrinol Metab 315(4):E662–e670. https://doi.org/10.1152/ajpendo.00201.2018
4. King AJ, Austin AL, Nandi M, Bowe JE (2017) Diabetes in rats is cured by islet transplantation...But only during daytime. Cell Transplant 26(1):171–172. https://doi.org/10.3727/096368916X692258
5. Hirsch IB, Brownlee M (2005) Should minimal blood glucose variability become the gold standard of glycemic control? J Diabetes Complications 19(3):178–181. https://doi.org/10.1016/j.jdiacomp.2004.10.001
6. Kovatchev B (2019) Glycemic variability: risk factors, assessment, and control. J Diabetes Sci Technol 13(4):627–635. https://doi.org/10.1177/1932296819826111
7. Wurbel H (2017) More than 3Rs: the importance of scientific validity for harm-benefit analysis of animal research. Lab Anim 46(4):164–166. https://doi.org/10.1038/laban.1220
8. Kappel S, Hawkins P, Mendl MT (2017) To group or not to group? Good practice for housing male laboratory mice. Animals 7(12). https://doi.org/10.3390/ani7120088

# Chapter 17

# Assessing Mouse Islet Function

**Patricio Atanes, Inmaculada Ruz-Maldonado, Oladapo E. Olaniru, and Shanta J. Persaud**

## Abstract

Islets of Langerhans are clusters of endocrine cells embedded within the exocrine pancreas. Islets constitute only approximately 1–2% of the total pancreas mass in all species, so methods have been developed to digest the pancreas and purify islets from the surrounding acinar cells. This chapter provides detailed protocols for isolation of mouse islets and their in vitro functional characterization in terms of assessments of islet viability, hormone content and secretion, second messenger generation and β-cell proliferation.

**Key words** Islets of Langerhans, β-cell, Mouse islet isolation, Islet viability assays, Islet hormone secretion, Second messengers

## 1 Introduction

Islets of Langerhans are small endocrine cell clusters distributed throughout the pancreas. They consist of five distinct cell types that secrete hormones involved in the regulation of fuel homeostasis: β-cells (insulin), α-cells (glucagon), δ-cells (somatostatin), PP-cells (pancreatic polypeptide) and ε-cells (ghrelin) [1, 2]. Islet size is remarkably consistent between species despite interspecies differences in the arrangement of the different islet cell types [3]. For example, the volume of a human pancreas is several thousand times bigger than a rodent's, but rather than scaling being achieved through increased islet size it occurs through increased islet number, such that a human pancreas contains approximately 1 million islets while a mouse pancreas contains only a few hundred [4]. Islets of all mammalian species are around 50–250 μm in diameter and each islet contains 1,000–3,000 cells [5]. In rodent islets 60–80% of the cells are β-cells, while α-, δ-, PP- and ε-cells account for 15–20%, <10%, <5% and <1% of the cell population, respectively [1, 6]. Human islets are the gold standard model for clinically relevant studies, but the majority of the physiological and pharmacological experimental assessments of islet function have been

Aileen J. F. King (ed.), *Animal Models of Diabetes: Methods and Protocols*, Methods in Molecular Biology, vol. 2128, https://doi.org/10.1007/978-1-0716-0385-7_17, © Springer Science+Business Media, LLC, part of Springer Nature 2020

carried out using isolated rodent islets due to the limited availability of human islets for research [7]. Recent studies have focused on comparative gene expression profiles between human and mouse islets. These have characterized conserved signaling pathways, establishing mouse islets as suitable surrogates for human islets, although some key differences between species have been identified [8–12].

Defining the processes by which islet function is normally regulated is important in understanding islet dysregulation that contributes to the pathogenesis of diabetes and in guiding the development of new therapies for diabetes [13]. As islets contribute only 1–2% of the total pancreas mass, robust protocols have been developed for isolation of high-purity islets [14] so that studies of gene and protein expression can be performed with confidence, and potential interactions of the exocrine acinar cells can be ruled out in functional studies. Obtaining a high yield of islets from the mouse pancreas, free from contaminating acinar cells, is one of the most important factors for functional islet biology experiments in vitro, and success depends on variables such as correct positioning of the clamp over the duodenal opening of the common bile duct, precise bile duct cannulation, careful removal of the pancreas, optimal enzymatic digestion time and removal of exocrine cell contamination during islet purification.

The protocols in this chapter describe methods for the isolation and purification of islets from the collagenase-perfused mouse pancreas, maintenance of isolated islets in culture, assessment of islet viability, quantification of islet hormone content and secretion, and of intracellular calcium and cAMP levels, β-arrestin assays using conditioned media from islets, and in vitro/in vivo BrdU incorporation into proliferating β-cells.

## 2 Materials (*See* Note 1)

### *2.1 Mouse Islet Isolation and Purification*

1. Small bulldog clamp.
2. Dissecting scissors, 110 mm.
3. Graefe forceps, 100 mm.
4. Small iris scissors for internal use, 90 mm.
5. 27- and 30-gauge sterile hypodermic needles.
6. Luer lock syringe, 2.5 mL.
7. Falcon™ 50 mL conical centrifuge tubes.
8. Scale balance AJ100.
9. Temperature-controlled water bath.
10. Petri dishes.
11. Tissue culture incubator.

12. Tissue culture laminar flow hood.
13. SZ61 stereo zoom dissecting microscope.
14. Collagenase type XI from *Clostridium histolyticum*.
15. Histopaque®-1077 sterile-filtered, density: 1.077 g/mL.
16. Medical wipes.
17. 70% ethanol solution.
18. 425 μm steel mesh.
19. Universal 320R centrifuge for Falcon™ 50 mL conical centrifuge tubes.
20. Minimum Essential Medium supplemented with 100 U/mL penicillin, 100 μg/mL streptomycin, 2 mM L-glutamine, and 10% newborn calf serum (complete MEM).
21. RPMI-1640 medium, supplemented with 100 U/mL penicillin, 100 μg/mL streptomycin, 2 mM L-glutamine, and 10% fetal bovine serum (complete RPMI-1640 medium).

### 2.2 Maintenance of Islets in Culture

1. Complete RPMI-1640 medium.
2. Petri dishes.
3. Tissue culture laminar flow hood.
4. Tissue culture incubator.

### 2.3 Viability Assays

#### 2.3.1 Trypan Blue Exclusion Test

1. 96-well microplate.
2. Trypan blue, diluted to 0.2% in phosphate-buffered saline.
3. Complete RPMI-1640 medium.
4. Tissue culture incubator.

#### 2.3.2 ATP Synthesis

1. 96-well microplate, white.
2. Complete RPMI-1640 medium.
3. CellTiter-Glo® 3D Cell Viability Assay.
4. Optoelectronically controlled shaker.
5. Luminometer plate reader.

#### 2.3.3 MTT Assay

1. 96-well microplate.
2. Complete RPMI-1640 medium.
3. 1 mg/mL 3-(4,5-dimethylthiazol-2-yl)-2,5-diphenyltetrazolium bromide (MTT) in complete RPMI-1640 medium; make up fresh when needed.
4. Tissue culture incubator.

5. Acidified isopropanol: 48 mL isopropanol plus 250 μL concentrated HCl.
6. Microplate reader.

#### 2.3.4 Caspase Assay

1. 96-well microplate, white.
2. 35 mm Petri dishes.
3. Caspase-Glo® 3/7 Assay.
4. Complete RPMI-1640 medium (2 or 10% FBS).
5. Recombinant murine TNF-α (1 U/μL), IFN-γ (1 U/μL), and IL-1β (0.25 U/μL).
6. Optoelectronically controlled shaker.
7. Luminometer plate reader.

#### 2.3.5 Quantification of Islet Hormone Content

1. 10 mM HCl, diluted as appropriate from concentrated HCl stock.
2. 0.5 M Monobasic sodium phosphate ($NaH_2PO_4$); adjust pH to 7.4 with 5 M NaOH and add 0.5 mg/mL bovine serum albumin (BSA).
3. Sonicator.
4. $^{125}$I-insulin, iodionated in-house.
5. Anti-insulin antibody, generated in-house in guinea pigs.
6. Insulin standard.
7. $^{125}$I-glucagon.
8. Anti-glucagon antibody, generated in-house in guinea pigs.
9. Glucagon standard.
10. Somatostatin radioimmunoassay kit.
11. Borate buffer: 133.4 mM boric acid, 10 mM ethylenediamine-tetraacetic acid (EDTA), 67.5 mM NaOH in distilled water; adjust pH to 8.0 with 10 mM HCl and then add 1 mg/mL BSA. Store at 4 °C.
12. 30% (w/v) polyethylene glycol (PEG): Dissolve 600 g PEG in 1 L distilled water and then make up to 2 L with distilled water.
13. Precipitant: Dissolve 0.1% γ-globulins in phosphate buffered saline, add an equal volume of 30% PEG, and supplement with 0.05% v/v polyoxyethylene (20) sorbitan monolaurate (Tween 20).
14. Wizard automatic gamma counter.
15. 0.5 mL Eppendorf® safe-lock microcentrifuge tubes.
16. Centrifuge for 0.5 mL Eppendorf® safe-lock microcentrifuge tubes.

## 2.4 Functional Assays

### 2.4.1 Quantification of Islet Hormone Secretion in Static Incubations

1. 1.5 mL Eppendorf® safe-lock microcentrifuge tubes.
2. Falcon™ 15 mL conical centrifuge tubes.
3. Physiological salt solution: 111 mM sodium chloride (NaCl), 27 mM sodium hydrogen carbonate ($NaHCO_3$), 5 mM potassium chloride (KCl), 1 mM magnesium chloride hexahydrate ($MgCl_2·6H_2O$), 0.28 mM magnesium sulfate heptahydrate ($MgSO_4·7H_2O$), 0.12 mM potassium dihydrogen orthophosphate ($KH_2PO_4$) in distilled water. Adjust pH to 7.4 with 5% $CO_2$ and then add 2 mM $CaCl_2$, 0.5 mg/mL BSA, and 2 or 20 mM glucose on the day of use.
4. Temperature-controlled water bath.
5. Tissue culture incubator.
6. Centrifuges for 15 mL Falcon™ tubes and 1.5 mL Eppendorf® safe-lock microcentrifuge tubes.
7. Radioimmunoassay reagents as for Subheading 2.3.5.

### 2.4.2 Quantification of Islet Dynamic Hormone Secretion

1. Perifusion chambers (Swinnex® 13 filter holders).
2. Physiological salt solution: 111 mM NaCl, 27 mM $NaHCO_3$, 5 mM KCl, 1 mM $MgCl_2·6H_2O$, 0.28 mM $MgSO_4·7H_2O$, 0.12 mM $KH_2PO_4$ in distilled water. Adjust pH to 7.4 with 5% $CO_2$ and then add 2 mM $CaCl_2$, 0.5 mg/mL BSA, and 2 or 20 mM glucose on the day of use.
3. Temperature-controlled water bath.
4. 1 μm Nylon mesh.
5. Peristaltic pumps and tubing.
6. Temperature-controlled chamber or room.
7. 2 mL 96-well plates.
8. Radioimmunoassay reagents as for Subheading 2.3.5.

### 2.4.3 cAMP Quantification

1. Physiological salt solution: 111 mM NaCl, 27 mM $NaHCO_3$, 5 mM KCl, 1 mM $MgCl_2·6H_2O$, 0.28 mM $MgSO_4·7H_2O$, 0.12 mM $KH_2PO_4$ in distilled water. Adjust pH to 7.4 with 5% $CO_2$ and then add 2 mM $CaCl_2$, 0.5 mg/mL BSA, and 2 mM glucose on the day of use.
2. Assay buffer: 10 mM HEPES, 0.2% BSA, 2 mM 3-isobutyl-1-methylxanthine (IBMX) in Hank's Balanced Salt Solution (HBSS). Adjust pH to 7.4 with 5% $CO_2$.
3. 96-well half-area microplate, white.
4. Tissue culture incubator.
5. Cisbio HiRange cAMP assay kit.
6. PHERAstar FS plate reader.

*2.4.4 Calcium Measurements*

1. Fura-2-acetoxymethyl ester (Fura-2AM), diluted to 1 mM with DMSO. Store at −20 °C.
2. Circular borosilicate cover glass, 22 mm diameter, thickness number 0.
3. Physiological salt solution: 111 mM NaCl, 27 mM $NaHCO_3$, 5 mM KCl, 1 mM $MgCl_2 \cdot 6H_2O$, 0.28 mM $MgSO_4 \cdot 7H_2O$, 0.12 mM $KH_2PO_4$ in distilled water. Adjust pH to 7.4 with 5% $CO_2$ and then add 2 mM $CaCl_2$, 10 mM 4-(2-hydroxyethyl)-1-piperazineethanesulfonic acid (HEPES), and 2 or 20 mM glucose on the day of use.
4. 6-well assay plate.
5. 0.1 mg/mL poly-D-lysine hydrobromide.
6. Tissue culture incubator.
7. Steel chamber mounted onto a temperature-controlled platform of an Axiovert 135 inverted microscope, with an ORCA-05G CCD camera.
8. Accutase® solution.
9. 0.02% w/v EDTA solution.
10. 2 N acetic acid.
11. 70% ethanol made by dilution of absolute ethanol with distilled water.
12. OptoFluor or MetaFluor software.
13. Temperature-controlled water bath.

*2.4.5 β-Arrestin Assays to Identify Islet-Derived G Protein-Coupled Receptor Regulatory Ligands*

1. Physiological salt solution: 111 mM NaCl, 27 mM $NaHCO_3$, 5 mM KCl, 1 mM $MgCl_2 \cdot 6H_2O$, 0.28 mM $MgSO_4 \cdot 7H_2O$, 0.12 mM $KH_2PO_4$ in distilled water. Adjust pH to 7.4 with 5% $CO_2$ and then add 2 mM $CaCl_2$, 0.5 mg/mL BSA, and 5.5 or 20 mM glucose on the day of use.
2. 96-well white culture plate, clear bottom.
3. PathHunter® eXpress β-Arrestin GPCR Assay.
4. Luminometer plate reader.
5. Tissue culture incubator.
6. Falcon™ 15 mL conical centrifuge tubes.
7. Centrifuges for 15 mL Falcon™ tubes.

### 2.5 Assessment of β-Cell Proliferation

*2.5.1 BrdU Incorporation into β-Cells of Isolated Islets In Vitro*

1. Cell culture Petri dishes, 60 mm, non-treated.
2. Tissue culture incubator.
3. Tissue culture laminar flow hood.
4. 1.5 mL Eppendorf® safe-lock microcentrifuge tubes.

5. Centrifuges for 1.5 mL Eppendorf® safe-lock microcentrifuge tubes.
6. Scalpel blade and handle.
7. Graefe forceps, 100 mm.
8. Small paintbrushes.
9. Small corkscrew reamer.
10. Tissue processing/embedding cassettes.
11. 15 × 15 × 7 mm stainless-steel base molds for embedding.
12. Paraffin-embedding station.
13. Gravity oven at 60 °C.
14. Complete RPMI-1640 medium (2% FBS).
15. 1 mg/mL 5-bromo-2′-deoxyuridine (BrdU) in RPMI-1640 medium supplemented as above; make up fresh when needed.
16. Paraformaldehyde, diluted to 4% in phosphate-buffered saline.
17. Absolute ethanol.
18. 95%, 80%, and 70% ethanol solutions made by dilution of absolute ethanol with distilled water.
19. 1% Aqueous eosin solution, diluted to 0.75% eosin in 70% ethanol.
20. Xylene.
21. Paraffin wax.

*2.5.2 BrdU Incorporation into β-Cells In Vivo*

1. Laboratory mice.
2. BrdU, diluted to 1 mg/mL in drinking water.
3. Sterile 1 mL syringes.
4. Sterile hypodermic 30-gauge needles.
5. Rodent water bottles covered with aluminum foil or black tape.
6. Portable balance.
7. 30 × 2 × 5 mm stainless-steel base molds for embedding.
8. Automated tissue processor.
9. Tissue processing/embedding cassettes.
10. Scalpel blade and handle.
11. Graefe forceps, 100 mm.
12. Small paintbrushes.
13. Paraffin wax.
14. Paraffin-embedding station.
15. Gravity oven at 60 °C.
16. Paraformaldehyde, diluted to 4% in phosphate-buffered saline.
17. Absolute ethanol.

18. 70% ethanol made by dilution of absolute ethanol with distilled water.
19. 15 mL Falcon™ centrifuge tubes.

*2.5.3 Fluorescence BrdU and Insulin Co-staining*

1. Microtome.
2. Water bath for paraffin sections.
3. Slide warmer for slides.
4. Hybridization oven at 37 °C.
5. Light microscope.
6. 25 × 75 × 1 mm glass microscope slides.
7. 22 × 40 mm cover glass.
8. Slide staining tray or humidifying chamber.
9. Glass Coplin staining jar.
10. Staining trough tray with handle.
11. Liquid blocker super PAP pen.
12. 10× TBS: 80 g NaCl and 60 g Tris base dissolved in 1 L of distilled water; dilute 100 mL in 900 mL of distilled water to make 1× working solution.
13. 1× Trypsin-EDTA solution.
14. 2 N HCl (1:5 dilution of concentrated HCl).
15. Fluorescence mounting medium.
16. Blocking buffer: 1% BSA, 5% goat serum, diluted in phosphate-buffered saline.
17. 75% and 95% ethanol made by dilution of absolute ethanol with distilled water.
18. Mouse monoclonal anti-BrdU antibody.
19. Polyclonal guinea pig anti-insulin antibody.
20. Anti-mouse FITC conjugated secondary antibody.
21. Anti-guinea pig Alexa Fluor 594-conjugated secondary antibody.
22. Fluorescence microscope.

## 3 Methods

The methods described below focus on isolation of islets from the mouse pancreas for subsequent experimental approaches. The primary goal of isolating islets for in vitro studies is to obtain viable, purified islets that respond in a manner consistent with their function in vivo. To achieve successful islet isolation, the following key points should be taken into consideration: (1) enzymatically digest exocrine tissue, (2) separate intact islets from non-islet tissue, and (3) maintain isolated islets in an environment that maintains their viability. Subheading 3.1 describes a protocol for mouse islet isolation and purification.

Islets may be used experimentally immediately after isolation, but overnight incubation allows recovery from proteolysis that may occur during exposure to collagenase, and from any stress experienced during Histopaque-mediated purification. Subheading 3.2 describes the appropriate procedures for islet maintenance in culture.

Some preliminary assessments can be carried out after isolation to determine islet health. For example, optical inspection of the islets under a light microscope can provide preliminary information about their health status and purity. The protocols described in Subheading 3.3 can be used to quantify viability of islet populations after isolation or after experimental approaches.

The primary function of islets is to release hormones in direct response to changes in blood glucose concentration or to other modulators, most of which bind to cell surface receptors. Glucose and receptor ligands transduce their effects by modifying levels of second messengers in islet cells, the most important of which are intracellular calcium levels ($[Ca^{2+}]_i$) and 3′,5′-cyclic adenosine monophosphate (cAMP). Protocols for assessment of islet function are described in Subheading 3.4.

Islet β-cells have a very limited proliferative capacity, but they can be stimulated under certain circumstances such as obesity or pregnancy to increase insulin secretion in the face of insulin resistance. Methods for quantifying β-cell proliferation in islets in vitro and in vivo are described in Subheading 3.5.

### 3.1 Mouse Islet Isolation and Purification

#### 3.1.1 Pancreas Perfusion

1. Weigh sufficient collagenase (1 mg/mL) into a 50 mL Falcon™ conical centrifuge tube to prepare 3 mL per mouse (*see* **Note 2**).
2. Add the required volume of cold complete MEM and gently invert the tube two to three times until the collagenase is fully dissolved (*see* **Note 3**).
3. Place 2.5 mL of collagenase solution into a 2.5 mL luer lock syringe and attach a 27- or 30-gauge needle (*see* **Note 4**). Keep filled syringes on ice.
4. Euthanize the mouse by cervical dislocation and sterilize the skin with 70% ethanol solution.
5. Open the abdominal cavity by laparotomy, using dissecting scissors to cut through the peritoneal wall horizontally following the rib line and then vertically.
6. Turn the mouse so that its nose is closest to the surgeon and the tail points away (*see* **Note 5**).

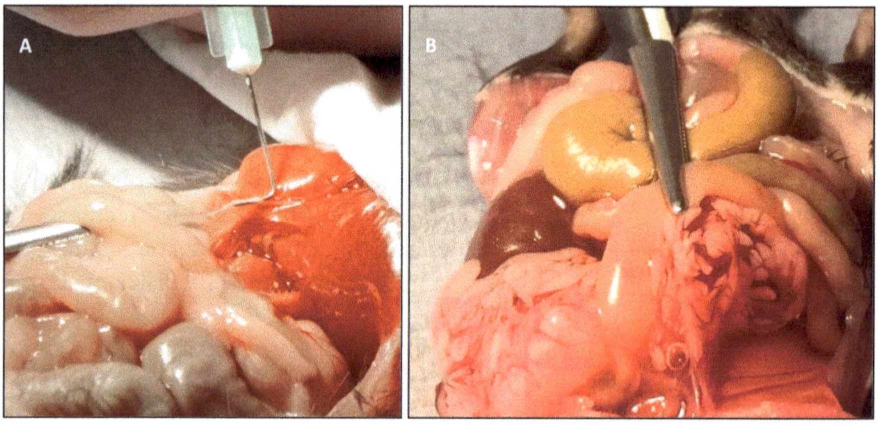


**Fig. 1** Mouse pancreas digestion. (**a**) Bile duct cannulation, (**b**) distension of the pancreas by perfusion with collagenase solution

7. Clip off the posterior sternum cartilage with a pair of scissors.
8. Gently press on either side of the body with your fingers to expose the liver. Move the liver over onto a medical wipe and fold the tissue over to cover it.
9. Place the mouse under the binocular dissecting microscope and expose the duodenum using forceps.
10. Clamp the common bile duct where it enters the duodenum using a small bulldog clamp.
11. Cannulate the bile duct with a needle secured to the 2.5 mL syringe containing the collagenase solution (*see* Fig. 1a).
12. Gently inject 2.5 mL collagenase solution into the bile duct to inflate the pancreas (*see* Fig. 1b) (*see* **Note 6**).
13. Use dissecting scissors to gently remove the pancreas from its connections with the intestines, liver, stomach, and spleen, and place it into a 50 mL conical tube (*see* **Notes** 7 and **8**).
14. If the pancreas does not inflate following injection of collagenase solution or if damage to the duct precludes cannulation, an alternative method involves direct injection of collagenase solution into the lobes of the pancreas after its removal from the body cavity (*see* **Note 9**).

#### 3.1.2 Pancreas Digestion

1. Transfer the 50 mL conical tube containing the distended pancreas to a water bath at 37 °C and incubate for 10 min (*see* **Note 10**).
2. Disrupt the digested pancreas by vigorously shaking the tube by hand for 1 min at a rate of three shakes per second.
3. Quickly place the tube on ice and fill to 25 mL with complete MEM to stop the digestive process.

4. Wash and pellet the digested pancreas three times in complete MEM (25 mL, 340 × *g*, 10 °C, 75 s).
5. Resuspend the digest in 25 mL of complete MEM.
6. Place a 425 μm mesh on top of a new 50 mL conical tube and pour the digested pancreas through the mesh (*see* **Note 11**).

#### 3.1.3 Islet Isolation and Purification

1. Adjust the volume of the digest to 25 mL by addition of complete MEM and pellet by centrifugation (365 × *g*, 10 °C, 90 s).
2. Decant the supernatant by gently pouring out the liquid and place the opening of the 50 mL conical tube onto a paper towel to remove as much liquid as possible (*see* **Note 12**).
3. Resuspend the digested pancreas in 15 mL Histopaque®-1077 and thoroughly mix by vortexing.
4. Gently add 10 mL of complete MEM to create a density interface.
5. Centrifuge (1900 × *g*, 10 °C, 24 min) with slow acceleration and no brake to avoid disturbing the density interface (*see* **Note 13**).
6. Collect the islets at the MEM-Histopaque interface and wash three times in complete MEM (50 mL, 365 × *g*, 10 °C, 90 s) (*see* **Note 14**).
7. To further purify the islets, carry out gravity sedimentation by resuspending the islets in 25 mL of complete MEM and allowing them to settle for 4 min on ice. Discard the top 10 mL containing the remaining small exocrine fragments.
8. Handpick islets under a light microscope in a tissue culture hood (*see* **Notes 15** and **16**) and transfer to 100 mm Petri dishes (*see* **Note 17**) at a density of approximately 50–100 islets/mL in 10 mL of complete RPMI-1640 medium.

### 3.2 Maintenance of Islets in Culture

1. After isolation and purification, incubate islets overnight for 16–20 h in complete RPMI-1640 medium in a tissue culture incubator at 37 °C with 5% $CO_2$ and humidified air.
2. If necessary, replace medium under aseptic conditions every 2–3 days for long-term maintenance of islets (*see* **Note 18**).

### 3.3 Viability Assays

#### 3.3.1 Trypan Blue Exclusion Test

1. Under aseptic conditions, transfer groups of ten islets in complete RPMI-1640 medium to four to five wells of a 96-well plate in a volume of 100 μL using a 100 μL Gilson pipette under a standard binocular microscope.
2. Add 100 μL 0.2% (w/v) trypan blue to the wells containing islets.

3. Incubate for 15 min in a tissue culture incubator (37 °C, 5% $CO_2$).
4. Gently remove the trypan blue solution from each well using a 100 μL Gilson pipette, and wash with 200 μL of phosphate-buffered saline, ensuring that the islets are not removed in the process (*see* **Note 19**).
5. Add 200 μL of phosphate-buffered saline to each well containing islets.
6. Visualize islets in the wells and take photographs as required.

#### 3.3.2 ATP Synthesis

1. Under aseptic conditions, transfer groups of three islets in complete RPMI-1640 medium to six to eight wells of a white-walled 96-well plate in a volume of 5 μL using a 10 μL Gilson pipette under a standard binocular microscope.
2. Add media containing agonists of interest at the desired concentrations in a volume of 30 μL using a 100 μL Gilson pipette.
3. Equilibrate the plate and its contents to 37 °C for approximately 1 h.
4. Prepare an ATP standard curve (0.1 nM to 10 μM) by diluting ATP in complete RPMI-1640 medium and add 35 μL of each ATP standard in duplicate to empty wells of the plate using a 100 μL Gilson pipette (*see* **Note 20**).
5. Add 35 μL of Promega CellTiter-Glo® 3D Reagent, reconstituted according to the manufacturer's recommendations, to each well.
6. Cover the plate and mix the contents by placing it on a shaker for 2 min.
7. Quantify ATP production after 15 min by measuring luminescence with a luminometer plate reader.

#### 3.3.3 MTT Assay (See Note 21)

1. Under aseptic conditions, transfer groups of five islets in complete RPMI-1640 medium to six to eight wells of a 96-well plate in a volume of 50 μL using a 100 μL Gilson pipette under a standard binocular microscope.
2. Add 50 μL of 1 mg/mL MTT to each of the wells containing islets.
3. Incubate for 4 h at 37 °C, after which time purple/brown formazan crystals will be generated by reduction of MTT by dehydrogenase enzymes in metabolically active islet cells.
4. Add 100 μL acidified isopropanol to each well containing islets.
5. Solubilize the formazan crystals by pipetting up and down several times.
6. Determine formazan production by quantifying absorbance at 570 nm using a Chameleon microplate reader.

#### 3.3.4 Caspase Assay

1. Under aseptic conditions, transfer groups of 100 islets in complete RPMI-1640 medium to 35 mm Petri dishes in a volume of 1.5 mL under a standard binocular microscope.
2. Agents under investigation for pro- or antiapoptotic effects may be added to some groups of islets at this stage (*see* **Note 22**), and islets should be incubated for 28 h in a tissue culture incubator at 37 °C with 5% $CO_2$ and humidified air.
3. Divide the islets from each Petri dish equally into two new Petri dishes, using complete RPMI-1640 medium (2% FBS) and containing the same test agents as used in **step 2**. Supplement half of the islets with inducers of apoptosis, most commonly a cytokine cocktail (1 U/μL TNF-α, 1 U/μL IFN-γ, and 0.25 U/μL IL-1β) or 500 μM sodium palmitate.
4. Pick five islets in 40 μL of complete RPMI-1640 medium (2% FBS) with or without cytokines/palmitate and transfer to six to eight wells of a white-walled 96-well white plate.
5. Incubate for 20 h at 37 °C, 5% $CO_2$, completing an incubation period of 48 h.
6. Equilibrate the plate at room temperature for approximately 30 min. Add 40 μL of Promega Caspase-Glo 3/7 reagent, reconstituted according to the manufacturer's recommendations, to each well.
7. Cover the plate and briefly mix the contents on a shaking platform.
8. Incubate for approximately 1 h at room temperature.
9. Determine caspase 3/7 activities by quantifying luminescence with a luminometer plate reader.

#### 3.3.5 Quantification of Islet Hormone Content

1. Transfer groups of ten islets to 0.5 mL Eppendorf tubes using a 200 μL Gilson pipette under a standard binocular microscope.
2. Centrifuge at room temperature for 30 s using a bench Eppendorf centrifuge at 1000 × *g*.
3. Carefully remove all supernatant using a 200 μL Gilson pipette under a standard binocular microscope.
4. Resuspend the islet pellet in 50 μL 10 mM HCl and leave at room temperature for approximately 15 min (*see* **Note 23**).
5. Neutralize the HCl by adding 200 μL 0.5 M $NaH_2PO4$ buffer supplemented with 0.5 mg/mL BSA.
6. Sonicate the samples on ice (*see* **Note 24**).
7. Serially dilute the islet extracts with borate buffer for immunoassay of islet hormones (*see* **Note 25**).
8. For the hormone radioimmunoassays, prepare standard curves in triplicate over 0.08–10 ng/mL (insulin and glucagon) or

3.9–250 pM (somatostatin) by performing 1:2 dilutions of standard stocks (*see* **Notes 26–28**). Borate buffer is used for diluting insulin and glucagon standards, and phosphate buffer (provided with the immunoassay kit) is used for the somatostatin assay. Each standard has a final volume of 100 μL.

9. For each hormone, assay dilutions of the islet extracts in duplicate samples of 100 μL. To quantify insulin and glucagon, follow **steps 10–14**. To quantify somatostatin, follow **steps 15–19**.
10. To quantify ***insulin and glucagon***, add 100 μL of hormone-specific antibody to all standards and samples except tubes quantifying nonspecific binding and total radioactivity (*see* **Note 29**).
11. Add $^{125}$I-labelled glucagon or insulin, diluted to approximately 10,000 cpm/100 μL, to all tubes.
12. Prepare control samples to determine the total amount of radioactivity added ($^{125}$I-labelled glucagon or insulin alone), non-specific binding (borate buffer plus $^{125}$I-labelled glucagon or insulin), and maximum binding (borate buffer plus $^{125}$I-labelled glucagon or insulin plus specific antibody).
13. Incubate all samples and standards at 4 °C for 48–72 h for antibody-hormone binding to reach equilibrium (*see* **Note 30**).
14. Precipitate hormone-antibody complexes by adding 1 mL of precipitant to all tubes except those quantifying total radioactivity.
15. To quantify ***somatostatin***, add 200 μL somatostatin antibody to all tubes except those quantifying total radioactivity.
16. Incubate all samples and standards at 4 °C for 24 h and then add $^{125}$I-labelled somatostatin, diluted to approximately 10,000 cpm/200 μL, to all tubes.
17. Prepare control samples as for subsection 12A.
18. Incubate all samples and standards at 4 °C for a further 24 h.
19. Precipitate hormone-antibody complexes by adding 100 μL double antibody solid phase (anti-rabbit IgG coupled to cellulose particles) to all tubes except those quantifying total radioactivity and incubate at 4 °C for 1 h.
20. For all assays, centrifuge tubes at 2000 × *g* for 15 min at 4 °C.
21. Aspirate the supernatant to leave a firm pellet.
22. Detect the γ-emissions from all pellets using a gamma counter and interpolate the hormone concentrations in the diluted islet extracts from the appropriate standard curves.

### 3.4 Functional Assays

#### 3.4.1 Quantification of Islet Hormone Secretion in Static Incubations

1. Transfer 100 islets to a 15 mL Falcon™ centrifuge tube under a standard binocular microscope.
2. Adjust the volume to 10 mL with physiological salt solution supplemented with 2 mM glucose and centrifuge at 200 × *g* for 1 min.
3. Remove the supernatant and add 10 mL of physiological salt solution supplemented with 2 mM glucose.
4. Preincubate islets for 1 h in a tissue culture incubator at 37 °C with 5% $CO_2$ infusion (*see* **Note 31**).
5. Centrifuge the islets at 200 × *g* for 1 min, remove the supernatant and replace with 5 mL of physiological salt solution supplemented with 2 mM glucose.
6. Transfer the islets to a 60 mm Petri dish.
7. Pick groups of ten islets on ice in 20 μL of physiological salt solution supplemented with 2 mM glucose to 6–8 1.5 mL Eppendorf tubes using a 20 μL Gilson pipette under a standard binocular microscope.
8. Add 480 μL of physiological salt solution supplemented with 2 mM or 20 mM glucose, containing test agents of interest at desired concentrations, to each tube.
9. Incubate the islets in a temperature-controlled water bath at 37 °C for 1 h.
10. Place the tubes on ice and centrifuge at 200 × *g* for 1 min at 4 °C.
11. Retrieve supernatants in 1.5 mL Eppendorf tubes and store at −20 °C (*see* **Note 32**).
12. Quantify hormone concentrations in supernatants according to the hormone immunoassay protocols described in Subheading 3.3.5.

#### 3.4.2 Dynamic Quantification of Islet Hormone Secretion

1. Pick groups of 40–150 islets using a 200 μL Gilson pipette under a standard binocular microscope (*see* **Note 33**).
2. Transfer the islets onto 1 μm filters in individual perifusion chambers connected to buffer reservoirs and peristaltic pumps by 2 mm diameter tubing.
3. Fill the chambers with physiological salt solution supplemented with 2 mM glucose and remove bubbles with a 5 mL syringe when connecting tubing to the chambers (*see* **Note 34**).
4. Use peristaltic pumps to perifuse the islets with the physiological salt solution at 0.5 mL/min for at least 1 h at 37 °C to allow a stable baseline of hormone release (*see* **Note 35**).

5. Select experimental buffers in a heated reservoir using a three-way tap system and use peristaltic pumps to transfer buffers to the perifusion chambers at a flow rate of 0.5 mL/min.
6. Every 2 min, collect 1 mL outflow fractions containing secreted islet hormones into labeled 2 mL 96-well plates.
7. Quantify hormone concentrations of perifusate fractions according to the hormone immunoassay protocols described in Subheading 3.3.5.

*3.4.3 cAMP Quantification (See* ***Note 36****)*

1. Transfer 100 islets to a 15 mL Falcon™ centrifuge tube under a standard binocular microscope.
2. Adjust the volume to 10 mL with physiological salt solution supplemented with 2 mM glucose and centrifuge at 200 × *g* for 1 min.
3. Remove the supernatant and add 10 mL of physiological salt solution supplemented with 2 mM glucose.
4. Preincubate islets for 1 h in a tissue culture incubator at 37 °C with 5% $CO_2$ infusion.
5. Centrifuge the islets at 200 × *g* for 1 min, remove the supernatant, and replace with 5 mL of physiological salt solution supplemented with 2 mM glucose.
6. Transfer the islets to a 60 mm Petri dish.
7. Pick groups of four islets in 10 μL of assay buffer to six to eight wells of a 96-well white-walled half-area plate using a 20 μL Gilson pipette under a standard binocular microscope.
8. Add 10 μL of assay buffer, containing test agents of interest at desired concentrations, to each well.
9. Incubate the plate at room temperature for 1 h.
10. Add 10 μL of serial dilutions of the cAMP standard (provided with kit).
11. Add 10 μL of D2 and cryptate working solutions (provided with kit).
12. Cover the plate and incubate at room temperature for 1 h.
13. Determine cAMP levels by quantifying homogeneous time-resolved fluorescence with a PHERAstar FS plate reader (*see* **Note 37**).

*3.4.4 Calcium Measurements*

1. Prepare circular glass coverslips by soaking in 2 N acetic acid for 1 h and then rinsing five times in distilled water (*see* **Note 38**).
2. Soak the acid-washed coverslips in 70% ethanol for 1 h. Remove the coverslips from ethanol, allow to dry, and sterilize by exposing to UV light for 15 min in a tissue culture hood (*see* **Note 39**).

3. Under aseptic conditions, transfer 400 islets to a 1.5 mL Eppendorf tube.
4. Pellet the islets by brief centrifugation at room temperature (200 × *g*, 1 min) and discard the supernatant.
5. Resuspend the islets in 200 μL 0.02% EDTA solution or 200 μL Accutase solution and incubate at 37 °C in a water bath for 4 min to partially disperse the islets.
6. Pellet by brief centrifugation at room temperature (200 × *g*, 1 min) and discard the supernatant.
7. Resuspend the dispersed islets in serum-free RPMI-1640 at a density of approximately 100,000 cells/50 μL.
8. Transfer 50 μL aliquots of islet cell suspension into the center of each coverslip placed in a 6-well plate using a 200 μL Gilson pipette.
9. Allow the islet cells to adhere to the coverslip by incubating for 2 h at 37 °C (*see* **Note 40**).
10. Add 2 mL of complete RPMI-1640 medium and maintain in culture overnight in a tissue culture incubator at 37 °C with 5% $CO_2$ and humidified air.
11. Incubate the cells with the calcium fluorophore Fura-2AM for 30 min at 37 °C (*see* **Note 41**).
12. Mount the coverslip onto a temperature-controlled platform of an Axiovert 135 inverted microscope (*see* **Note 42**).
13. Perifuse cells on coverslips with a physiological solution containing 2 mM glucose or 20 mM glucose in the absence or presence of test agents (*see* **Note 43**).
14. Illuminate cells alternately at 340 and 380 nm, and filter emitted light at 510 nm (*see* **Notes 44** and **45**).
15. Collect data from selected cells of interest every 3 s using the OptoFluor or MetaFluor software (*see* **Note 46**).
16. Visualize the dynamic profile of intracellular calcium ($[Ca^{2+}]_i$) using a software package such as Excel to plot ratiometric emission at 510 nm after excitation at 340 nm and 380 nm (*see* Fig. 2).

#### 3.4.5 β-Arrestin Assays to Identify Islet-Derived G Protein-Coupled Receptor Regulatory Ligands

1. Under aseptic conditions, transfer 10,000 GPCR-specific β-arrestin cells (provided with the kit) in 100 μL of culture medium to a 96-well plate using a 100 μL Gilson pipette (*see* **Note 47**).
2. Incubate the cells for 48 h in a tissue culture incubator at 37 °C, with 5% $CO_2$ infusion and humidified air.
3. Transfer 100 islets to a 15 mL Falcon™ centrifuge tube under a standard binocular microscope.

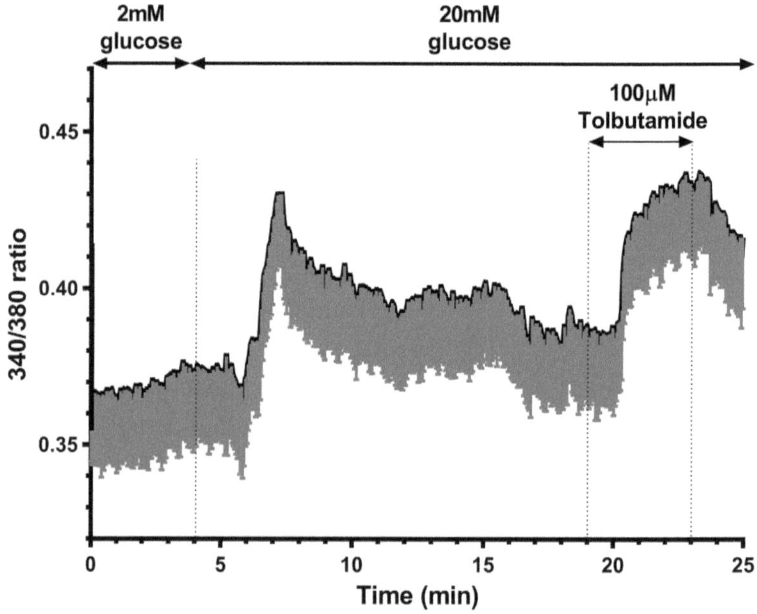


**Fig. 2** Changes in $[Ca^{2+}]_i$ measured using the calcium fluorophore Fura-2AM in mouse dispersed islets. Dynamic profiles of $[Ca^{2+}]_i$ in Fura-2-loaded dispersed C57BL6/J mouse islets. The horizontal arrows indicate the duration of exposure to 2 mM and 20 mM glucose and the period of perifusion with the positive control (100 μM tolbutamide). Data of the 340/380 fluorescence ratios are expressed as mean-SEM of 28 β-cells

4. Adjust the volume to 10 mL with physiological salt solution supplemented with 2 mM glucose and centrifuge at 200 × *g* for 1 min.
5. Remove the supernatant and add 10 mL of physiological salt solution supplemented with 2 mM glucose.
6. Preincubate the islets for 1 h in a tissue culture incubator at 37 °C with 5% $CO_2$ infusion.
7. Centrifuge the islets at 200 × *g* for 1 min, remove the supernatant, and replace with 5 mL of physiological salt solution supplemented with 2 mM glucose.
8. Transfer the islets to a 60 mm Petri dish.
9. Pick groups of 25 islets in 150 μL physiological salt solution supplemented with 2 mM or 20 mM glucose, containing test agents of interest at desired concentrations, to 6–8 1.5 mL Eppendorf tubes using a 200 μL Gilson pipette under a standard binocular microscope.
10. Incubate the islets in a temperature-controlled water bath at 37 °C for 1 h.
11. Place tubes on ice and centrifuge at 200 × *g* for 1 min at 4 °C.

12. Retrieve 125 μL of supernatant in 0.5 mL Eppendorf tubes and store at −20 °C (*see* **Note 48**).
13. Remove culture medium from GPCR-specific β-arrestin cells and replace with 100 μL of thawed islet supernatant using a 100 μL Gilson pipette.
14. Incubate GPCR-specific β-arrestin cells with islet supernatant for 90 min at 37 °C.
15. Add 55 μL of detection reagent to each well (provided with kit) (*see* **Note 49**).
16. Cover the plate and incubate for 1 h at room temperature.
17. Quantify luminescence with a luminometer plate reader on an unfiltered full-range setting.

### 3.5 Assessment of β-Cell Proliferation

#### 3.5.1 BrdU Incorporation into β-Cells of Isolated Islets In Vitro

1. Under aseptic conditions, transfer groups of 250 islets in complete RPMI-1640 medium (2% FBS) to 60 mm Petri dishes in a volume of 5 mL under a standard binocular microscope.
2. Supplement the culture medium with 1 mg/mL BrdU and test agents of interest (*see* **Note 50**).
3. Incubate the islets for 5 days in a tissue culture incubator at 37 °C with 5% $CO_2$ and humidified air, with changes of BrdU-supplemented medium every 48 h (*see* **Note 51**).
4. Transfer islets into 1.5 mL Eppendorf tubes and add 1 mL 4% paraformaldehyde for 2 h.
5. Centrifuge at 135 × *g* for 2 min, discard paraformaldehyde, and add 1 mL phosphate-buffered saline.
6. Centrifuge at 135 × *g* for 2 min, discard phosphate-buffered saline, and add 1 mL 70% ethanol.
7. Replace 70% ethanol with 1 mL 80% ethanol, and add three drops of 0.75% eosin in 70% ethanol (*see* **Note 52**).
8. Incubate for 5 min and then replace 80% ethanol with 1 mL 95% ethanol.
9. Incubate for 5 min and then replace 95% ethanol with 1 mL 100% ethanol.
10. Incubate for 5 min and then replace 100% ethanol with 1 mL xylene (*see* **Note 53**).
11. Incubate for 3 min and then remove all of the xylene (*see* **Note 54**).
12. Carefully fill the Eppendorf tube to the top with paraffin wax heated to 60 °C.
13. Leave the Eppendorf tube in the gravity oven at 60 °C for 4 h (*see* **Note 55**).
14. Store the wax-embedded islets in Eppendorf tubes at −20 °C.

15. Retrieve the Eppendorf tubes containing islets and place a corkscrew reamer in the wax.
16. Run cold water over the tube exterior for 2 min and then place the tube in boiling water for 3 s.
17. Carefully pull out the paraffin plug and screw out the corkscrew.
18. Heat a scalpel blade and cut away the paraffin tip containing islets.
19. Heat the scalpel blade again and place it between the stainless-steel base mold and the paraffin tip and remove the scalpel blade. Melt around the joint. Add more melted paraffin from the paraffin-embedding station.
20. Mark a tissue processing/embedding cassette with a pencil.
21. Place the tissue processing/embedding cassette over the immersed islets in paraffin and add more melted paraffin to the cassette.
22. Leave the steel base mold with the cassette in the paraffin-embedding station until the paraffin has fully solidified.
23. Store the paraffin blocks at −20 °C overnight.
24. Cool the paraffin-embedded tissue blocks on ice before sectioning (*see* **Note 56**).
25. Place the blade in the microtome blade holder, ensure it is secure, and determine the clearance angle (*see* **Note 57**).
26. Insert the paraffin block and orientate so that the blade will cut straight across the block.
27. Cut a few thin sections to ensure the positioning is correct. Adjust if necessary.
28. Cut three to four sections at a thickness of 5 μm.
29. Use a small paintbrush to pick up the ribbons of sections and leave them on the surface of the water in a water bath at 45 °C until they flatten out.
30. Pick the sections out of the water bath using microscope slides.
31. Check the sections under a light microscope to examine islet integrity.
32. Store the microscope slides upright in a slide rack.
33. Place the slide rack into an oven and allow the sections to dry overnight at 37 °C.

#### 3.5.2 BrdU Incorporation into β-Cells In Vivo

1. Provide mice with 1 mg/mL BrdU in drinking water *ad libitum* for up to 2 weeks using rodent water bottles covered with aluminum foil or black tape (*see* **Note 58**).

2. Weigh the mice to determine the volumes of specific concentrations of test agents of interest or vehicle to be delivered and load the appropriate volume into 1 mL syringes.
3. Deliver the agents of interest to the mice by intraperitoneal injections every 2 days for 2 weeks or according to the protocol requirements (*see* **Note 59**).
4. Euthanize the mice, retrieve the pancreases, and fix them in 4% paraformaldehyde for a minimum of 48 h at room temperature.
5. Transfer the pancreases into 8 mL 70% ethanol for storage at 4 °C in a 15 mL Falcon™ centrifuge tube.
6. Place the pancreases into embedding cassettes using forceps and label them with a pencil.
7. Place the embedding cassettes into an automated tissue processor using a standard program of 20 h (2 h in 70% ethanol, 2 h in 90% ethanol, 3 × 2 h in 100% ethanol, 2 h in 1:1 ethanol: xylene, 2 × 2 h in xylene, 2 × 2 h in paraffin wax).
8. Embed the tissues in paraffin blocks.
9. Select a mold that best corresponds to the size of the pancreas and discard the cassette lid.
10. Put a small amount of melted paraffin in the mold using the paraffin-embedding station.
11. Transfer the pancreas into the mold using warm forceps, placing the cut side down, as it was placed in the cassette.
12. Transfer the mold to a cold plate, and gently press the pancreas flat. Paraffin will solidify in a thin layer that holds the tissue in position.
13. Add the labeled tissue cassette on top of the mold as a backing. Press firmly.
14. Hot paraffin is added to the mold from the paraffin-embedding station. Make sure there is enough paraffin to cover the face of the plastic cassette. Paraffin should solidify in 30 min. When the paraffin is completely cooled and compacted, the paraffin block can be easily removed from the mold.
15. Repeat **steps 24–33** of Subheading 3.5.1.

#### 3.5.3 Fluorescence BrdU and Insulin Co-staining

1. Place the slides containing islet or pancreas sections on a hot plate until the wax has melted.
2. Quickly place the slides into a rack before the wax resolidifies, and wash sections in xylene for 2 × 5 min to remove the wax.
3. Rehydrate the sections in a series of ethanol washes: 2 min in 100% ethanol, 2 min in 95% ethanol, 2 min in 75% ethanol.
4. Wash sections in running tap water for 10 min.

5. While the sections are washing in tap water, warm the 2 N HCl and trypsin-EDTA solution in a 37 °C water bath.
6. Wash the sections in 1× TBS for 5 min.
7. Transfer the slides from the rack to a humidifying chamber hybridization oven that contains water at the bottom.
8. Draw around the sections carefully with a wax pen.
9. Pipette a drop of 2 N HCl onto each section, making sure that the tissue is covered.
10. Transfer the slides in the humidifying chamber into an oven at 37 °C and incubate for 20 min.
11. Place the slides back in the rack and wash in distilled water for 5 min.
12. Wash the sections in 1× TBS for 5 min.
13. Put the slides back in the humidifying chamber and pipette 50–100 μL of the pre-warmed trypsin-EDTA solution onto each section (*see* **Note 60**).
14. Incubate the slides with trypsin for 15 min at 37 °C in the humidifying chamber.
15. Place the slides back in the rack and wash 3 × 5 min in 1× TBS.
16. Transfer the slides to the humidifying chamber and add a drop of blocking buffer to each section to minimize nonspecific binding of antibodies.
17. Leave for 15 min at room temperature, during which time the appropriate dilutions of primary antibodies should be prepared in blocking buffer.
18. Tap the slides on a piece of tissue paper to remove the blocking buffer.
19. Add 50–100 μL of a mixture of monoclonal mouse anti-BrdU antibody (1:100) and polyclonal guinea pig anti-insulin antibody (1:150) and incubate slides for 3 h in the humidifying chamber in the oven at 37 °C (*see* **Note 61**).
20. Wash 3 × 5 min in 1× TBS, during which time the appropriate dilutions of secondary antibodies should be prepared in blocking buffer (*see* **Note 62**).
21. Add 50–100 μL of a mixture of FITC anti-mouse (1:50; green fluorescence) and Alexa 594 anti-guinea pig (1:50; red fluorescence) and incubate slides for 1 h at room temperature in the dark (*see* **Notes 63** and **64**).
22. Wash 3 × 5 min in 1× TBS.
23. Place coverslips on a piece of blue roll and add two to three drops of fluorescent mounting medium onto each coverslip.

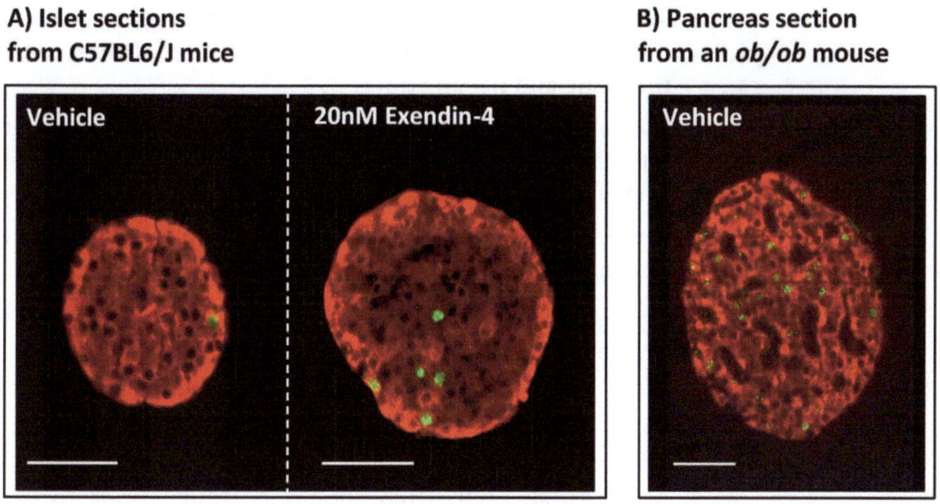


**Fig. 3** Representative images of double immunofluorescence for insulin and BrdU of paraffin-embedded sections. (**a**) Islets from C57BL6/J mice after 5 days of exposure to 1 mg/mL BrdU in vitro in the presence of vehicle (left) or 20 nM of the GLP-1 analogue exendin-4 (right). (**b**) Pancreas from an *ob/ob* mouse that was administered 1 mg/mL BrdU for 2 weeks
Antibodies in all panels were directed against insulin (red cytoplasm) and BrdU (green nuclei). Scale bar: 50 μm

Lower the slides down onto the coverslips so that the sections are covered by the coverslip.

24. Visualize the insulin and BrdU co-staining using a fluorescence microscope (*see* Fig. 3).

## 4 Notes

1. The equipment described for the methods (centrifuges, microscopes, plate readers, etc.) may be replaced by other suitable apparatus.
2. It is advisable to include an extra volume of collagenase solution in case injections are unsuccessful and the pancreas must be distended outside of the body cavity.
3. Keep the collagenase solution on ice from this point onward. This will allow maintenance of uniform conditions while isolating islets from multiple mice.
4. The choice of hypodermic needle size will depend on the mouse strain or size: C57BL6/J and small ICR mice may require a 30-gauge needle, while medium and large ICR mice and *ob/ob* mice usually require a 27-gauge needle. However, a 25-gauge needle can be used if the duct is large as a tight fit allows better distention. The needle should be bent to 90° to facilitate entry into the bile duct.

5. This position allows for easy cannulation into the common bile duct from the rostral end closest to the liver at the junction of the cystic and hepatic ducts.
6. Care should be taken when clamping and inserting the needle into the duct to minimize unwanted perforation or destruction, and the collagenase solution should be injected slowly.
7. It is very important to avoid intestinal rupture to minimize risk of contamination by intestinal bacteria.
8. It is possible to fit two to three pancreases in one 50 mL conical tube.
9. A few drops of collagenase solution should be injected into as many of the lobes of the pancreas as possible.
10. Different batches of collagenase have different proteolytic activities, so each batch should be tested prior to use for isolation of islets from multiple pancreases, and incubation time should be adjusted accordingly.
11. This will allow removal of fat, large pieces of exocrine tissue and undigested pancreas.
12. It is important to remove all medium so the Histopaque is not diluted.
13. This will result in accumulation of the islets at the density interface and exocrine fragments will be centrifuged into the pellet.
14. To maximize islet yield, the Histopaque at the interface and the waste washes can be viewed under a light microscope and any residual islets can be collected.
15. From this stage onward, procedures should be carried out under aseptic conditions in a tissue culture hood to minimize islet contamination.
16. The islets appear as solid white spheres under a light microscope and are readily distinguishable from any grayish contaminating acinar cells.
17. Islets are maintained in culture in uncharged Petri dishes to minimize cell adhesion.
18. Islets are most often used experimentally soon after isolation, but they may be maintained in culture for extended periods for assessment of effects of experimental agents on gene expression or viability.
19. It is useful to use a microscope to ensure that islets are not lost during medium removal and islet washing steps.
20. The ATP standards may be prepared in advance and stored frozen for use in multiple assays.

21. MTT may be replaced by MTS, which has the advantage of producing a colored formazan product that is soluble in culture media, removing the requirement of repeated pipetting to solubilize the formazan crystals obtained with MTT.
22. Exendin-4, a glucagon-like peptide-1 receptor agonist, can be used as positive control since it protects islets from apoptosis induced by cytokines [15] and palmitate [16].
23. HCl is used to disrupt islets if insulin, glucagon, and somatostatin contents are to be quantified in the same samples. If measurements are to be made of insulin content alone, then acidified ethanol (ethanol:water:HCl, 52:17:1) rather than $HCl/NaH_2PO_4$ can be used to solubilize the islet cells, and extracts must be diluted at least 1/500 for immunoassay.
24. Any suitable method may be used to release the hormones from secretory granules.
25. The required dilutions of islet extracts can be estimated from mouse islet contents of approximately 80 ng insulin, 20 ng glucagon, and 5 ng somatostatin. In this way, the diluted samples should fall on the appropriate hormone assay standard curve.
26. Ionizing radiation local rules should be followed for handling radioactivity and disposing of radioactive waste.
27. Islet hormones may also be quantified by enzyme-linked immunosorbent assays (ELISAs) or homogeneous time-resolved fluorescence (HTRF) assays.
28. Assays that are sensitive over different standard curve ranges may also be used.
29. The antibody dilutions used for the assays will depend on the titer of the antibodies being used.
30. Binding of hormones to antibodies can be accelerated by incubation at room temperature, but this may reduce sensitivity of the assays.
31. Islets may be preincubated in a water bath at 37 °C.
32. Islet hormone secretion may be quantified immediately without freezing the samples beforehand.
33. Forty islets are sufficient for dynamic quantification of insulin secretion, but 150 islets should be used for profiling somatostatin secretion. The perifusion flow rate can be reduced if a low-abundance hormone is being quantified and/or if islets are rate-limiting.
34. The presence of bubbles in the closed perifusion system may cause shear stress on the islets, which can induce unregulated hormone release.

35. Islet metabolism and hormone secretion are temperature-sensitive, so perifusion experiments should take place in a cabinet or room maintained at 37 °C.
36. The Cisbio cAMP HiRange assay is described here, but other assays may be used for quantification of islet cAMP levels.
37. The following settings should be used: delay time = 100 μs, integration time = 100 μs, number of flashes = 50.
38. This removes a thin film of grease or dust that may be present on the surface of the coverslips. A clean coverslip allows cells to adhere.
39. The dry, acid-washed coverslips should be kept in a clean airtight container until use.
40. To increase cell adherence, coverslips may be coated with poly-D-lysine or Cell-Tak solutions before seeding the cells.
41. During the incubation period, Fura-2AM enters the cells where it is cleaved by cellular esterases to the active, impermeant polycarboxylate form of the dye.
42. Glucose metabolism by islet β-cells is temperature-dependent. The heating chamber should be maintained at 37 °C.
43. 100 μM ATP and 50 μM tolbutamide may be used as positive controls, to stimulate $Ca^{2+}$ mobilization and $Ca^{2+}$ influx, respectively.
44. A flywheel or rotary chopper to rapidly alternate between the two excitation wavelengths and wavelength-specific filters are required.
45. When bound to $Ca^{2+}$, Fura-2 changes its emission characteristics at 340 nm and 380 nm excitation wavelengths, allowing measurements of $Ca^{2+}$-free and $Ca^{2+}$-bound forms of the dye to be made. Expressing the level of $[Ca^{2+}]_i$ as a ratio of emitted florescence at these excitation wavelengths circumvents changes in fluorescence that may occur independently of changes in $Ca^{2+}$.
46. Other suitable software packages may be used.
47. These cell lines are genetically manipulated to only generate a reporter signal when the recombinantly expressed GPCR of interest is activated by its cognate ligand.
48. For labile ligands, the supernatant should be added to the β-arrestin cells immediately to preclude loss of secreted products.
49. Activation of the GPCR by a secreted ligand stimulates binding of β-arrestin to the receptor leading to activation of β-galactosidase, which cleaves a luminescent substrate.

50. Exendin-4 may be employed as a positive control as it is known to induce mouse β-cell proliferation [17–19].
51. BrdU is light-sensitive and must be replaced regularly to maintain its concentration.
52. Eosin is a negatively charged acidic dye that stains basic structures red, so the islets become visible in the pellet.
53. Xylene is a volatile irritant, so should be used in a vented fume hood.
54. Remove as much xylene as possible since this agent can make the islets brittle.
55. The islets will sink to the bottom of the Eppendorf tube so that they accumulate in the paraffin tip. The sedimentation time may be increased from 4 h to overnight, but longer is not recommended.
56. Cold wax allows thinner sections to be obtained by providing support for harder elements within the tissue specimen. The small amount of moisture that penetrates the block from the melting ice will also make the tissue easier to cut.
57. The clearance angle prevents contact between the knife facet and the face of the block. Follow the microtome manufacturer's instructions for guidance on setting the clearance angle. For Leica blades, this is normally between 1° and 5°.
58. This solution should be protected from light and should be refreshed at least every 2 days. BrdU can be toxic, so mice should be closely monitored for signs of toxicity if they receive BrdU for more than 2 weeks.
59. Reduce the risk of injection site infection by using a new needle for each mouse and minimize discomfort by injecting fluid that is at body temperature. If repeated injections are required, consider alternatives such as the use of minipumps.
60. Trypsin breaks protein cross-links that form following tissue fixation, so is used as an antigen retrieval step to reveal epitopes to the primary antibodies.
61. Primary antibodies can be mixed if they recognize different antigens and are raised in different species. The volume of the antibody cocktail added depends on the size of the tissue section.
62. Fluorescent secondary antibodies are light-sensitive, so minimize their exposure to the light.
63. All subsequent steps should avoid exposure to light where possible.
64. Similar to primary antibodies, secondary antibodies can also be mixed and applied as a cocktail.

## References


1. Arrojo E, Drigo R, Ali Y, Diez J et al (2015) New insights into the architecture of the islet of Langerhans: a focused cross-species assessment. Diabetologia 58:2218–2228
2. Steiner DJ, Kim A, Miller K et al (2010) Pancreatic islet plasticity: interspecies comparison of islet architecture and composition. Islets 2:135–145
3. Cabrera O, Berman DM, Kenyon NS et al (2006) The unique cytoarchitecture of human pancreatic islets has implications for islet cell function. Proc Natl Acad Sci U S A 103:2334–2339
4. Dolensek J, Rupnik MS, Stozer A (2015) Structural similarities and differences between the human and the mouse pancreas. Islets 7: e1024405
5. Soria B, Tuduri E, Gonzalez A et al (2010) Pancreatic islet cells: a model for calcium-dependent peptide release. HFSP J 4:52–60
6. Roscioni SS, Migliorini A, Gegg M et al (2016) Impact of islet architecture on beta-cell heterogeneity, plasticity and function. Nat Rev Endocrinol 12:695–709
7. Kaddis JS, Olack BJ, Sowinski J et al (2009) Human pancreatic islets and diabetes research. JAMA 301:1580–1587
8. Amisten S, Atanes P, Hawkes R et al (2017) A comparative analysis of human and mouse islet G-protein coupled receptor expression. Sci Rep 7:46600
9. Atanes P, Ruz-Maldonado I, Hawkes R et al (2018) Defining G protein-coupled receptor peptide ligand expressomes and signalomes in human and mouse islets. Cell Mol Life Sci. https://doi.org/10.1007/s00018-018-2778-z
10. Benner C, Van Der Meulen T, Caceres E et al (2014) The transcriptional landscape of mouse beta cells compared to human beta cells reveals notable species differences in long non-coding RNA and protein-coding gene expression. BMC Genomics 15:620
11. Dai C, Brissova M, Hang Y et al (2012) Islet-enriched gene expression and glucose-induced insulin secretion in human and mouse islets. Diabetologia 55:707–718
12. Kutlu B, Burdick D, Baxter D et al (2009) Detailed transcriptome atlas of the pancreatic beta cell. BMC Med Genet 2:3
13. Cnop M, Welsh N, Jonas JC et al (2005) Mechanisms of pancreatic beta-cell death in type 1 and type 2 diabetes: many differences, few similarities. Diabetes 54(Suppl 2): S97–S107
14. Li DS, Yuan YH, Tu HJ et al (2009) A protocol for islet isolation from mouse pancreas. Nat Protoc 4:1649–1652
15. Velmurugan K, Balamurugan AN, Loganathan G et al (2012) Antiapoptotic actions of exendin-4 against hypoxia and cytokines are augmented by CREB. Endocrinology 153:1116–1128
16. Natalicchio A, Labarbuta R, Tortosa F et al (2013) Exendin-4 protects pancreatic beta cells from palmitate-induced apoptosis by interfering with GPR40 and the MKK4/7 stress kinase signalling pathway. Diabetologia 56:2456–2466
17. Fusco J, Xiao X, Prasadan K et al (2017) GLP-1/Exendin-4 induces beta-cell proliferation via the epidermal growth factor receptor. Sci Rep 7:9100
18. Wang C, Chen X, Ding X et al (2015) Exendin-4 promotes beta cell proliferation via PI3k/Akt signalling pathway. Cell Physiol Biochem 35:2223–2232
19. Wu X, Liang W, Guan H et al (2017) Exendin-4 promotes pancreatic beta-cell proliferation via inhibiting the expression of Wnt5a. Endocrine 55:398–409

# Chapter 18

# Assessing Immune Responses in the Nonobese Diabetic Mouse Model of Type 1 Diabetes

**Terri C. Thayer, Dimitri Kakabadse, Joanne Boldison, and F. Susan Wong**

## Abstract

Type 1 diabetes is an autoimmune disease resulting in the loss of insulin production and, consequently, hyperglycemia. The nonobese diabetic (NOD) mouse develops spontaneous diabetes with considerable similarity to the disease in humans. Immunological studies using the NOD mouse model allow for the investigation of the natural history of the disease and leukocyte and lymphocyte pathogenic and regulatory functions, as well as testing potential therapies for intervention. The analyses of the cellular events leading up to diabetes may utilize different in vitro cellular assays, immunohistochemistry, and in vivo adoptive transfer, to study mechanisms of the disease and the effects of therapeutic intervention. In this chapter, we describe some common techniques for phenotyping and mechanistic analyses of function, particularly of CD8+ T cells.

**Key words** Type 1 diabetes, Autoimmune, Nonobese diabetic mouse, T cells, Flow cytometry, Immunohistochemistry, Autoantigens

## 1 Introduction

Type 1 diabetes (T1D) is a result of the targeted attack and destruction of the pancreatic beta cells. Leukocytic infiltration of islets contains macrophages, dendritic cells (DCs), T cell subsets, and B cells, all contributing to the autoimmune process in varying capacities (reviewed in refs. 1–4). Macrophages and DCs drive early inflammatory responses and provide antigen-specific activation of lymphocytes. T cells that respond to islet autoantigens are implicated as the final effectors mediating beta cell death. CD4+ T cells produce pro-inflammatory cytokines, including IFN-γ, and are important for the recruitment of CD8+ T cells and B cells. CD8+ T cells contribute directly to beta cell killing via cytotoxic mechanisms, and B cells serve as antigen-presenting cells, as well as producers of autoantibodies. Many factors that include surface marker expression, signaling cascades, and cytokine synthesis contribute to

Aileen J. F. King (ed.), *Animal Models of Diabetes: Methods and Protocols*, Methods in Molecular Biology, vol. 2128,
https://doi.org/10.1007/978-1-0716-0385-7_18, © Springer Science+Business Media, LLC, part of Springer Nature 2020

the complex interaction of leukocytes and combine to direct the immune response to promote tolerance or induce pathogenicity. The nonobese diabetic (NOD) mouse has been an invaluable tool in defining autoimmune processes. The main genetic susceptibility factor, the major histocompatibility complex (MHC) class II gene ($IA^{g7}$), has features very similar to the most important genetic susceptibility factor in humans (HLA-DQ8). The mice are highly susceptible to the environment, and immune processes are very important in the development of autoimmune diabetes. The mice may be used for studying ways to intervene in the disease process.

## 2 Materials

### 2.1 Cell Phenotyping

#### 2.1.1 In Vitro Flow Cytometry Methods for Assessing Leukocyte Phenotype and Function

1. Cells (which can be isolated from various sources, ranging from blood, thymus, and spleen to tissue-specific locations including draining lymph nodes and islets).
2. Petri dishes, sterile.
3. 30 g needles.
4. Glass tissue homogenizer.
5. 1 mL syringe with 27 g needle.
6. Medium: RPMI with 5% FBS, 2 mM L-glutamine, 100 U/mL penicillin, 100 μg/mL streptomycin, and 50 μM β-2-mercaptoethanol.
7. Deionized water.
8. Phosphate-buffered saline (10×).
9. Phosphate-buffered saline (1×).
10. Buffer for blood samples: 1× PBS with 0.5% BSA, 4 mM EDTA, 2% sodium azide.
11. Lysis buffer.
12. Foxp3/Transcription Factor Staining Buffer.
13. Antibodies (*see* Tables 1 and 2 for routine markers for T cell subsets and Table 3 for B cell identification) (reviewed in ref. 5).
14. FACS buffer, 1× PBS with 0.5% BSA, stored at 4 °C.
15. FACS tubes.

#### 2.1.2 Labeling Cells with MHC Class I Tetramers [6]

1. Cells can be isolated from various sources, ranging from the blood, thymus, and spleen to tissue-specific locations including draining lymph nodes and islets (*see* **Note 1**).
2. 15 mL non-sterile tubes for tissue collection.
3. Glass tissue homogenizer.
4. Phosphate-buffered saline (PBS) 1×, sterile.

**Table 1**
**Identifying T cell subsets**

| Classification | Surface markers | Cytokines/ chemokines | Transcription factors |
|---|---|---|---|
| Regulatory T cells | CD4+ CD25+ CD127+ CTLA4 | TGF-β, IL10, IL12 | FoxP3, STAT5 |
| Cytotoxic T lymphocytes (CTLs) | CD8+ CD107a | IFN-γ, TNF-α, | EOMES, T-bet |
| T-helper 1 | CD4+ CXCR3, | IFN-γ, IL2, IL12, IL18, | STAT4, STAT1 |
| T-helper 2 | CD4+ CCR4 | IL2, IL4 | GATA3 |

**Table 2**
**T cell activation status**

| | |
|---|---|
| Naïve | CD44− CD62L+ |
| Central memory | CD44+ CD62L+ |
| Effector memory | CD44+ CD62L− |
| Tissue-resident memory | CD69+ CD103+ |
| Anergic | PD-1+, CD7+, CD28lo |

**Table 3**
**Identifying B cell subsets**

| | Surface markers | Intracellular factors | Location |
|---|---|---|---|
| Activated B cells | CD19 CD86 CD80, MHC I, MHC II, CD40, PD-1 | IL-6 | Peripheral tissues |
| Regulatory B cells | CD5 + CD1dhi CD38hi CD24hi | IL-10, TGF-β | Marginal zone, transitional zone |
| Marginal zone | CD19+ CD21hi CD23lo | | Spleen |
| Follicular zone | CD19 CD21lo CD23hi | | Spleen |
| Transitional B cells | CD19 CD21hi CD23hi | | Spleen |
| B1 cells | CD5 CD11b PD-L2 | | Peritoneum and pleural cavity |
| Immature naive B cell | IgM+ IgD+ | | Bone marrow |
| Mature naive B cell | IgM+ | | Peripheral lymphoid organs |
| Plasma cells | CD138+ MHC II- IgD- CD19- | Blimp1 | Bone marrow |
| Memory B cells | CD73 PD-L2 | | Peripheral tissues |

5. PBS, 10×.
6. Deionized water.
7. Wash buffer: 1× PBS with 2% FBS.
8. 10 mg dasatinib diluted in DMSO to 1 mM (*see* **Note 2**).
9. Fc block.
10. MHC class I peptide tetramers.
11. Viability dye.
12. FACS tubes.

#### 2.1.3 T Lymphocyte Proliferation

1. Vybrant® CFDA SE Cell Tracer Kit or other suitable dyes (*see* **Note 3**).
2. Phosphate-buffered saline (1×).
3. Medium: RPMI with 5% FBS.
4. Splenocytes, whole lymphocyte populations, or T lymphocytes isolated via magnetic separation.

#### 2.1.4 In Vitro Assessment of CD8+ T Cell Cytotoxicity

1. Medium: RPMI with 5% FBS, 2 mM L-glutamine, 100 U/mL penicillin, 100 μg/mL streptomycin, and 50 μM β-2-mercaptoethanol.
2. PKH-26 labeling kit or other suitable cell-labeling kits.
3. Serum-free medium, stored at 4 °C.
4. Fetal bovine serum, stored at −20 °C.
5. Insulin B chain15–23 peptide (or other known autoantigenic peptides), stored at −20 °C.
6. CD8a Microbeads with suitable magnetic columns.
7. TOPRO-3 or other suitable viability dyes.
8. Permeabilization buffer, stored at 4 °C.

### 2.2 In Vivo Adoptive Transfer of Cells for Diabetes Induction

1. Splenocyte/lymphocyte source: NOD prediabetic or diabetic donor mice.
2. NOD-Scid 5–6-week-old female recipient mice.
3. Phosphate-buffered saline (1×), sterile.
4. Deionized water.
5. Phosphate-buffered saline (10×).
6. Saline, sterile.
7. Mouse restrainer.
8. 1 mL syringe and 27 g needle.

### 2.3 Immunohistochemistry

#### 2.3.1 Preparation of Frozen Tissue for Sectioning and Staining

1. Tissues may include the spleen, pancreas, thymus, salivary gland, and lymph nodes.
2. 0.1 M Phosphate buffer, pH 7.6: Combine monobasic sodium phosphate ($NaH_2PO_4$) and dibasic sodium phosphate ($Na_2HPO_4$) to adjust pH accordingly (*see* **Note 4**).
3. Paraformaldehyde-lysine-periodate (PLP) fixative: 0.1 M phosphate buffer with 1% paraformaldehyde, 75 mM lysine. Just before use, add sodium m-periodate at 10 mM (*see* **Note 5**).
4. Sucrose 10% and 20% solutions made in 0.1 M phosphate buffer.
5. OCT compound.
6. Isopentane (2-methyl-butane).
7. Dry ice.
8. Molds for embedding.

#### 2.3.2 Sectioning of Frozen Blocks

1. Microscope slides, SuperFrost Plus 25 × 75 × 1 mm.
2. Blades, MB35 Premier microtome blade 34°/80 mm.
3. OCT embedding matrix.

#### 2.3.3 Staining Using a Biotinylated Antibody for Frozen Sections

1. 0.1 M Tris buffer, 0.9% NaCl (TBS), pH 7.4.
2. Wash buffer: TBS with 0.01% Triton X-100 (TX100).
3. 0.1 M Tris buffer, pH 8.2–8.5.
4. Serum for blocking, either using the same species in which the antibodies were generated or BSA.
5. Antigen-specific antibodies or cell type-specific antibodies.
6. Secondary antibodies, if needed.
7. Avidin/biotin kit.
8. Streptavidin-alkaline phosphatase kit.
9. Vector Red.
10. Levamisole solution.
11. 0.01% Triton X-100 (TX100).
12. Hematoxylin.
13. Hydromount.
14. Moisture chamber: large dish with grid to hold up slides and lid covered in aluminum foil.
15. PAP pen.
16. Glass coverslip.

*2.3.4 Preparation of Formalin-Fixed Paraffin-Embedded Tissue for Immunohistochemistry*

1. Tissues may include the spleen, pancreas, thymus, salivary gland, and lymph nodes.
2. 10% formalin buffer solution
3. Phosphate-buffered saline (PBS), 1×.
4. Histology cassettes.
5. Ethanol, 70%, 95%, and 100%.
6. Xylene.
7. Paraffin.

*2.3.5 Sectioning of Paraffin-Embedded Tissue*

1. Microscope slides, SuperFrost Plus 25 × 75 × 1 mm.
2. Blades, MB35 Premier microtome blade 34°/80 mm.
3. Water bath, 35–40 °C.
4. Oven, 65 °C.

*2.3.6 Staining of Paraffin-Embedded Tissues*

1. Ethanol, 100%, 95%, 70%, and 50%.
2. Deionized water.
3. 0.1 M Tris buffer, 0.9% NaCl (TBS), pH 7.4.
4. Wash buffer: TBS with 0.01% Triton X-100 (TX100).
5. 0.1 M Tris buffer, pH 8.2–8.5.
6. Serum for blocking, either using the same species in which the antibodies were generated or BSA.
7. Antigen-specific antibodies or cell type-specific antibodies.
8. Secondary antibodies, if needed.
9. Avidin/biotin kit.
10. Streptavidin-alkaline phosphatase kit.
11. Vector Red.
12. Hematoxylin.
13. Hydromount.
14. Moisture chamber: large dish with grid to hold up slides and lid covered in foil.
15. PAP pen.
16. Glass coverslip.

## 3 Methods

### 3.1 Cell Phenotyping

Phenotyping lymphocytes is an important step in describing and understanding an immune response. It is common now to evaluate >15 surface and intracellular markers, at a time with the advancement of lasers, detectors, and fluorophores, using flow cytometry.

*3.1.1 In Vitro Flow Cytometry Methods for Assessing Leukocyte Phenotype and Function*

1. Remove desired tissues from the mouse after culling by approved methods.
   (a) For lymph nodes: Place in a petri dish with 1 mL of medium. Tease apart lymph nodes using a 30 g needle to release lymphocytes. Transfer cell suspension to a 15 mL tube containing 5 mL of medium, rinsing the petri dish to collect all cells.
   (b) For the spleen or thymus: Collect tissue in a 15 mL tube containing 5 mL of medium. Homogenize tissue with the glass homogenizer. Lyse red blood cells from the spleen by resuspending cell pellet in 900 μL of deionized water, followed by the quick addition of 100 μL of 10× PBS.
   (c) For peripheral blood: With a 1 mL needle and 27 g needle, perform cardiac puncture to collect blood sample. Transfer to a FACS tube with buffer for blood samples to prevent clotting.
   (d) For cultured cells in vitro: Collect cells from the tissue culture plate, and transfer to a FACS tube.
2. Pellet cells by centrifugation at 300 × *g* for 5 min, at room temperature.
3. Resuspend in 1× PBS in a suitable volume for counting.
   (a) Lymph nodes: 5 mL, 1:2 dilution with trypan blue.
   (b) Spleen and thymus: 10 mL, 1:40 dilution with trypan blue.
4. Centrifuge cells again.
5. Place sample, $1 \times 10^6$ cells from the spleen, thymus, and lymph nodes or whole sample of peripheral blood or cultured cells in a FACS tube with desired antibodies for surface expression (Tables 1–3).
6. Label for 30 min, aluminum foil covered to protect from the light, at 4 °C.
7. For blood samples only:
   (a) Add 1 mL diluted lysis buffer, and mix well.
   (b) Incubate for 10 min, at room temperature and protected from the light.
8. Wash with 1 mL FACS buffer, centrifuge, decant buffer, and resuspend cell pellet in minimal residual FACS buffer.
9. For intracellular staining: Foxp3/Transcription Factor Staining Buffer. Follow manufacturer's protocol for nuclear staining.
10. Wash with 1 mL FACS buffer, centrifuge, decant buffer, and resuspend cell pellet in minimal residual FACS buffer.
11. Resuspend in 200 μL FACS buffer.
12. Collect and analyze on a flow cytometer.

*3.1.2 Labeling Cells with MHC Class I Tetramers [6]*

1. Precool centrifuge to 4 °C.
2. Collect the spleen/thymus in a 15 mL tube containing 5 mL wash buffer. For lymph nodes, dissect them and transfer them into a petri dish containing 1 mL wash buffer. Use a 30 g needle to tease open the lymph nodes, and then transfer the cell suspension into a non-sterile FACs tube containing 3 mL PBS on ice.
3. Homogenize the spleen/thymus in PBS using glass tissue homogenizer.
4. Pellet cells (400 × *g*, 5 min, 4 °C).
5. Pour off supernatant into a waste beaker and resuspend pellet in remaining liquid.
6. **For spleen cells only**: Add 900 μL $dH_2O$, immediately followed by 100 μL 10× PBS (to lyse red blood cells). **For thymus/lymph nodes**: Resuspend cells in 10 mL (thymus) or 1 mL (lymph nodes) of wash buffer and proceed to **step 10**.
7. Top up with 9 mL 1× PBS.
8. Pellet cells (400 × *g*, 5 min, 4 °C).
9. Pour off supernatant and resuspend cell pellet in 10 mL wash buffer.
10. Count cells.
11. Aliquot cells at 0.5–1.0 × $10^6$ cells per sample; top up to 1 mL with wash buffer.
12. Pellet cells (400 × *g*, 5 min, 4 °C).
13. While cells are pelleting, defrost 5 μL 1 mM aliquot of dasatinib and dilute 1:10,000 (100 nM) in wash buffer.
14. Pour off supernatant and remove any additional supernatant using a pipette.
15. Resuspend cell pellet in 50 μL wash buffer followed by 50 μL 100 nM dasatinib solution (made in **step 13**) to give the final concentration: 50 nM. Dasatinib is a protein kinase inhibitor, so it will prevent the T cell receptor from being recycled from the cell surface upon tetramer binding.
16. Incubate at 37 °C, 5% $CO_2$ for 30 min.
17. While cells are incubating, change temperature on centrifuge to room temperature (~22 °C). This prevents any sudden temperature shock to the cells and improves viability.
18. Post-incubation, pellet cells (400 × *g*, 5 min, 22 °C) and discard the supernatant.
19. Resuspend cells in 25 μL wash buffer containing 2 μL Fc block and incubate at room temperature for 5 min.

20. Add 100 μL wash buffer and pellet cells (400 × *g*, 5 min, 22 °C) and discard the supernatant.
21. Resuspend cell pellet in 100 μL wash buffer containing tetramer-BV421 (usually at 1 μg per sample; *see* **Note 6**).
22. Incubate at 37 °C, 5% $CO_2$ for 15 min.
23. Wash cells with 1 mL wash buffer and pellet (400 × *g*, 5 min, 22 °C), and discard supernatant.
24. Resuspend pellet in 100 μL wash buffer containing antibodies for surface staining at optimal concentrations (e.g., CD8, CD4, CD19, CD11b) and viability dye (*see* **Note 7**).
25. Incubate at 4 °C for 30 min.
26. While cells are incubating, change centrifuge back to 4 °C.
27. Wash cells with 1 mL wash buffer and pellet (400 × *g*, 5 min, 4 °C), and discard the supernatant.
28. Resuspend pellet in 100 μL wash buffer.
29. Store samples in the dark at 4 °C until ready to analyze on a flow cytometer (*see* **Note 8**).
30. Ensure that relevant controls are used (*see* **Notes 9–11**).

### 3.1.3 T Lymphocyte Proliferation

Cytoplasmic cell tracers are a convenient and reliable way to track cells in vivo and assess cell proliferation via flow cytometry [7].

1. Prepare a 10 mM CFDA SE stock solution immediately prior to use by dissolving the contents of one vial (component A) in 90 μL of the high-quality DMSO provided (component B). Aliquot this in 5 μL aliquots and store at −20 °C for future use.
2. Dilute the stock solution in phosphate-buffered saline (PBS) or other suitable buffers to the desired working concentration (0.5–25 μM); *see* **Note 12**.
   (a) Splenocytes for in vivo transfer: 2 μM
   (b) Lymphocytes for in vitro culture: 0.5 μM
3. Centrifuge at 300 × *g* for 5 min to obtain a cell pellet and aspirate the supernatant.
4. Resuspend the cells gently in pre-warmed (37 °C) PBS containing the probe.
   (a) Splenocytes for in vivo transfer: $5^{e6}$/mL
   (b) Lymphocytes for in vitro culture: $1^{e6}$/mL
5. Incubate the cells for 15 min at 37 °C.
6. Re-pellet the cells by centrifugation and resuspend in fresh pre-warmed medium. Incubate the cells for another 30 min to ensure complete modification of the probe and then wash the cells again.

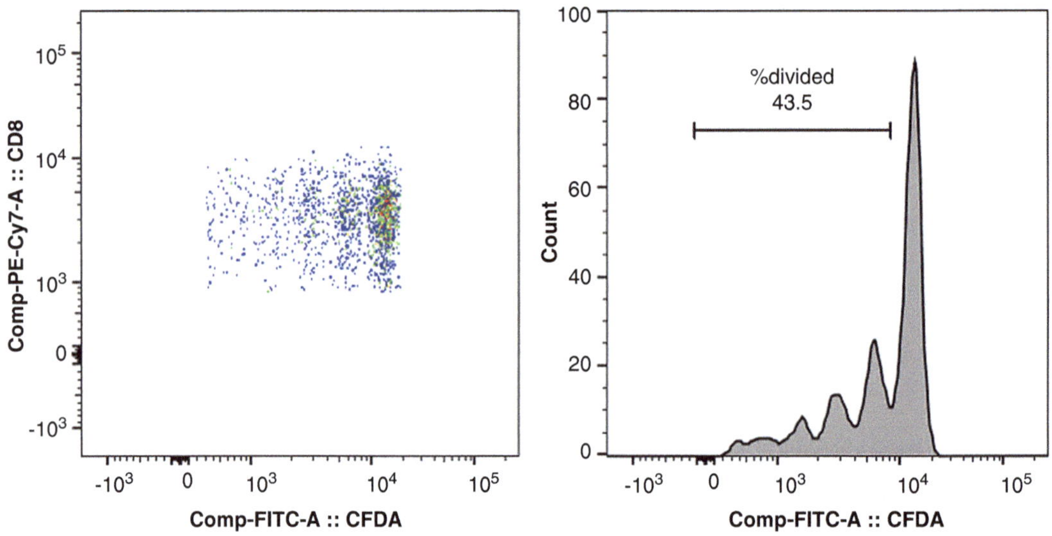


**Fig. 1** In vivo proliferation of autoimmune effectors. Transferred G9 CD8+ T cells (left panel) were analyzed for dilution of CFDA cell tracer, and the percentage of cells divided was determined (right panel, histogram)

7. Labeled cells can be used for in vivo or in vitro analysis.
   (a) Splenocytes for in vivo transfer: analyzed 2, 3, 4, and 5 days post-transfer
   (b) Lymphocytes for in vitro culture: assay duration of 4 days
8. Proliferation analysis can be combined with phenotyping by using the above protocol in Subheading 3.1.1.
9. The approximate excitation and emission peaks of this product after hydrolysis are 492 nm and 517 nm, respectively.
10. Each round of proliferation reduces the fluorescence intensity of the dye (Fig. 1). Proliferation generations can be analyzed using appropriate software, for example, FlowJo.

*3.1.4 In Vitro Assessment of CD8+ T Cell Cytotoxicity*

The primary function of CD8+ T cells is the elimination of infected or cancerous cells via cytotoxic factors. The gold standard of measuring cytotoxicity has previously been $^{51}$chromium release assays. However, this comes with the potential risks and precautions needed for handling radioactive material. A flow-based method is favored because it eliminates the need for radioactive isotopes and provides the opportunity to combine cytotoxicity measures with other flow cytometric-based readouts, such as proliferation and expression of activation markers [8]. P815, a mastocytoma cell line, is a suitable target for cytotoxicity as these cells express the NOD MHC class I $K^d$ haplotype.

1. P815 target cells are grown following standard sterile tissue culture guidelines.

2. Day 0: Split cells (*see* **Note 13**).
   (a) Culture cells overnight in medium at 37 °C, 5% $CO_2$.
3. Day 1: Prepare P815 target cells.
   (a) Prepare cells in serum-free medium to prevent interference with PKH labeling.
   (b) Collect cells from the flask using a cell scraper, transferring to a 50 mL conical tube, and centrifuge at 300 × *g* for 5 min.
   (c) Decant, resuspend in 5 mL serum-free medium, and count cells.
   (d) Wash $1 \times 10^6$ cells in 10 mL serum-free medium, and centrifuge again.
   (e) Volumes below are for labeling $1 \times 10^6$ cells; scale up as needed.
   (f) Save unlabeled cells for compensation controls.
   (g) Dilute dye by adding 1 μL PKH-26 dye to 250 μL diluent C in a sterile microfuge tube, protected from light.
   (h) Resuspend cells in 250 μL diluent C of the PKH-26 labeling kit.
   (i) Mix diluted dye with cells; incubate for 90 s at room temperature, covered in aluminum foil to protect from light.
   (j) Add 500 μL FBS and incubate for 1 min at room temp, aluminum foil covered, to stop labeling.
   (k) Wash cells twice with complete medium (3–5 mL), and centrifuge 300 × *g* for 5 min to remove excess dye.
   (l) Resuspend in 1 mL medium for counting.
   (m) Re-pellet cells if needed and resuspend at $2 \times 10^5$ cells/mL in complete medium.
   (n) Take 50 μL of cell suspension, $1 \times 10^4$ cells, into flow tubes (set up all in duplicate or triplicate), keeping cells covered to protect dye from light for
      - Targets alone-negative control
      - Targets/peptide alone-negative control
      - Targets/peptide alone-positive control (with added permeabilization buffer, to provide gating control for dead cells)
      - Two to three tubes for each experimental condition of peptide concentration and effector cell replicates

4. Day 1: Prepare peptide.
   (a) Dilute autoantigenic peptide in complete medium at a range of concentrations covering five- to tenfold dilutions (e.g., for insulin B chain 15–23 peptide, the concentrations 20, 4, 0.8, and 0.16 μg/mL are used).
   (b) Add 50 μL of peptide or medium to P815 experimental tubes (for final concentrations, 5, 1, 0.2, and 0.04 μg/mL and no peptide) in a final volume of 200 μL.
5. Day 1: Isolate and prepare cytotoxic CD8+ T cell effectors (e.g., we describe insulin-specific CD8+ T cell effectors) [9].
   (a) Splenic CD8+ T cells are isolated using CD8a Microbeads following the manufacturer's instructions.
   (b) Count cells and prepare at $2 \times 10^6$ cells/mL in complete medium.
   (c) Take 100 μL of CD8+ T cells ($2 \times 10^5$ cells for 20:1, dilute as needed for other ratios) and add to peptide-loaded P815 targets in flow tubes, ratio E:T of 5:1, 10:1, and 20:1.
   (d) Final volume per tube is 200 μL.
   (e) Recap tubes loosely to allow air exchange. Incubate (37 °C, 5% $CO_2$) overnight for 16 h.
6. Day 2: Flow cytometry.
   (a) Add a drop of permeabilization buffer to the targets/peptide alone-positive control tube as a gating control for dead cells.
   (b) Dilute viability dye 1:50, or suitable concentration for dye used, in medium.
   (c) Add 10 μL diluted viability dye to experimental tubes, for final 1:1000 dilution; mix well.
   (d) Set appropriate cytometer settings: for PKH26 emission 567 nm and viability emission 780 nm.
   (e) Gate on PKH-26+ P815 population, collecting 3000–5000 events in this gate; assess % viable (780−) and dead (780+) (Fig. 2).

### 3.2 In Vivo Adoptive Transfer of Cells for Diabetes Induction

Adoptive transfer of cells is a useful technique to determine the potency of polyclonal diabetogenic cells or oligoclonal/monoclonal T cells, such as those derived from antigen-specific TCR transgenic mice. Furthermore, this technique can be used for tracking antigen-specific cells in vivo. It is also useful for testing cellular strategies that could potentially delay and/or prevent the onset of diabetes, in principle. Adoptive transfer is also useful for assessing in vivo cytotoxicity for antigen-specific T cells that kill their targets in the context of the NOD MHC class I $K^d$ or $D^b$ alleles. The

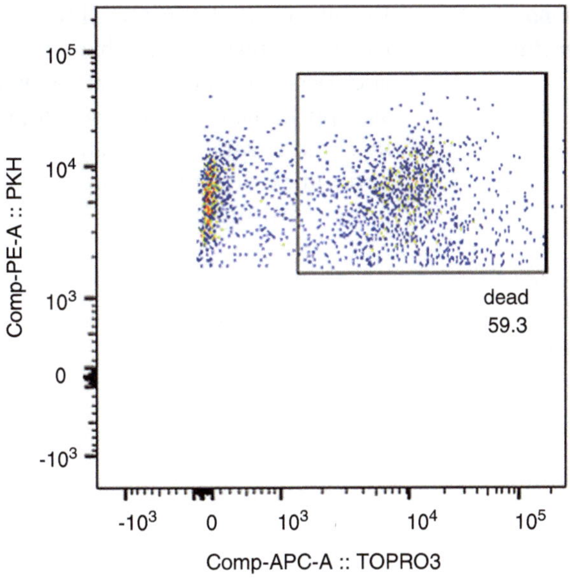


**Fig. 2** Antigen-specific cytotoxicity. Insulin peptide-presenting P815 targets (PKH+) were analyzed for cell death (TOPRO3+) to determine direct killing by G9 CD8+ T cell effectors

following will describe induction of diabetes following adoptive transfer of splenocytes from a polyclonal NOD mouse.

1. For polyclonal diabetogenic T cells, remove the spleen from the diabetic mouse after culling by approved methods.
2. Collect tissue in a 15 mL tube containing 5 mL of 1× PBS.
3. Homogenize tissue with glass homogenizer.
4. Lysis of red blood cells from the spleen:
   (a) Pellet cells via centrifugation at 300 × *g* for 5 min.
   (b) Resuspend cells in 900 μL of deionized water, followed by the quick addition of 100 μL of 10× PBS.
5. Wash cells in 10 mL 1× PBS, centrifuge, and decant.
6. Resuspend cells in 10 mL 1× PBS for counting.
7. Wash cells with sterile saline, centrifuge, and decant.
8. Resuspend at 10–20 million cells per 100 μL saline.
9. Inject 100 μL cell suspension in the tail vein of the recipient mouse (*see* **Note 14**).
10. Recipient mice can be monitored regularly by testing urine for glycosuria, and diabetes is confirmed by testing blood glucose (*see* **Note 15**) to determine onset of type 1 diabetes (*see* **Note 16**).

### 3.3 Immunohistochemistry

Immunohistochemistry (IHC) is a technique used for morphological characterization, which reveals the abundance, distribution, and localization of biomarkers within the tissue. These techniques are applicable for basic research, in addition to being indispensable in clinical tests. Target antigens may be evaluated using specific antibodies directly conjugated with enzyme or fluorophores or indirectly using similarly labeled secondary antibodies and reagents.

#### 3.3.1 Preparation of Frozen Tissue for Sectioning and Staining

1. Collect pancreatic tissue in a glass bottle containing 3–5 mL PLP buffer, stored on ice.
2. Incubate overnight at 4 °C. Do not leave for longer than 24 h as this may result in over-fixing and fractured tissue.
3. Decant PLP buffer and wash with 5 mL 0.1 M phosphate buffer.
4. Decant and replace with 5 mL 10% sucrose solution.
5. Incubate for a minimum of 30 min at 4 °C.
6. Decant and replace with 20% sucrose solution.
7. Incubate for a minimum of 30 min at 4 °C.
8. Label mold with experimental details.
9. Prepare a bath for freezing the tissue blocks by filling up a tray with a thin layer of isopentane and some dry ice leaving enough room for mold to float on isopentane.
10. Store bath on top of dry ice in a polystyrene box, and cover the box with lid.
11. Carefully remove tissue from the bottle using forceps.
12. Blot tissue on tissue paper to remove excess liquid.
13. Put a layer of OCT in mold, being careful not to overfill the mold.
14. Remove any air bubbles with fine forceps or P200 tip.
15. Arrange tissue in OCT, surface of interest facing down (e.g., for pancreas, vein facing down).
16. Cover tissue with OCT and remove air bubbles.
17. Transfer mold to the isopentane bath without any liquid spilling over the edges of mold.
18. Leave in bath until fully frozen.
19. Store blocks at −80 °C.

#### 3.3.2 Sectioning of Frozen Blocks

1. Prepare cryostat with the following parameters for pancreatic tissue:
   (a) Chamber temp: −25 °C
   (b) Sample temp: −15 to −20 °C
   (c) Blade angle: 0°

2. Lock in the blade and place the block in the cryostat chamber and allow to acclimatize to the temperature.
3. Mount the block using OCT to sample arm.
4. Section tissue at 5–10 μm thickness, collecting on microscope slide.
5. Confirm presence of islets from pancreas sections.
6. Label slides with pencil and store at −80 °C (up to 1–2 years) or −20 °C freezer for the short term (up to 3 months).

*3.3.3 Staining Using a Biotinylated Antibody for Frozen Sections*

1. Defrost slides, and allow to air-dry for 30 min.
2. Circle the tissue section with a PAP pen to create a barrier. All incubations are performed in the moisture chamber to prevent slides from drying and to protect from light.
3. Rehydrate in wash buffer (TBS with 0.01% Triton X-100) for 30–40 min.
4. Remove liquid by tipping the slide and dabbing on a paper towel after each wash.

   Block with 10% (or 5% rat + 5% goat) normal serum (same species as primary antibody) or BSA 1% in wash buffer for 20 min (*see* **Note 17**).
5. Rinse with wash buffer.
6. Block with avidin (kit) one to two drops per section for 15 min.
7. Rinse with wash buffer.
8. Block with biotin (kit) one to two drops per section for 15 min.
9. Incubate with wash buffer for 5 min and then 1% $H_2O_2$ (*see* **Note 18**).
10. Add primary biotinylated antibody at the appropriate dilution (determined by serial dilution experiments). Antibodies are diluted in wash buffer plus 1% normal serum (same species as primary or BSA). Incubate for 90 min at room temperature or overnight at 4 °C (*see* **Note 19**).
11. Wash three times, 5 min each in wash buffer.
12. Incubate sections with streptavidin-alkaline phosphatase for 30 min. SAP (kit) is diluted 1/500 in wash buffer plus 1% normal serum or BSA.
13. Wash three times, 5 min each in wash buffer.
14. Vector Red staining (following manufacturer's instruction):
    (a) To 5 mL of the 10 mM Tris–HCl, pH 8.2–8.5:
    - Add a drop of levamisole solution, and mix well.
    - Add two drops of Reagent 1, and mix well.

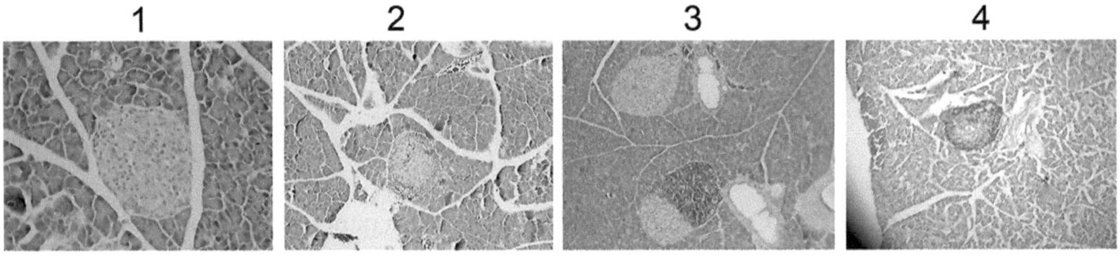


**Fig. 3** Leukocytic infiltration of pancreatic islets

- Add two drops of Reagent 2, and mix well.
- Add two drops of Reagent 3, and mix well.

(b) Apply mixture to the sample and incubate for 20 min.

15. Wash in TBS for 5 min (*see* **Note 20**).
16. Rinse in tap water.
17. Counter stain with hematoxylin for 7–15 s or longer if darker color is desired.
18. Wash with tap water, remove excess liquid, and let dry.
19. Mount coverslip to the slide following directions and let dry overnight.
20. Sections can be visualized with a light microscope.
21. Islets can be assessed for insulitis severity with the following scoring system (Fig. 3):
    (a) 0 – no insulitis
    (b) 1 – peri-insulitis, accumulation around islet perimeter
    (c) 2 – <50 of islet infiltrated
    (d) 3 – >50 of islet infiltrated
22. Infiltrate composition can be assessed by analysis of CD4, CD8, B220, and insulin staining.

*3.3.4 Formalin Fixation and Paraffin Embedding for Immunohistochemistry*

The purpose of this step is to preserve the tissues with formalin so that they remain as biologically similar to when they were removed as possible.

1. Collect the pancreas into formalin in a glass bottle within 5 min of harvesting (or other tissues), arranging in a tissue cassette covered by formalin, ensuring that the tissue cassette has been prelabeled with a pencil.
2. Incubate overnight to allow for complete fixation of tissue at 4 °C. Do not exceed 24 h.
3. Decant formalin, rinse tissues in cold water for about 5 min, and place tissues in PBS.

4. Store at 4 °C until ready for processing.
5. Place the tissues, while still in the cassettes, into 70% ethanol for 20 min (*see* **Note 21**).
6. Transfer the tissues into 95% ethanol for 20 min. Repeat incubation in fresh 95% ethanol.
7. Transfer the tissues into 100% ethanol for 20 min. Repeat incubation in fresh 100% ethanol.
8. "Clear" the tissues by placing the tissues into a xylene bath for 20 min to remove the ethanol; ensure that tissue is placed with surface of interest facing down (e.g., for pancreas, vein facing down)
9. Open the cassettes and embed the tissues in molten paraffin for 30 min to replace the xylene in the tissues. Fill the mold with paraffin to make a "block" (*see* **Note 22**).
10. Allow paraffin to cool and harden (*see* **Note 23**).

#### 3.3.5 Sectioning of Paraffin-Embedded Tissue

The purpose of this step is to make cuts or slices of the block, which contains the tissue embedded in paraffin, and mount these sections on glass microscope slides.

1. Warm water bath to 35–40 °C.
2. Place the tissue block in a microtome.
3. Cut the tissue into slices between 4 and 5 μm in thickness.
4. Place the cut slices of tissue onto the surface of the water.
5. Place a glass microscope slide under the floating tissue slice and lift the slide up to catch the tissue section on the slide.
6. Place the slides in an oven at 65 °C for 10–20 min until the paraffin starts to melt and the tissue is mounted to the slide.
7. Store samples at room temperature until ready for staining.

#### 3.3.6 Staining of Paraffin-Embedded Tissue

1. Immerse the slides in xylene (mixed isomers) two times for 10 min each.
2. Immerse the slides in 100% ethanol two times for 10 min each.
3. Immerse the slides in 95% ethanol for 5 min.
4. Immerse the slides in 70% ethanol for 5 min.
5. Immerse the slides in 50% ethanol for 5 min.
6. Rinse the slides with deionized $H_2O$.
7. Rehydrate the slides with wash buffer for 10 min.
8. Circle the tissue section with a PAP pen to create a barrier. All incubations are performed in the moisture chamber to prevent slides from drying and to protect from light.

9. Block with 10% (or 5% rat + 5% goat) normal serum (same species as primary antibody) or BSA 1% in wash buffer for 20 min.
10. Rinse with wash buffer.
11. Block with avidin (kit) one to two drops per section for 15 min.
12. Rinse with wash buffer.
13. Block with biotin (kit) one to two drops per section for 15 min.
14. Incubate with wash buffer for 5 min.
15. Add primary biotinylated antibody at the appropriate dilution (determined by serial dilution experiments). Antibodies are diluted in wash buffer plus 1% normal serum (same species as primary or BSA). Incubate for 90 min at room temperature or overnight at 4 °C.
16. Wash three times, 5 min each in wash buffer.
17. Incubate sections with streptavidin-alkaline phosphatase for 30 min. SAP (kit) is diluted 1/500 in wash buffer plus 1% normal serum or BSA.
18. Wash three times, 5 min each in wash buffer.
19. Vector Red staining (following manufacturer's instruction):
    (a) To 5 mL of the 10 mM Tris–HCl, pH 8.2–8.5:
        - Add a drop of levamisole solution, and mix well.
        - Add two drops of Reagent 1, and mix well.
        - Add two drops of Reagent 2, and mix well.
        - Add two drops of Reagent 3, and mix well.

    (b) Apply mixture to the sample and incubate for 20 min.
20. Wash in TBS for 5 min.
21. Rinse in tap water.
22. Counter stain with hematoxylin for 7–15 s or longer if darker color is desired.
23. Wash with tap water, remove excess liquid, and let dry.
24. Mount coverslip to the slide following directions and let dry overnight.
25. Sections can be visualized with a light microscope.
26. Islets can be assessed for insulitis severity with the following scoring system (Fig. 3):
    (a) 0 – no insulitis
    (b) 1 – peri-insulitis, accumulation around perimeter of the islet

(c) 2 – <50 of islet infiltrated

(d) 3 – >50 of islet infiltrated

27. Infiltrate composition can be assessed by analysis of CD4, CD8, B220, and insulin staining.

## 4 Notes

1. Where possible, ensure you include a **positive** control, i.e., a cell clone or line known to respond to the specific MHC class I peptide.
2. We store stocks in 1 mL or 5 μL aliquots, stored at −20 °C. Aliquots should be defrosted **only once**, and a fresh aliquot is used for every experiment even in the next day.
3. The CFDA SE dye reacts with amine groups and should not be used with amine-containing buffers or lysine-coated slides.
4. Can be stored at room temperature.
5. Store at 4 °C for up to 1 week.
6. Initial tetramer titrations should be done to ensure the correct concentration to use (normal range: 0.25 μg, 0.5 μg, 1 μg, 1.5 μg, 2 μg).
7. Antibody panels should be devised to ensure **no spectral overlap** of any antibody with the tetramer.
8. Tetramer-stained samples can be fixed if immediate analysis on the flow cytometer is not possible, but ensure all antibodies/tetramers are compatible with fixation.
9. Include single-antibody stained samples for each fluorochrome by staining either beads or cells with only one antibody or viability dye.
10. As a control for the tetramer-BV421, use any BV421-labeled antibody.
11. As a control for the live/dead marker, use cells that have been treated with 100 μL permeabilization buffer mixed 1:1 with non-treated cells prior to addition of live/dead marker.
12. Different cell types may be more sensitive to dye uptake. Titrating the concentration and checking cell viability is recommended. Concentrations listed have been suitable for our work with murine CD8 T cells. Dye concentration should also be titrated to determine stability for the course of the assay. In vivo assays typically require a higher concentration due to cell metabolism in vivo. Long-term assays may require higher concentrations to account for loss and more generations of divi-

sion. These concentrations should also be validated for any adverse effect on cell viability.

13. Cells should be in dividing phase and healthy. Splitting the day before the experiment allows room for expansion and removes dead cells. A 1:10 split from a 75–80% confluent T75 culture flask is sufficient to obtain $1 \times 10^6$ cells.
14. Place the mouse in a restrainer. To dilate the vein for easier injection, rub the tail between your fingers to warm the tail. Gently squeeze the base of the tail to enlarge the vein and then release.
15. When the mouse is scruffed, it will usually urinate spontaneously. If not, press on the bladder gently. For males it is important that semen is not tested. We use blood glucose concentrations >13.9 mmol/l as indication of diabetes onset.
16. T1D onset typically occurs 8–12 weeks post-transfer with splenocytes from prediabetic donors and within 4 weeks after splenocyte transfer from diabetic donors.
17. The use of normal serum or BSA before the application of the primary antibody also eliminates Fc receptor binding of both the primary and secondary antibodies.
18. $H_2O_2$ suppresses endogenous peroxidase activity and therefore reduces background staining. Incubating pancreatic sections with peroxide after the primary incubation avoids this problem. Peroxide can be diluted in TBS or water.
19. Overnight incubation allows antibodies of lower titer or affinity to be used by simply allowing more time for the antibodies to bind.
20. The use of TBS to give a cleaner background than PBS. The use of 0.01% Triton X-100 (TX100) in the TBS helps to reduce surface tension, allowing reagents to cover the whole tissue section.
21. The incubations in ethanol and xylene can be done with the tissues still in the cassettes. Many automated machines will do these steps for you overnight.
22. Be careful with molten paraffin; it has a melting point of around 55–65 °C.
23. Paraffin blocks can be stored at room temperature in addition to the antibody-stained slides that have been cut from the blocks.

## References


1. Thayer TC, Wilson SB, Mathews CE (2010) Use of nonobese diabetic mice to understand human type 1 diabetes. Endocrinol Metab Clin N Am 39(3):541–561. https://doi.org/10.1016/j.ecl.2010.05.001
2. Chaparro RJ, Dilorenzo TP (2010) An update on the use of NOD mice to study autoimmune (Type 1) diabetes. Expert Rev Clin Immunol 6 (6):939–955. https://doi.org/10.1586/eci.10.68
3. Pearson JA, Wong FS, Wen L (2016) The importance of the Non Obese Diabetic (NOD) mouse model in autoimmune diabetes. J Autoimmun 66:76–88. https://doi.org/10.1016/j.jaut.2015.08.019
4. Lehuen A, Diana J, Zaccone P, Cooke A (2010) Immune cell crosstalk in type 1 diabetes. Nat Rev Immunol 10(7):501–513. https://doi.org/10.1038/nri2787
5. Raphael I, Nalawade S, Eagar TN, Forsthuber TG (2015) T cell subsets and their signature cytokines in autoimmune and inflammatory diseases. Cytokine 74(1):5–17. https://doi.org/10.1016/j.cyto.2014.09.011
6. Pearson JA, Wong FS (2016) Identification of islet antigen-specific CD8 T cells using MHCI-peptide tetramer reagents in the non obese diabetic (NOD) mouse model of type 1 diabetes. Methods Mol Biol 1433:119–125. https://doi.org/10.1007/7651_2015_295
7. Thayer TC, Wong FS (2016) Tracking immunological responses of islet antigen-specific T cells in the nonobese diabetic (NOD) mouse model of type 1 diabetes. Methods Mol Biol 1433:127–134. https://doi.org/10.1007/7651_2015_293
8. Höppner M, Luhm J, Schlenke P, Koritke P, Frohn C (2002) A flow-cytometry based cytotoxicity assay using stained effector cells in combination with native target cells. J Immunol Methods 267(2):157–163
9. Wong FS, Siew LK, Scott G, Thomas IJ, Chapman S, Viret C, Wen L (2009) Activation of insulin-reactive CD8 T-cells for development of autoimmune diabetes. Diabetes 58 (5):1156–1164. https://doi.org/10.2337/db08-0800

# Chapter 19

# Assessment of Insulin Tolerance Ex Vivo


## Irene Cózar-Castellano and Germán Perdomo



## Abstract

Insulin is a hormone produced and secreted by the β-cells of the pancreatic islets of Langerhans in response to increased blood glucose levels after a meal. The hormone binds to its receptor located on the plasma membrane triggering an intracellular signaling cascade. This signaling pathway is responsible for the pleiotropic actions of insulin on different tissues, such as regulation of glucose and lipid metabolism, proliferation, and differentiation. Although considerable efforts have been made to understand the molecular mechanism linking the action of the hormone to biological processes, our knowledge is incomplete. Of note, under certain conditions, physiological circulating levels of the hormone are insufficient to properly regulate these processes, a term coined as insulin resistance. The ex vivo analysis of insulin action provides valuable information to decipher intracellular signaling events downstream of the insulin receptor under physiological and pathophysiological conditions. In this chapter, we focus on the analysis of intracellular insulin action ex vivo.



**Key words** Insulin sensitivity, Insulin resistance, Insulin signaling, Insulin receptor, Liver, Glucose metabolism, Lipid metabolism


## 1 Introduction

Insulin is a hormone produced by pancreatic β-cells as preproinsulin, which is processed to proinsulin. Then, proinsulin is cleaved to insulin and C-peptide and stored in secretary granules. In response to elevated blood glucose levels, insulin is secreted to circulation [1]. Through the portal vein, circulating insulin reaches the liver binding to the insulin receptor located in hepatocytes [2]. The pleiotropic actions of insulin through its receptor regulate hepatic glucose and lipid metabolism, including metabolic processes such as glucose uptake, glycogen synthesis, gluconeogenesis, de novo fatty acid synthesis, and VLDL assemblage and secretion [3–5].

Upon the extracellular insulin binding, the kinase domains of the insulin receptor (IR) are activated by autophosphorylation on tyrosine residues, resulting in tyrosine phosphorylation of insulin receptor substrate (IRS) proteins. Among the six IRS identified proteins, IRS1 and IRS2 are important in the integration of insulin


Aileen J. F. King (ed.), *Animal Models of Diabetes: Methods and Protocols*, Methods in Molecular Biology, vol. 2128,
https://doi.org/10.1007/978-1-0716-0385-7_19,

signaling network of the hepatocytes. Of note, IRS2 depletion in hepatocytes results in dysregulation of gluconeogenic gene expression and hepatic glucose production [6]. Phosphorylated IRS proteins allow the association of other proteins containing src homology region 2 (SH2) domains, such as the p85 regulatory subunit of phosphatidylinositol 3-kinase (PI3K). Activation of PI3K catalytic subunit p110 promotes phosphorylation of phosphoinositide at the 3'position of the inositol ring and the production of phosphatidylinositol-3,4,5-triphosphate (PIP3), a lipid second messenger located on the plasma membrane [7]. PIP3 allows the recruitment and activation of proteins containing the pleckstrin homology (PH) domain, such as the 3-phosphoinositide-dependent protein kinase 1 (PDK1) and the serine/threonine protein kinase AKT [AKT, also known as protein kinase B (PKB)] [8]. The relocation of AKT/PKB from the cytoplasm to the plasma membrane goes along with the phosphorylation on Thr308 by PDK1. However, full activation of AKT/PKB requires phosphorylation on Ser473, which is mediated by the rictor-mammalian target of rapamaycin (mTOR) complex (mTORC2) [9, 10]. Among the three AKT/PKB known isoforms (AKT1, AKT2, AKT3), AKT2 is a major mediator of the metabolic effects of insulin in the liver [11], whereas AKT1 is more related to cellular growth and angiogenesis [12]. It has been proposed that AKT/PKB is an essential mediator or critical node of the insulin signaling pathway [13]. The downstream targets of AKT are the Rab-GTPase-activating protein (AS160), involved in the translocation of the glucose transporter GLUT4 to the plasma membrane; tuberin (tuberous sclerosis complex-2, TSC2), which regulates protein synthesis through mTOR complexes; the forkhead (FOX) family of transcriptional factors (FOXO1 and FOXA2), which regulate glucose and lipid metabolism; and the glycogen synthase kinase-3 (GSK3) which regulates glycogen synthesis [13].

This chapter focuses on the assessment of the insulin signaling pathway in liver tissues ex vivo. The methodology presented can also be applied to other tissues such as skeletal muscle and adipose tissue.

## 2 Materials

1. Mice (we use C57Bl/6).
2. Insulin solution, rapid action (we use Humulin®R [100 U/mL]).
3. Blood glucose monitor and test strips for glucose measurement (we use Ascensia Contour glucometer, Bayer).
4. Sterile saline buffer.

5. Animal scale.
6. 1 mL syringe and 30 G × 1/2 in needles.
7. Timer.
8. Clean mouse cages.
9. Surgical material: scissors, forceps, etc.
10. Refrigerated centrifuge.
11. Tissue homogenizer (we use OMNI TH).
12. Dounce homogenizer.
13. Electrophoresis and transfer chamber systems.
14. Electrophoresis power supply (we use PowerPac™).
15. Cell lysis buffer 10×.
16. Protease and Phosphatase Inhibitor Cocktail.
17. Laemmli sample buffer (LSB) 4×: 200 mM Tris–HCl (pH 6.8), 40% (v/v) glycerol, 4% (v/v) SDS, 20% (v/v) β-mercaptoethanol, and 0.4% (w/v) bromophenol blue.
18. Pierce BCA protein assay kit.
19. Plus Membrane Protein Extraction Kit.
20. NE-PER Nuclear and Cytoplasmic Extraction Reagents Kit.
21. Criterion Tris–HCl protein gel.
22. Immune-Blot PVDF membranes.
23. Blocking buffer (1× PBS, 0.1% Tween-20 with 5% w/v non-fat dry milk).
24. Stripping buffer (2% SDS, 62.5 mM Tris–HCl, pH 6.8, and 100 mM β-mercaptoethanol).
25. Running and transfer buffers for protein electrophoresis.
26. Clarity Western ECL Substrate kit.
27. Imaging software (we use ImageJ).

## 3 Methods

### 3.1 Insulin Injection and Organ Retrieval

1. The day before, fast mice (between 5:00 and 6:00 p.m.) for a period of 14–16 h (overnight).
2. The next day (between 8:00 and 9:00 a.m.), place mice in new cages with clean bedding and free access to water (*see* **Note 1**). A period of acclimatization is recommended between placing mice in the new cages and the beginning of the experiment (e.g., 15 min).

3. Prepare an experimental record table to register body weight and blood glucose levels at basal ($t = 0$) and at the end of the experiment ($t = 10$ min).
4. Weigh mice and measure baseline glucose levels ($t = 0$) with the glucometer. Record both values in the table.
5. Using body weight data, calculate the volume of insulin that is necessary to inject 0.75 U/kg of body weight for each mouse (*see* **Note 2**). Record the value in the table.
6. Split mice into two groups: one group will receive an intraperitoneal injection of saline buffer (control group), and the other group will receive an intraperitoneal insulin injection (*see* **Note 3**).
7. Dilute the stock solution of insulin (typically for a mouse of 25 g of body weight, insulin will be diluted to 1:1000) and prepare a sterile solution of insulin in the saline buffer (0.9% NaCl).
8. Preload 1 mL syringes with insulin solution (usually the volume to inject per mouse is ~100–200 μL) and remove air from syringes.
9. Pick the mouse up with gloves and restrain it with the other hand. Proceed with the intraperitoneal insulin (or saline) injection with a 10 min interval between each animal.
10. Start the timer when the first mouse is injected. It is important to maintain the same time course from one injection to the other (*see* **Note 4**).
11. At the end of the experiment ($t = 10$ min), measure and record blood glucose levels. It is advisable that no more than eight to ten mice should be assessed during the same experimental setting. Ideally, two to three trained people should participate during the experiment, each one of them performing specific roles in the experiment (i.e., one person injecting mice and the other dissecting tissues).
12. Sacrifice mice in the order established at the beginning of the experiment by cervical dislocation or $CO_2$ inhalation.
13. Dissect the liver as quickly as possible to prevent degradation by proteases.
14. Rinse the liver with ice-cold PBS to remove excess of blood.
15. Dry tissues in paper to remove excess PBS.
16. Snap freeze tissues in liquid nitrogen to avoid protein degradation. Store tissues at −80 °C until analysis or keep them on ice for immediate homogenization (*see* **Note 5**).

### 3.2 Extracting Cytosolic, Membrane, or Nuclear Protein Fractions to Analyze the Intracellular Insulin Signaling Pathway

Insulin triggers a cascade of phosphorylation events on proteins located in the plasma membrane (IR) and the cytosol (PI3K, PDK1, IRS1/2, AKT1/2, GSK3, mTOR, etc.) or promotes the translocation of transcriptional factors to the nucleus (FOXO1). To dissect the effects of insulin on these proteins, it is necessary to separate subcellular cell fractions (i.e., plasma membrane, cytosol, and nucleus). Below, we describe how to proceed to obtain protein lysates from liver tissues.

#### 3.2.1 Using Cytosolic Protein Lysates to Analyze Phosphorylated Proteins in the Cytoplasm (e.g., AKT/PKB) in Response to Insulin

1. Using a tissue homogenizer, homogenize 20–40 mg of frozen liver tissue in ~200 μL of ice-cold lysis buffer (1×) supplemented with protease and phosphatase cocktail inhibitors (*see* **Notes 6–8**).
2. Sonicate lysates on ice and centrifuge them at 18,500 × *g* for 10 min at 4 °C (*see* **Note 9**).
3. Keep the supernatant fraction and discard insoluble materials.
4. Quantify protein content in supernatants with BCA protein assay kit.
5. Afterward, mix an aliquot of tissue lysate with LSB 1× and boil for 5 min at 95 °C. Store the lysates at −80 °C until analysis.

#### 3.2.2 Using Membrane Protein Lysates to Analyze Phosphorylated Proteins in Membranes (e.g., IR) in Response to Insulin

*We currently use the Mem-PER Plus Membrane Protein Extraction Kit following manufacturer's instructions (see* ***Note 10****). The main steps are outlined as follows*:

1. Wash 30 mg of liver tissue with 4 mL of Cell Wash Solution in a 10 mL glass tube.
2. Discard the wash buffer and transfer the tissue to a 2 mL glass tissue grinder. Homogenize the tissue in 1 mL of permeabilization buffer, containing protease and phosphatase inhibitors, by six to ten strokes.
3. Add another 1 mL of permeabilization buffer, transfer the lysate to a new tube, and incubate 10 min at 4 °C with constant mixing.
4. Centrifuge lysates at 18,500 × *g* for 15 min at 4 °C to separate cytosolic (supernatant) and membrane (pellet) fractions.
5. At this point, discard supernatants containing cytosolic proteins (or store at −80 °C for further downstream analyses if you are interested in cytoplasmic fraction).
6. Resuspend the pellet in 1 mL of solubilization buffer and incubate it 40 min at 4 °C with constant mixing.
7. Centrifuge samples at 18,500 × *g* for 15 min at 4 °C.
8. Remove and save the supernatant fraction containing solubilized membrane and membrane-associated proteins. Discard the pellet fractions.

9. Quantify protein content with the BCA protein assay kit. Afterward, mix an aliquot of tissue lysate with LSB 1× and heat for 5 min at 30 °C (*see* **Note 11**). Membrane fractions can be used immediately or stored at −80 °C until further analyses.

#### *3.2.3 Using Nuclear Fractions to Analyze Translocation of Nuclear Factors to the Nucleus (e. g., FOXO1) in Response to Insulin*

*We currently use the N-PER Plus Membrane Protein Extraction Kit following manufacturer's instructions (see* ***Note 12****). The main steps are outlined as follows:*

1. Wash 20 mg of liver tissue with 4 mL of PBS in a 10 mL glass tube.
2. Discard the PBS and transfer the tissue to a 2 mL glass tissue grinder. Homogenize tissue in 200 μL of ice-cold CER I buffer containing protease and phosphatase inhibitors using a Dounce homogenizer by six to ten strokes.
3. Transfer samples to a new Eppendorf tube, vortex vigorously for 15 s, and incubate on ice for 10 min.
4. Add 11 μL of ice-cold CER II buffer, vortex vigorously for 5 s, and incubate on ice for 1 min.
5. Vortex again for 5 s and centrifuge samples at 18,500 × *g* for 5 min at 4 °C.
6. Transfer supernatants (cytoplasmic extracts) to a prechilled tube. Mix the insoluble fraction (pellet), which contains nuclei, with 100 μL of ice-cold NER I buffer, and incubate for a total of 40 min on ice.
7. Centrifuge samples at 18,500 × *g* for 10 min at 4 °C and transfer the supernatant (nuclear extract) to a new tube.
8. Quantify protein content with the BCA protein assay kit.
9. Mix an aliquot of tissue lysate with LSB 1× and boil for 5 min at 95 °C. Nuclear fractions can be used immediately or stored at −80 °C until further analyses.

### *3.3 Analysis of Proteins by Western Blotting*

1. Use 10% SDS-PAGE precast polyacrylamide gel (Criterion Tris–HCl protein gel) to separate protein samples from cytosolic, membrane, or nuclear fractions.
2. Electro-transfer onto polyvinylidene difluoride filters (PVDF) for immunoblotting by conventional means.
3. Electro-transferred PVDF membranes are blocked for 1 h at room temperature using the blocking buffer.
4. Probe with a specific antibody of interest (against the phosphorylated protein).
5. Strip the membrane using the stripping buffer for 30 min at 50 °C.

6. Wash and reprobe with a second antibody to detect the total protein.
7. Detect signal using chemiluminescence (Clarity Western ECL Substrate kit) (*see* **Notes 13–15**).
8. Use the ImageJ software to quantify the intensity of the bands.
9. Plot the data as the ratio of phosphorylated vs. total protein for cytoplasmic or membrane fractions and the ratio of nuclear vs. cytoplasmic proteins translocated to the nucleus (*see* **Notes 16** and **17**).

## 4 Notes

1. For animal fasting, place mice into a new cage with clean bedding (without feces and food pellets) and free access to water. A shorter period of fasting (5–6 h, e.g., 9:00–14:00) is more physiological and reduces metabolic stress of mice. However, to obtain more consistent baseline plasma insulin levels, overnight fasting is recommended. Whenever possible, randomize mice and exclude mice with extreme body weight, avoiding grouping bias during the experimental setting.
2. The dose of insulin may vary from 0.5 to 7.5 U/kg of body weight depending on genetic background of mice, body weight, age, nutritional status, and diabetic status. Insulin-resistant mice will require higher amounts of insulin.
3. Insert the needle with an angle of ~25° to the abdominal wall in the lower right or ~45° in the left quadrant of the abdomen trying to avoid hitting the bladder, liver, or other internal organs. Likewise, avoid subcutaneous injection. Of note, in obese mice, be aware to avoid fat pad areas. Insulin solution should be injected at body temperature to further reduce mice discomfort.
4. It is important to record blood glucose levels ($t = 0$ and $t = 10$ min) in order to check whether intraperitoneal insulin injections cause a fall in blood glucose levels. In control mice (saline injections), blood glucose levels must remain unchanged. If insulin injections fail to decrease blood glucose levels, place food in animal cages and repeat the experiment 3 days later.
5. We recommend dissecting the whole liver and wrapping the tissue with aluminum foil paper and immersing it in liquid nitrogen to snap freeze. Each sample must be correctly labeled (e.g., +/− insulin treatment). Alternatively, small tissue pieces can be placed in round-bottom microcentrifuge tubes (2 mL) and immersed in liquid nitrogen.

6. The literature describes several lysis buffers for obtaining protein lysates. You may use them, but the key point is to avoid protein degradation while maintaining protein phosphorylation inhibiting phosphatases in protein lysates. Usually, phenylmethylsulfonyl fluoride (PMSF) is added to the protease cocktail. Stock solutions of PMSF (100 mM) are prepared in 100% ethanol and stored at −20 °C. Due to the short half-life of PMSF, fresh solution (1 mM) must be added to the protease inhibitor cocktail. Likewise, stock solutions (100 mM, 100×) of orthovanadate, an inhibitor of alkaline and acid phosphatases and tyrosine phosphatases, are prepared in water (pH ~ 10) and stored in aliquots at −20 °C. Do not reuse sodium orthovanadate once defrosted.
7. Volume of the lysis buffer is determined in relation to tissue weight to avoid loss of proteins and large volumes of samples to be loaded onto gels. Optimal concentration is 1–5 mg/mL.
8. Homogenization of liver tissues with the tissue homogenizer must be performed in short periods of time (less than a minute) to avoid sample heating and denaturation of proteins in the lysate. Repeat the procedure for three to five times (or until complete homogenization of the tissue). During and between each cycle, samples must be kept on ice.
9. Sonication of the samples is important for disruption of inclusion bodies and membranes and fragmentation of genomic DNA. If nucleic acids are not correctly fragmented (seen as a transparent jelly in lysates), the density of the sample would affect the quality of protein electrophoresis. Specific conditions of the duration, wattage, and frequency used for ultrasonic homogenization depend on the characteristics of the equipment. Depending on the equipment (ultrasonic bath or probe sonicators), the time of sonication varies between 5 and 20 min and 15 and 45 s, respectively. In the case of probe sonicators, the procedure should be performed using short periods of time (15 s, two to three cycles). During sonication and between each cycle, samples must be kept on ice to avoid protein denaturation.
10. The Mem-PER Plus Membrane Protein Extraction Kit allows the isolation of membrane proteins and cytosolic proteins. Cross-contamination of cytosolic proteins in the membrane fraction is less than 10%. Therefore, this kit can be used for isolation of both membrane and cytosolic fractions.
11. To avoid protein aggregation and precipitation, membrane fractions are mixed with LSB 1× and heated for 5 min at 30 °C.
12. The N-PER Plus Membrane Protein Extraction Kit allows the isolation of nuclear from cytoplasmic fractions from different tissues. Cytoplasmic content is released by the addition of two

buffers, and after centrifugation, nuclei are isolated. Then, nuclear proteins are extracted by a third buffer. Cross-contamination of cytosolic proteins in the nuclear fraction is less than 10%. Therefore, this kit can be used for isolation of both nuclear and cytosolic fractions.

13. Western blot detection of target proteins depends on the specificity of antibodies. Use antibody dilution as recommended by the manufacturer. If the signal of the protein is too low, we recommend increasing the amount of protein loaded into the gel, rather than to increase the amount of primary antibody for its detection.
14. As a loading control in the Western blot, we recommend using actin or tubulin for cytosolic proteins and α1 subunit of the Na/K-ATPase for membrane proteins. To confirm similar loading in your samples, check the ratio of total protein vs. actin or total protein vs. Na/K-ATPase of your target protein. If you do not perform this, you may misinterpret data regarding phosphorylation of your target protein.
15. To analyze phosphorylation of your target protein, we recommend using the phospho-antibody first, and after stripping, reprobe the membrane with the total antibody.
16. To quantify the total and phosphorylated protein levels, use ImageJ software. The ratio of phosphorylated vs. non-phosphorylated protein levels in arbitrary units is plotted. Samples treated with the saline buffer are normalized to 100% or 1, whereas samples treated with insulin are referred to saline samples.
17. To quantify proteins that are translocated to nuclei in response to insulin, data are plotted as the ratio of nuclear vs. cytosolic target protein. Likewise, samples treated with the saline buffer are normalized to 100% or 1, whereas samples treated with insulin are referred to saline samples.

## Acknowledgments

This work was supported by grants from the Ministerio de Economía, Industria y Competitividad: SAF2014-58702-C2-1-R and SAF2016-77871-C2-1-R to IC and SAF2014-58702-C2-2-R and SAF2016-77871-C2-2-R to GP. The EFSD European Research Programme on New Targets for Type 2 Diabetes is supported by an educational research grant from MSD to ICC and GP.

## References


1. Fu Z, Gilbert ER, Liu D (2013) Regulation of insulin synthesis and secretion and pancreatic Beta-cell dysfunction in diabetes. Curr Diabetes Rev 9(1):25–53
2. Bergman RN (2000) Non-esterified fatty acids and the liver: why is insulin secreted into the portal vein? Diabetologia 43(7):946–952
3. Otero YF, Stafford JM, McGuinness OP (2014) Pathway-selective insulin resistance and metabolic disease: the importance of nutrient flux. J Biol Chem 289 (30):20462–20469
4. Kamagate A, Qu S, Perdomo G, Su D, Kim DH, Slusher S, Meseck M, Dong HH (2008) FoxO1 mediates insulin-dependent regulation of hepatic VLDL production in mice. J Clin Invest 118(6):2347–2364
5. Horton JD, Goldstein JL, Brown MS (2002) SREBPs: activators of the complete program of cholesterol and fatty acid synthesis in the liver. J Clin Invest 109(9):1125–1131
6. Valverde AM, Gonzalez-Rodriguez A (2011) IRS2 and PTP1B: two opposite modulators of hepatic insulin signalling. Arch Physiol Biochem 117(3):105–115
7. Shepherd PR, Withers DJ, Siddle K (1998) Phosphoinositide 3-kinase: the key switch mechanism in insulin signalling. Biochem J 333(Pt 3):471–490
8. Li H, Marshall AJ (2015) Phosphatidylinositol (3,4) bisphosphate-specific phosphatases and effector proteins: a distinct branch of PI3K signaling. Cell Signal 27(9):1789–1798
9. Sarbassov DD, Guertin DA, Ali SM, Sabatini DM (2005) Phosphorylation and regulation of Akt/PKB by the rictor-mTOR complex. Science 307(5712):1098–1101
10. Toker A, Marmiroli S (2014) Signaling specificity in the Akt pathway in biology and disease. Adv Biol Regul 55:28–38
11. Whiteman EL, Cho H, Birnbaum MJ (2002) Role of Akt/protein kinase B in metabolism. Trends Endocrinol Metab 13(10):444–451
12. Gonzalez E, McGraw TE (2009) The Akt kinases: isoform specificity in metabolism and cancer. Cell Cycle 8(16):2502–2508
13. Taniguchi CM, Emanuelli B, Kahn CR (2006) Critical nodes in signalling pathways: insights into insulin action. Nat Rev Mol Cell Biol 7 (2):85–96

# Chapter 20

# Using Pancreas Tissue Slices for the Study of Islet Physiology

**Julia K. Panzer, Christian M. Cohrs, and Stephan Speier**

## Abstract

Studies on islet of Langerhans physiology are crucial to understand the role of the endocrine pancreas in diabetes pathogenesis and the development of new therapeutic approaches. However, so far most research addressing islet of Langerhans biology relies on islets obtained via enzymatic isolation from the pancreas, which is known to cause mechanical and chemical stress, thus having a major impact on islet cell physiology. To circumvent the limitations of islet isolation, we have pioneered a platform for the study of islet physiology using the pancreas tissue slice technique. This approach allows to explore the detailed three-dimensional morphology of intact pancreatic tissue at a cellular level and to investigate islet cell function under near-physiological conditions. The described procedure is less damaging and faster than alternative approaches and particularly advantageous for studying infiltrated and structurally damaged islets. Furthermore, pancreas tissue slices have proven valuable for acute studies of endocrine as well as exocrine cell physiology in their conserved natural environment. We here provide a detailed protocol for the preparation of mouse pancreas tissue slices, the assessment of slice viability, and the study of pancreas cell physiology by hormone secretion and immunofluorescence staining.

**Key words** Tissue slices, In situ islet cell function, Insulin secretion, Immunofluorescence staining, Live/dead staining

## 1 Introduction

Functional studies on islet physiology have been successfully performed on islets obtained by isolation for many years. However, the isolation process involves enzymatic and mechanical stress, causing considerable changes to cell physiology. This involves, e.g., upregulation of stress genes and cytokines, production of reactive oxygen species, induction of inflammatory responses, and apoptosis [1–6]. In addition, isolated islets are disconnected from their endogenous environment, which precludes the investigation of processes that involve cells or factors in the surrounding tissue. Thus, the isolation process has major influence on islet physiology that may not reflect the in vivo response.

Aileen J. F. King (ed.), *Animal Models of Diabetes: Methods and Protocols*, Methods in Molecular Biology, vol. 2128, https://doi.org/10.1007/978-1-0716-0385-7_20, 

To circumvent the abovementioned limitations, Speier and Rupnik have developed a platform for the study of islet physiology using ex vivo tissue slice preparations from the rodent pancreas [7]. The technique is based on stabilizing the viable pancreatic tissue by common bile duct injection of and embedding in agarose to enable the cutting of 150 μm-thick slices by vibratome sectioning. This allows the study of intact islets preserved in their natural environment without enzymatic disturbance and minimal mechanical stress. Additionally, the slicing procedure requires no other chemical substances than agarose, which does not affect pancreatic cell function. Finally, the preparation time is substantially shorter compared to isolation of acini or islets, and slices do not require a recovery culture period like isolated islets [8, 9].

The obtained slices with their preserved anatomy allow in-depth analysis of the three-dimensional tissue architecture as well as assessment of cellular function of the different compartments within the pancreas under near-physiological conditions. Slices are therefore superior to studies on dispersed cells or isolated units (islets, acini) that lack connectivity resulting in altered function, gene expression, and identity [2, 3, 10]. Moreover, pancreas tissue slices enable the investigation of structurally altered islets, which isolation by enzymatic digestion would hardly allow. These benefits have already been shown to be of major value for islet research in rodent animal models [7, 11].

Since the establishment of the technique, several groups have shown the diverse application of living pancreas tissue slices in regard to their potential native in vivo state. This includes the simultaneous observation of endocrine and exocrine cells, but also the influence of neuronal, vascular endothelial, and adipose tissue. These advantageous features of the tissue slice technique enable various functional approaches like confocal microscopy of living pancreas tissue slices using transgenic mice and measurements of hormone release in perifusion experiments [12, 13]. Additionally, electrophysiological characterization of endocrine and exocrine cells can be accomplished in deeper cell layers of the tissue by using the patch clamp technology [7, 14–18]. Moreover, fixed tissue slices can be used to investigate islet morphology in three dimensions by immunostaining for various cell types including hormones, the islet vascular network, tissue-resident macrophages, and the pattern of sensory innervation [19, 20]. Furthermore, the successful establishment of an organotypic culture platform for rodent tissue slices facilitates mid- to long-term in situ studies of pancreas cell biology, enabling experimental manipulation of the cells and the evaluation of pharmaceutical drugs [12].

Notably, this technique has recently been optimized by our group for the preparation of human pancreas tissue slices from samples procured from patients undergoing partial pancreatectomy [13]. As a result, this platform now serves as a valuable platform for the study of human islet and exocrine physiology and enables to gather novel insights into the pathogenesis of diabetes.

## 2 Materials

### 2.1 Agarose Injection

1. Surgical scissors, forceps, and clamp.
2. 5-mL syringe.
3. 27-gauge needle.
4. Petri dish 35 mm, sterile; and petri dish 60 mm, sterile.
5. Low-melting-point agarose.
6. Microwave oven.
7. Extracellular solution (ECS): 125 mM NaCl, 2.5 mM KCl, 26 mM $NaHCO_3$, 1.25 mM $NaH_2PO_4$, 1 mM $MgCl_2$, 2 mM $CaCl_2$, 10 mM HEPES, 3 mM glucose, pH 7.4.

### 2.2 Slice Procedure

1. Surgical scissors and surgical forceps.
2. Curved forceps.
3. Scalpel.
4. Petri dish 35 mm, sterile; and petri dish 60 mm, sterile.
5. Multiwell plate for cell suspension culture, 24-well, sterile.
6. Superglue (we use 90–120 CPS).
7. Vibratome VT1200 S (we use Leica).
8. Extracellular solution (ECS): 125 mM NaCl, 2.5 mM KCl, 26 mM $NaHCO_3$, 1.25 mM $NaH_2PO_4$, 1 mM $MgCl_2$, 2 mM $CaCl_2$, 10 mM HEPES, 3 mM glucose, pH 7.4.
9. Dulbecco's phosphate-buffered saline.
10. Krebs-Ringer bicarbonate HEPES (KRBH) buffer: 137 mM NaCl, 5.36 mM KCl, 0.34 mM $Na_2HPO_4$, 0.81 mM $MgSO_4$, 4.17 mM $NaHCO_3$, 1.26 mM $CaCl_2$, 0.44 mM $KH_2PO_4$, 10 mM HEPES, 0.1% BSA, 3 mM glucose, pH 7.3.
11. Soybean trypsin inhibitor. Prepare a stock solution of 10 mg/mL in deionized water and store the aliquots at −20 °C. Use a final concentration of 0.1 mg/mL.

### 2.3 Viability Assessment

1. Hoechst 33342 is diluted in deionized water to a concentration of 10 mg/mL and stored at 4 °C.
2. Propidium iodide (PI) is diluted in PBS to a concentration of 0.5 mg/mL and stored at 4 °C for up to 12 months.

3. Fluorescein diacetate (FDA) is dissolved in acetone to a concentration of 5 mg/mL and stored at −20 °C for up to 3 months.
4. Dulbecco's phosphate-buffered saline (DPBS).

Prepare stock solutions and store them in aliquots. All reagents have to be stored light protected.

### 2.4 Slice Perifusion

1. Small hairbrush.
2. Silicone Grease Kit.
3. Krebs-Ringer bicarbonate HEPES buffer (137 mM NaCl, 5.36 mM KCl, 0.34 mM $Na_2HPO_4$, 0.81 mM $MgSO_4$, 4.17 mM $NaHCO_3$, 1.26 mM $CaCl_2$, 0.44 mM $KH_2PO_4$, 10 mM HEPES, 0.1% BSA, 3 mM glucose, pH 7.3) + aprotinin (25 KIU/mL).
4. Acid ethanol: 2% HCl [37%, 12 M] in absolute ethanol.
5. Perifusion system with automated tray handling.
6. Perifusion Two-Stop Tubing Sets.
7. Small-volume imaging chamber—70 μL Series 20 chamber.
8. Chamber platform—P-5 platform for Series 20 chambers, non-heater.
9. Polyethylene tubing, PE-160, 10 ft.
10. Silicone tubing.
11. Round cover glass, #1 thickness, 15 mm, 100 pack.

### 2.5 Staining

1. 24-well plate.
2. Dulbecco's phosphate-buffered saline.
3. GSDB 0.6%: For 24 mL of 0.6% in 2× GSDB stock solution: 8 mL of goat serum (30% (vol/vol) final conc.), 5.4 mL of 4 M NaCl (900 mM final conc.), 4 mL of 240 M sodium phosphate buffer (pH 7.4; 40 mM final conc.), 288 μL of 100% Triton X-100 (0.6% final conc.), and 6.46 mL of purified, autoclaved water. Store at −20 °C for 3 months. Freshly prepare the final 0.6% Triton X-100 in 1× GSDB solution for antibody staining by diluting the stock 1:1 with purified, autoclaved water.
4. Primary antibodies:
   (a) Insulin.
   (b) Glucagon.
5. Secondary antibodies:
   (a) Goat-anti-guinea pig 488.
   (b) Goat-anti-mouse 633.
   (c) DyLight 594-labeled *Lycopersicon esculentum* lectin.
   (d) DAPI.

## 3 Methods

### 3.1 Agarose Injection

1. For each pancreas, prepare 10 mL of low-melting-point agarose solution (1.25%, in ECS), heat it up in a microwave oven, and keep it in a water bath at 37 °C (*see* **Note 1**).
2. Euthanize the mouse by cervical dislocation and place it on its back to open the abdominal cavity using surgical scissors (*see* **Note 2**).
3. Place the mouse under a stereomicroscope and clamp off the common bile duct at the ampulla of Vater (*see* **Note 3**).
4. Use a 5-mL syringe and a 27-gauge needle to inject steadily 1.5–2.5 mL agarose solution at 37 °C into the distally clamped bile duct (*see* **Note 4**).
5. Wait 2–3 min to let the agarose solidify.
6. Extract the hardened pancreas using forceps and scissors and transfer it into a 60-mm petri dish filled with ECS (3 mM glucose, room temperature) (Fig. 1a) and place it under a stereomicroscope.
7. Remove connective, fibrotic, and adipose tissue using sterilized surgical instruments and cut tissue into smaller blocks of approximately 3 × 3 × mm size (*see* **Note 5**).
8. Remove the tissue pieces from the solution and blot-dry them by placing them on tissue paper.
9. Transfer four to five tissue pieces into a 35 mm petri dish and fill the dish with the remaining pre-heated agarose until it is fully submerged.
10. Allow the agarose to solidify by placing the dish shortly in the fridge (2–3 min).
11. Once the agarose is solid, cut small cubes containing tissue (Fig. 1b) by using a scalpel and remove the blocks from the dish.

### 3.2 Slicing Procedure

#### 3.2.1 Preparations

1. Prepare the vibratome by placing a blade and calibrate it using the VibroCheck (VC) according to the manual. Calibration is necessary to check the optimal position of the blade and minimize vertical vibration.
2. Prepare 12 mL of KRBH (3 mM) and add soybean trypsin inhibitor (0.1 mg/mL). Fill two 60-mm petri dishes with each 6 mL solution to collect slices in.

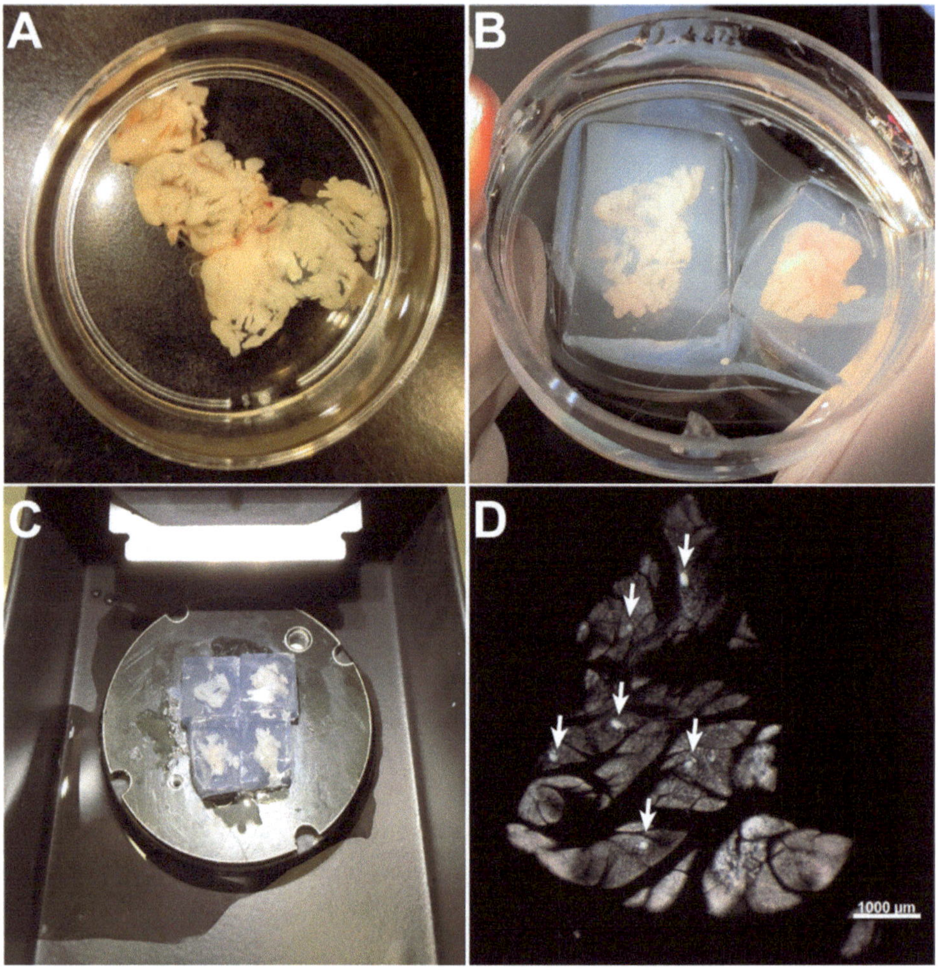


**Fig. 1** Mouse tissue slice preparation and vibratome slicing. (**a**) Extracted mouse pancreas after agarose injection. (**b**) Pancreas tissue pieces embedded in low-melting-point agarose. (**c**) Tissue slicing of mounted tissue blocks using a vibratome. (**d**) Mouse pancreas tissue slice with islets of Langerhans (white arrows) in reflected light. Scale bars, 1000 μm

*3.2.2 Vibratome Slicing*

To ensure tissue viability, do not exceed a cutting time of 3 h.

1. Apply superglue to the metal plate and mount the tissue blocks upside down (Fig. 1c). Wait a few seconds for the glue to solidify.
2. Mount the plate to the tray and fill it with ECS (3 mM glucose) until tissue blocks are entirely covered in solution.
3. Move the blade to an angle of 18° (upper mark) and set the amplitude to 1.0 mm and the speed to 0.1 mm/s. Use a step thickness of 150 μm.
4. Set the programmable sectioning window by moving the blade close to the edges of the tissue blocks.
5. Before slicing, make sure the blade is also covered in solution.

6. Start slicing by pressing the start button (*see* **Notes 6** and 7).
7. Collect the slices (Fig. 1d) by lifting them gently using a brush or curved forceps and transfer them into a 60-mm petri dish containing KRBH (3 mM glucose) and soybean trypsin inhibitor (0.1 mg/mL). Place a maximum number of 30 slices per dish and keep the slices at room temperature on a shaker until further use. Do not exceed a resting time of 4 h for functional assessments.

### 3.3 Assessment of Tissue Viability

#### 3.3.1 Preparations

Prepare staining solutions in a 24-well plate by mixing the stock solutions in the described amount of solution. Do not stain more than two slices in one well to ensure proper staining of the entire slice.

Prepare four wells with different staining solutions in advance:

1. PBS 1000 μL for washing.
2. Hoechst staining solution: 2 μL Hoechst stock solution + 998 μL KRBH 3 mM glucose.
3. PI staining solution: 100 μL PI stock + 900 μL PBS.
4. FDA staining solution: 10 μL FDA stock + 990 μL PBS.

Solutions should be protected from light and have to be used within 30 min after preparation.

#### 3.3.2 Staining Procedure

1. Start the staining by transferring the tissue slices from KRBH 3 mM glucose into the well filled with Hoechst staining solution and incubate for 15 min, shaking at RT and protecting it from light (*see* **Note 8**).
2. After staining, transfer slices into PBS and wash for 1 min, shaking at room temperature.
3. Transfer slices into a well filled with PI staining solution and add the FDA staining solution to the PI staining solution containing the slices.
4. Let the slices incubate for 2 min at RT, protected from light.
5. Transfer slices into a 60 mm petri dish with PBS for subsequent imaging using a laser scanning microscope.

### 3.4 Assessment of Hormone Secretion Using a Perifusion Machine

#### 3.4.1 Preparations

Prepare KRBH stock solutions prior to the experiments and store them without BSA and glucose for up to a month at 4 °C.

1. Mix necessary amount of buffers by adding glucose, BSA, and soybean trypsin inhibitor (0.1 mg/mL) and place them in a water bath at 37 °C to heat up.
2. Prepare closed perifusion chambers: Grease the upper and lower rim of the chamber and attach a coverslip to the lower part. Place the chamber into the metal platform and tighten the screws (Fig. 2a, c).

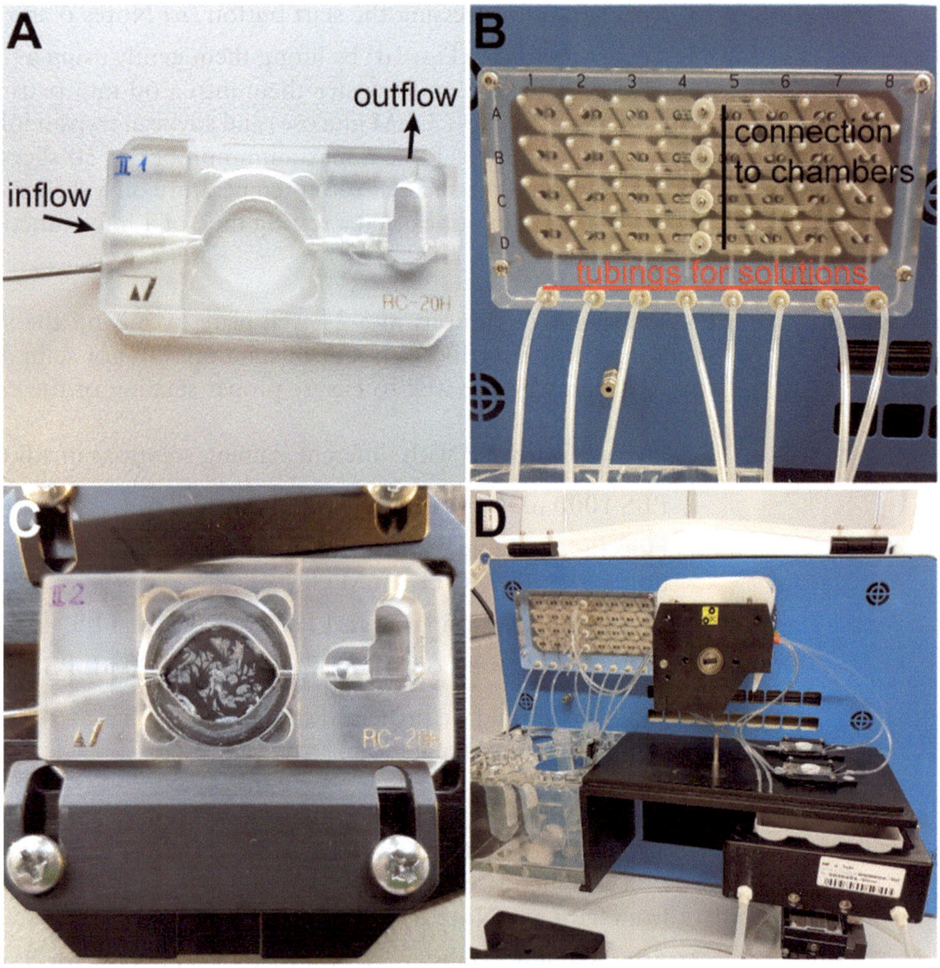


**Fig. 2** Experimental setup for the assessment of hormone secretion from mouse pancreas tissue slices. (**a**) Perifusion chamber with indicated in- and outflow (black arrowheads). (**b**) Manifold of the Biorep® perifusion machine with eight tubings for different solutions and four outflow connections. (**c**) Mouse tissue slices embedded in a perifusion chamber and mounted on a metal platform. (**d**) Closed perifusion chambers connected to a Biorep® perifusion machine

3. Prepare the perifusion machine by attaching all tubings necessary for running your experiment (*see* **Note 9**) (Fig. 2b).
4. Program your individual protocol. The protocol should include 1 h flushing in resting buffer prior to the actual protocol.
5. Put all necessary solutions in the tube holder and place the tubings into the solutions.
6. Open the perifusion protocol and prime the system.
7. After priming, proceed one step further and start the heater. Attach tubings and cooled water to the tray pump and start the pump. The machine is now ready to attach the chamber and to start the protocol.

#### 3.4.2 Perifusion Procedure

Prior to secretion, slices should be rested in KRBH (3 mM glucose + soybean trypsin inhibitor (0.1 mg/mL)) for 1 h on a shaker at room temperature. During resting time, slices can be preselected and trimmed under a stereomicroscope using forceps and a scalpel.

1. After 1 h resting time is over, place tissue slices into prepared perifusion chambers (*see* **Note 10**). Pipette 50 μL of the resting buffer in the middle of the chamber and place the slices one by one. Make sure slices are not folded or overlaying by rearranging them using a brush.
2. Close the chamber by placing another coverslip and a white o-ring.
3. Place the chamber and platform onto the stage of the perifusion machine (*see* **Note 11**) (Fig. 2c, d).
4. Attach the in and out tubings by using connectors (Fig. 2d) and place a 96-well collecting plate into the tray holder.
5. Start the protocol.
6. Monitor the chambers throughout the perifusion to see if bubbles or leaks appear (*see* **Note 12**).
7. Once the protocol is finished, switch off the heater and cooling pump and generate a report.
8. Disconnect all chambers and remove them from the platform.
9. Open the chambers and transfer slices into 1.5 mL tubes containing 500 μL acid ethanol for protein extraction. Keep the tubes at −20 °C for the assessment of insulin content afterward (at least overnight incubation).
10. Proceed with the cleaning protocol for the machine and follow the on-screen instructions.
11. Meanwhile, clean the chambers by applying a three-step process using trisodium phosphate, 0.1% HCl, and deionized water in this order to flush all in- and outlets. Use a cotton tip to remove remaining grease.
12. Once the cleaning protocol is finished, exit the protocol, park the tray holder, and shut down the machine.
13. Plates and lysates can be stored at −20 °C for later measurements.
14. Determine insulin concentrations of the perfusates and lysates by using a mouse ultra-sensitive insulin ELISA in accordance with the instructions provided with the kit.

### 3.5 Staining Protocol

1. Prepare a 24-well plate and fill every well with 500 μL PBS.
2. Transfer slices into the plate. Do not place more than two slices to each well.
3. Remove PBS and add 500 μL of 4% PFA.

4. Incubate the slices for 30 min at 4 °C, shaking.
5. Once incubation time is over, remove the fixative and add 1 mL PBS. Slices can be stored that way for up to 3 months at 4 °C.
6. Prepare the first antibody solution for insulin (1:500) and glucagon (1:2000) in GSDB 0.6% (400 μL/well).
7. Remove PBS and add the first antibody solution and incubate overnight at 4 °C, shaking.
8. The next day, remove the staining solution with a pipette and wash slices three times for 10 min with PBS containing 0.3% Triton X on a shaker at room temperature.
9. Meanwhile, prepare the second antibody solution (400 μL/well) and protect it from light:

   Alexa Fluor 488 goat-anti-guinea pig (1:200) → insulin

   Alexa Fluor 633 goat-anti-mouse (1:200) → glucagon

   DyLight 594-labeled *Lycopersicon esculentum* (tomato) lectin (1:100)

   DAPI (1:2000) in GSDB 0.6%
10. Once washing steps are finished, remove PBS and add 400 μL second antibody solution and incubate overnight at 4 °C, shaking. Cover the plate to protect it from light.
11. Remove the second antibody and wash three times for 10 min with PBS.
12. Keep slices protected from light in PBS until imaging at 4 °C.

## 4 Notes

1. Heat the agarose 30 min prior to the procedure and place it in a water bath to avoid bubbles in the solution and ensure body temperature.
2. Wet the abdominal fur with 80% ethanol prior to opening the abdomen, in order to avoid hair contaminations during the injection procedure.
3. Make sure that the clamp is positioned correctly to avoid inflation of the small intestine or the liver.
4. Try to be as quick as possible during cannulation or agarose infusion, as agarose solidifies very fast.
5. Difficulties during the slicing procedure can be reduced by proper processing. Try to remove all structures that might disturb the process and cut tissue pieces smaller if necessary.
6. Start the sectioning very slow 0.05 mm/s and only increase the speed if applicable (up to 0.3 mm/s). In case the tissue is very

stiff, it helps to increase the amplitude from 1.0 mm up to 1.3 mm.

7. Tissue pieces might break out of the agarose if there are ducts or adipose tissue inside. Try to process the tissue again, re-embed it in agarose, and mount it fresh on the plate. Make sure the razor blade is sharp or renew it.
8. Preselect the slices used for experiments by identifying slices rich in islets with transmitted light microscopy. If you have difficulties recognizing islets, apply dithizone staining to a selected number of slices and monitor before and after staining to gain experience in islet identification. Important: Do not use the slices after dithizone staining, as this has effects on islet biology.
9. Seal all unneeded tubings and outlets by clamping its tubing or attaching a dead-end tubing (knot on a short perifusion tubing ending). Otherwise, air bubbles might occur during perifusion.
10. Trim the slices before embedding into the chamber to fit more tissue without overlaying.
11. Tighten screws leaving some distance to the reservoir outlet for easier tubing access.
12. Air bubbles during perifusion can be sucked out using a small syringe entering via the metal tube very slowly.

## Acknowledgment

This work was supported with funds from the Paul Langerhans Institute Dresden (PLID) of Helmholtz Zentrum München at the University Clinic Carl Gustav Carus of Technische Universität Dresden, the German Ministry of Education and Research (BMBF) to the German Centre for Diabetes Research (DZD), the DFG SFB/Transregio 127, the European Foundation for the Study of Diabetes (EFSD)/Boehringer Ingelheim Basic Research Programme, and the Helmsley Charitable Trust George S. Eisenbarth nPOD Award for Team Science.

### References

1. Bhagat L et al (2000) Heat shock protein 70 prevents secretagogue-induced cell injury in the pancreas by preventing intracellular trypsinogen activation. J Clin Invest 106:81–89
2. Blinman TA et al (2000) Activation of pancreatic acinar cells on isolation from tissue: cytokine upregulation via p38 MAP kinase. Am J Physiol Cell Physiol 279:C1993–C2003
3. Irving-Rodgers HF et al (2012) Pancreatic islet basement membrane loss and remodeling after mouse islet isolation and transplantation: impact for allograft rejection. Cell Transplant 23:59–72
4. Negi S et al (2012) Analysis of beta cell gene expression reveals inflammatory signaling and evidence of dedifferentiation following human islet isolation and culture. PLoS One 7:e30415

5. Paraskevas S et al (1999) Activation and expression of ERK, JNK, and p38 MAP-kinases in isolated islets of Langerhans: implications for cultured islet survival. FEBS Lett 455:203–208
6. Raposo do Amaral AS et al (2013) Glutathione ethyl ester supplementation during pancreatic islet isolation improves viability and transplant outcomes in a murine marginal islet mass model. PLoS One 8:e55288
7. Speier S, Rupnik M (2003) A novel approach to in situ characterization of pancreatic beta-cells. Pflugers Arch 446:553–558
8. Lacy PE, Kostianovsky M (1967) Method for the isolation of intact islets of Langerhans from the rat pancreas. Diabetes 16:35–39
9. Li DS, Yuan YH, Tu HJ, Liang QL, Dai LJ (2009) A protocol for islet isolation from mouse pancreas. Nat Protoc 4:1649–1652
10. Logsdon CD, Williams JA (1983) Pancreatic acini in short-term culture: regulation by EGF, carbachol, insulin, and corticosterone. Am J Phys 244:G675–G682
11. Ivanova A et al (2013) Age-dependent labeling and imaging of insulin secretory granules. Diabetes 62(11):3687–3696
12. Marciniak A, Selck C, Friedrich B, Speier S (2013) Mouse pancreas tissue slice culture facilitates long-term studies of exocrine and endocrine cell physiology in situ. PLoS One 8:e78706
13. Marciniak A, Cohrs C et al (2014) Using pancreas tissue slices for in situ studies of islet of Langerhans and acinar cell biology. Nat Protoc 9(12):2809–2822
14. Speier S, Yang SB, Sroka K, Rose T, Rupnik M (2005) KATP-channels in beta cells in tissue slices are directly modulated by millimolar ATP. Mol Cell Endocrinol 230:51–58
15. Speier S, Gjinovci A, Charollais A, Meda P, Rupnik M (2007) Cx36-mediated coupling reduces beta cell heterogeneity, confines the stimulating glucose concentration range, and affects insulin release kinetics. Diabetes 56:1078–1086
16. Huang YC et al (2013) In situ electrophysiological examination of pancreatic alpha cells in the streptozotocin-induced diabetes model, revealing the cellular basis of glucagon hypersecretion. Diabetes 62:519–530
17. Rose T, Efendic S, Rupnik M (2007) Ca2+–secretion coupling is impaired in diabetic Goto Kakizaki rats. J Gen Physiol 129:493–508
18. Huang YC, Rupnik M, Gaisano HY (2011) Unperturbed islet alpha cell function examined in mouse pancreas tissue slices. J Physiol 589:395–408
19. Cohrs C, Chen C et al (2017) Vessel network architecture of adult human islets promotes distinct cell-cell interactions in situ and is altered after transplantation. Endocrinology 158(5):1373–1385
20. Weitz J et al (2018) Mouse pancreatic islet macrophages use locally released ATP to monitor beta cell activity. Diabetologia 61(1):182–192

# Chapter 21

# Determining Beta Cell Mass, Apoptosis, Proliferation, and Individual Beta Cell Size in Pancreatic Sections

Noèlia Téllez and Eduard Montanya

## Abstract

Pancreatic beta cells have a significant remodeling capacity which plays an essential role in the maintenance of glucose homeostasis. Beta cell apoptosis, replication, size, dedifferentiation, and (neo)generation contribute to the beta cell mass regulation. However, the extent of their respective contribution varies significantly depending on the specific condition, and it is the balance among them that determines the eventual change in beta cell mass. Thus, the study of the pancreatic beta cell mass regulation requires the determination of all these factors. In this chapter, we describe the quantification of beta cell replication based on the incorporation of thymidine analogs into replicated DNA strands and on the expression of Ki67 antigen and phosphorylation of histone H3. Beta cell apoptosis is analyzed by the TUNEL technique, and beta cell mass and cross-sectional area of individual beta cells are determined by computerized image processing methods.

**Key words** BrdU, EdU, PHH3, Ki67 antigen, Replication, TUNEL, Apoptosis, Beta cell mass, Cross-sectional individual beta cell area, Pancreatic islets

## 1 Introduction

The endocrine pancreas has a significant remodeling capacity that plays an essential role in the maintenance of glucose homeostasis [1]. Beta cell mass is dynamic and adjusts to meet the changes in metabolic demand, both in physiologic and pathologic conditions [2, 3]. The failure to increase beta cell mass in response to increased metabolic demand results in hyperglycemia.

Beta cell mass reduction is a central event in the development of type 1 and type 2 diabetes mellitus. In type 1 diabetes, there is an extensive reduction in β-cell mass due to autoimmune destruction of β-cells [4], while in type 2 diabetes there is an inadequately low β-cell mass that cannot meet the increased demands encountered by insulin resistance [5]. The remodeling capacity of the endocrine pancreas could be used to design strategies aimed at the expansion of the beta cell mass.

Aileen J. F. King (ed.), *Animal Models of Diabetes: Methods and Protocols*, Methods in Molecular Biology, vol. 2128, https://doi.org/10.1007/978-1-0716-0385-7_21, 

Quantification of the beta cell mass in vivo has significantly evolved over the last years. Noninvasive techniques involving radiotracers and positron emission tomography (PET) or single photon emission computed tomography (SPECT) have shown promising results [6, 7]. Nevertheless, reliable in vivo quantification of beta cell mass is not yet well established, and current estimation of beta cell mass is only possible based on tissue sections.

Consistent measurement of β-cell mass in histological specimens depends on six fundamental steps: (1) accurate dissection of the pancreas, (2) pancreas processing for histology, (3) sampling, (4) insulin labeling, (5) image acquisition, and (6) image analysis. In this chapter, the six steps will be methodically described.

Depending on the specific condition analyzed, beta cell apoptosis, replication, size, neogenesis, transdifferentiation, and dedifferentiation contribute differently to the remodeling of the endocrine pancreas, and the balance among them determines whether beta cell mass is eventually increased or reduced. Thus, the study of the pancreatic remodeling requires the determination of all these factors for a proper understanding of the events taking place.

Beta cell apoptosis, replication, and mass and cross-sectional individual beta cell area can be detected and quantified directly, and the methods that will be presented in this chapter are well established.

On the other hand, available methods to evaluate beta cell neogenesis, transdifferentiation, and dedifferentiation are indirect due to the current absence of specific markers (i.e., dual labeling of keratin 20 and insulin used for the identification of neogenesis [8]; dual labeling of insulin and glucagon, for alpha-to-beta conversion [9]; or dual labeling of chromogranin A and islet hormones, for dedifferentiation [10]). However, these markers do not provide direct proof of the aforementioned events, and it is currently established that cell tracing experiments are required for a direct proof of cell conversion. Genetic cell tracing is based on the expression of tracer proteins in a particular cell type to allow recognition of the ancestral cell identity in eventual transdifferentiated, dedifferentiated, or newly generated cells. This type of analysis requires gene manipulation, and therefore the use of transgenic rodents or recombinant virus for transgene delivery into target cells is mandatory [11–14].

The techniques that will be described in this chapter for measuring beta cell mass, beta cell apoptosis, beta cell replication, and individual beta cell size have been used mostly in pancreatic sections. However, they can be also used to measure these parameters in transplanted islets [15, 16] and in cultured islets [17], essentially with no modifications.

The measurement of beta cell apoptosis or proliferation requires a double staining to identify both the beta cells and the cells that are apoptotic or actively replicating. The identification of beta cells may be performed using an anti-insulin antibody to stain the cytosol of beta cells or an anti-NKX6.1 antibody to stain the nucleus of beta cells. Staining with insulin provides optimal results; nevertheless, nuclear labeling of beta cells may be more accurate when assessing beta cell replication in conditions of beta cell degranulation where some beta cell nuclei lack adjacent insulin staining.

There are several reliable commercial kits available to measure apoptosis and replication. The principle of studying apoptosis is that it is a form of cell death that eliminates compromised or superfluous cells. It is controlled by multiple signaling and effector pathways that mediate active responses to external growth, survival, or death factors.

Of all the aspects of apoptosis, the defining characteristic is a complete change in cellular morphology. The cell undergoes shrinkage, chromatin margination, membrane blebbing, DNA fragmentation, nuclear condensation and then segmentation, and division into apoptotic bodies which may be phagocytosed. In this chapter, we describe the TUNEL (TdT- mediated dUTP nick-end labeling) method to determine beta cell apoptosis in histological specimens. This technique is based on the detection of the 180 bp internucleosomal DNA multimers that result from DNA fragmentation in apoptotic cells. These DNA strand breaks are detected by enzymatically labeling the free 3′-OH termini with modified nucleotides. Labeled DNA ends are typically localized in morphologically identifiable condensed nuclei and apoptotic bodies. Nevertheless, apoptosis measurement can be difficult to assess, and it may be advisable to use a second technique. We have previously used as confirmatory measurement of beta cell apoptosis cell cytometry [18].

In this chapter, we also describe several different methods to measure beta cell proliferation including thymidine analog incorporation and staining for the proliferation markers Ki67 and phosphohistone H3 (PHH3).

Thymidine analogs get incorporated into the DNA of cells in the S phase of the cell cycle. During cell proliferation, DNA replicates during the S phase of the cell cycle before the cell is divided into two daughter cells. This close association between DNA synthesis and cell doubling makes the measurement of DNA synthesis very attractive for assessing cell proliferation. If labeled DNA precursors are administered to the experimental animal or added to the cell culture, cells that are about to divide incorporate the labeled nucleotide into their DNA. BrdU and EdU are thymidine analogs that are incorporated into the DNA of proliferating cells in the S phase of the cell cycle. Therefore, they can be subsequently detected by immunohistochemistry.

The technique requires the administration of the nucleoside some time before harvesting the organ to allow its incorporation into the DNA of the replicating cells.

5-Ethynyl-2′-deoxyuridine (EdU) is structurally similar to the natural nucleoside in which a terminal alkyne group replaces the methyl group in the fifth position. EdU detection is based on a copper-catalyzed covalent reaction between a dye-conjugated azide and the alkyne group of the EdU, known as click chemistry [20]. The small-sized dye-azide complex allows for efficient EdU detection avoiding harsh conditions that degrade the structure of the cells [21, 22]. The main limitation for a broader use of EdU, replacing BrdU, is the higher cost of EdU reagents (up to fivefold).

If nucleoside injection is not possible (for instance, with human organs) or has not been done, the staining with Ki67 or PHH3 antibodies offers an alternative. Ki67 is a nuclear antigen associated with cell proliferation and is present throughout the active cell cycle ($G_1$, S, $G_2$, and M phases), but absent in resting cells ($G_0$). Compared with BrdU staining, the advantage of Ki67 is that no antigen must be injected to identify replication. This makes Ki67 particularly useful when beta cell replication is measured in the human pancreas. Ki67 antibody stains the cells that are in $G_1$, S, $G_2$, and M phases of the cell cycle, but not resting cells in $G_0$. A caveat of the technique is that cells that may not progress beyond the $G_1$ phase, and that will not replicate, are also stained. Thus, Ki67 staining may be less accurate than thymidine analog staining in the identification of cells that really replicate. Conversely, PHH3 antibody labels cells that passed the $G_2$ checkpoint of the cell cycle and that are committed to undergo cell division. Phosphorylation of histone H3 in serine 10 is negligible during interphase but strongly increases in late $G_2$ and M phases of the cell cycle, which is associated with mitotic chromatin condensation. Thus, PHH3 labels cells that undergo true mitosis after $G_2$ checkpoint. The main limitation of the use of this marker for the determination of beta cell replication is the low proportion of cells that become labeled due to the short duration of the M phase in these cells ($\approx$0.5 h) [19] combined with the slow turnover of these cells. Therefore, the use of PHH3 labeling is recommended to confirm the results obtained with thymidine analogs and Ki67.

Reliable in vivo quantification of beta cell mass is not yet well established despite recent progress [6], and current estimation of beta cell mass is only possible based on tissue sections. The estimation of pancreatic beta cell mass (or any other tissue) based on microscopic study of tissue sections has been a complex issue. Flat, two-dimensional sections must be related to the three-dimensional parameters that define the volume of interest, and the tissue sections provide restricted information about the distribution of the tissues of interest in the whole structure (s***tructure*** being defined as "something made up of interdependent parts in a

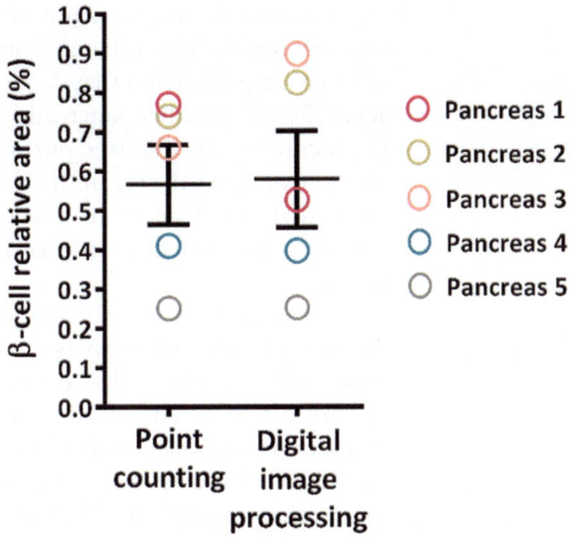


**Fig. 1** Comparison of traditional point counting morphometry vs. computerized image processing analysis of beta cell relative area in control pancreases. Values are individually represented, and mean with the standard error has been included to help in comparison between methodologies

definite pattern of organization"). Stereology, defined as "the body of mathematical methods relating three dimensional parameters defining the structure to two dimensional measurements obtainable on sections of the structure," was born to solve the difficulties of the spatial interpretation of sections. It provides methods by which to obtain reliable quantitative data on structure despite the restrictive nature of the information gathered by the study of tissue sections. To quantify the relative cell volume of the tissue of interest, several methods can be used, such as point counting, planimetry, or linear scanning. Systematic point counting has been largely considered to be the best method for estimating the volume density by stereological methods. However, this method is labor intensive and time consuming, and over the last decade it has been increasingly replaced by computerized image processing methods. Currently, digital image processing has become a reliable alternative to point counting morphometry for pancreatic beta cell mass analysis [10] (*see* Fig. 1); thus, this method will be discussed in the chapter.

To estimate the total beta cell mass of the pancreas, it is indispensable to have the weight of the pancreas and to determine the relative beta cell area. The pancreas weight is obtained as described in Subheading 3.1. The relative beta cell area is obtained by digital image processing. With these data, the absolute mass of the pancreatic beta cells can be estimated with the following formula:

$$\text{Beta cell mass} = \text{Relative beta cell area} \times \text{Pancreatic weight}$$

If pancreatic weight is not available, then the relative beta cell area, but not the beta cell mass, can be estimated. Since relative beta cell area is a proportion that depends on the tissue area occupied by beta cells and by other pancreatic cells (basically exocrine pancreas), a reduction in the exocrine portion of the pancreas would increase the relative beta cell area in the absence of changes in total beta cell mass (that could be even reduced). Thus, increments in relative beta cell area do not necessarily correlate with increased beta cell mass.

The contribution of beta cell hypertrophy or atrophy to changes in beta cell mass is often overlooked, despite the very important role that cell size has in order to determine the total mass of a tissue. The method that we will discuss quantifies the mean cross-sectional area of individual beta cells, a measure of individual beta cell size. In essence, the beta cell area of the pancreatic islets, measured by digital image processing, is divided by the number of beta cell nuclei in those islets. It must be taken into account that this method overestimates the size of the beta cells because the actual number of beta cells in the islets is higher than that counted since not all beta cells are sectioned across their nuclei and therefore some beta cell nuclei are not identified.

The protocols below detail the procedures which can be used to determine rates of beta cell apoptosis, beta cell proliferation, beta cell size, and calculations of beta cell mass in human and rodent tissues.

## 2 Materials

### 2.1 Pancreas Extraction and Processing

1. Dissecting scissors, straight, blunt/blunt, 10.5 cm long. 2 units.
2. Splinter forceps, curved, 10.5 cm long. 2 units.
3. Cold light shadowless surgery lamp.
4. Isoflurane vaporizer connected to a scavenger system with exhalation hose.
5. 70% ethanol in a squeeze bottle.
6. 10% iodopovidone in a squeeze bottle.
7. Sterile surgical drapes and gauze pads.
8. Petri dishes, 3.5 cm in diameter.
9. 10 ml syringes connected to a 20-gauge needle.
10. Sterile saline (0.9% (w/v) NaCl).
11. 3.7% Formaldehyde solution
12. Plastic cassettes with lid and metal molds for paraffin embedding.

13. Grease-proof paper pieces (3 × 2 cm).
14. Analytical balance.
15. Phosphate buffer (PB) 0.1 M (500 mL): Prepare 250 ml of buffer A (0.2 M $NaH_2PO_4 \cdot H_2O$) and 250 mL of buffer B (0.2 M $Na_2HPO_4$). Mix 169 mL of buffer A with 81 mL of buffer B and bring this final solution up to 500 mL by adding 250 mL of double distilled water ($ddH_2O$). pH of the solution: 7.2.
16. Cold plate.
17. Positively charged microscope slides 76 × 26 mm and microscope cover slips 24 × 32 mm.

### 2.2 Beta Cell Identification

1. Xylene, absolute ethanol, 90% ethanol, and 70% ethanol.
2. Phosphate-buffered saline (PBS): 160 mM NaCl, 7 mM $Na_2HPO_4$, 3 mM $NaH_2PO_4 \cdot H_2O$ (adjust to pH 7.4 with HCl if necessary).
3. Citrate buffer: 9.9 mM citric acid, 43 mM trisodium citrate dihydrate. Adjust to pH 6 with citric acid (when higher) or NaOH (when lower).
4. DNase I (10 μg/mL).
5. Streptavidin/biotin blocking kit.
6. Blocking solution: 5% (v/v) horse serum, 1% (v/v) Triton X-100 in PBS.
7. Identification of beta cells, cytosolic. Rabbit anti-human insulin (dilution 1/100) (*see* **Note 1**).
8. Identification of beta cells, nuclear. NKX6.1 antibody: mouse anti-NKX6.1 (dilution 1/100) (*see* **Note 1**).
9. Secondary antibody for immunofluorescence: Alexa Fluor® 555-labeled anti-rabbit IgG (dilution 1/400) (*see* **Note 1**).
10. Biotin-conjugated secondary antibody (dilution 1/200) and Alexa Fluor® 555-conjugated streptavidin (dilution 1/200) (*see* **Note 1**).
11. 4′-6-Diamidino-2-phenylindole (DAPI 300 nM).
12. Mounting medium for immunofluorescence.

### 2.3 Apoptosis

1. Proteinase K (20 μg/mL in PBS).
2. ApopTag® Plus Fluorescein in situ apoptosis detection kit (*see* **Note 2**).

### 2.4 Proliferation: BrdU

1. BrdU (20 mg/mL solution in PBS) (*see* **Note 3**).
2. HCl: 2 N HCl.
3. 0.1 M Borax (in $ddH_2O$).

4. BrdU mouse monoclonal antibody (dilution 1/100) (*see* **Note 4**).
5. Secondary antibody for immunofluorescence: Alexa Fluor® 488-labeled anti-rabbit IgG (dilution 1/400).

#### 2.5 Proliferation: EdU

1. EdU (20 mg/mL solution in PBS).
2. Click-iT® EdU Alexa Fluor® 594 Imaging Kit.

#### 2.6 Proliferation: PHH3

1. Trypsin (0.5 mg/mL in PBS).
2. Washing solution: 1% v/v horse serum, 1% v/v Triton X-100 in PBS.
3. Rabbit anti-PHH3 antibody (dilution 1/250).
4. Secondary antibody for immunofluorescence: Alexa Fluor® 488-labelld anti-rabbit IgG (dilution 1/400).

#### 2.7 Proliferation: Ki67

1. Domestic stainless-steel pressure cooker.
2. Hot plate.
3. Vessel with plastic slide rack to hold approximately 400–500 mL.
4. Ki67 antibody: rabbit anti-human Ki67 (diluted 1/100 in PBS) (*see* **Note 5**).
5. Secondary antibody for immunofluorescence: Alexa Fluor® 488-labeled anti-rabbit IgG (we use dilution 1/400).

## 3 Methods

### 3.1 Dissection of the Rodent Pancreas

#### 3.1.1 Set the Area

1. Stage a work area with the analytical balance and the formaldehyde-containing vessel. Label plastic cassettes according to mouse identification numbers and weigh them individually without lid and with the grease-proof piece of paper accommodated in the cassette (*see* **Note 6**).
2. Stage the operation table with all the equipment and supplies in a manner that allows comfortable performance of the procedure. Cover the surgical area with a sterile drape and place the following supplies: cold light shadowless surgery lamp, two pairs of scissors, two pairs of forceps, enough sterile surgical gauze pads, 10 ml syringe with sterile saline solution, 1 petri dish, 1 plastic cassette, and the grease-proof paper piece (pre-weighed).

#### 3.1.2 Anesthesia

1. Place the mouse in the induction chamber for anesthesia and adjust the oxygen flowmeter to approximately 0.5–1.0 L/min and the isoflurane vaporizer to 3–5%. Supervise the mouse while undergoing anesthesia. After the induction of anesthesia,

adjust the isoflurane vaporizer to 1–3% for maintenance and place the mouse on the operation table in supine position with the head pointed ahead from the surgeon and connected to the anesthesia hose.

#### 3.1.3 Pancreas Dissection

1. Perform the stimulus test by a foot pinch ensuring that the mouse is unresponsive to stimuli.
2. Pinch the skin of the mouse at the urethral opening and pull upward to make an incision with the surgical scissors.
3. Expose the abdominal cavity by midline laparotomy. If blood collection is required for metabolic analysis, perform exsanguination of the mouse at this point and finally carry out euthanasia by cutting the diaphragm and heart with scissors.
4. Pinch the cecum of the large intestine, pull slightly upward, and slide the intestines to your left-hand side to expose the pancreas. Start pancreas dissection from the distal colon by detaching the pancreas from the intestines and go upward until the pylorus. At this point, proceed with the pancreas dissection from the stomach toward the pylorus and duodenum to completely resect the pancreas. Keep the spleen attached to the pancreas in order to identify the tail of the pancreas. Hold and lift up the pancreas and spleen and detach the tissue from mesenteric connective tissue (*see* **Note 7**).
5. Spread out the pancreas on the petri dish preserving the spatial orientation of the tissue (*see* Fig. 2a) and remove contaminating tissues that may have been also excised (i.e., fat, duodenum, lymph nodes) (*see* Fig. 2b). If necessary, sprinkle the pancreas with sterile saline solution to clean from any residual blood, and carefully blot the pancreas with a sterile gauze pad (*see* **Note 8**). At this point, remove the spleen and accommodate the pancreas in the previously weighed cassette with the head facing the beveled end and spread it flat so that the tail faces the opposite side of the cassette (*see* Fig. 3).
6. Weigh the pancreas placed in the cassette in a scientific scale, close the lid, and immerse the cassette in the formaldehyde-containing vessel (*see* **Note 9**). Calculate pancreas weight by subtracting the cassette weight from the total weight (*see* **Note 10**).

### 3.2 Pancreas Processing for Immunohistochemistry

1. Leave the tissue in fixation solution overnight at 4 °C.
2. Wash the cassette containing the pancreas with phosphate buffer (PB) by changing the solution three times for 5 min each.
3. Process for paraffin embedding. If the pancreas is not embedded immediately, it can be kept in phosphate buffer solution for 2 weeks at 4 °C.

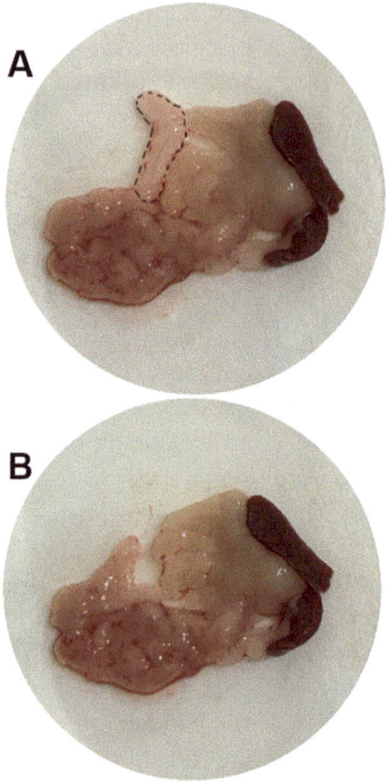


**Fig. 2** Dissected pancreas with spleen before (**a**) and after (**b**) removal of contaminating fat tissue. (**a**) Fat tissue is highlighted by black dashed lines. (**b**) Pancreas cleaned from fat. Pancreas dissection was performed as described in Subheading 3.1

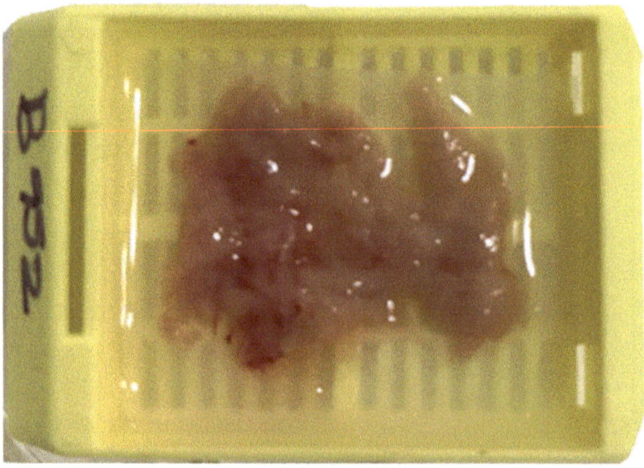

**Fig. 3** Oriented mouse pancreas in the histology cassette. Left to right: head, body, and tail of the pancreas. Pancreas dissection was performed as described in Subheading 3.1

### 3.3 Paraffin Embedding

1. Take the cassettes through a series of graded EtOH baths to dehydrate the tissues and then into xylene. Hot paraffin can then permeate the tissues.
2. Wash the specimen in cold running water for 5 min.
3. Dehydration: 70% ethanol 1 h (×1), 90% ethanol 1 h (×2), 100% ethanol 1 h (×3), xylene 30 min (×2).
4. Paraffin (65 °C) 1 h (×2) (*see* **Note 11**).
5. Turn the heat block on to melt the paraffin 1 h before adding the tissue cassettes. Also warm metal block molds on the hot plate.
6. Add hot paraffin to the mold. Use heated forceps to orient the tissue in the mold. When the tissue is in the desired orientation, add the labeled tissue cassette on top of the mold as a backing. Be sure there is enough paraffin to cover the face of the plastic cassette.
7. Slide the mold off the hot plate onto a cold plate. When the wax is completely cooled and hardened (~20 min), the paraffin block can be popped out of the mold. If the wax cracks or the tissues are not aligned well, melt them again and start over.

### 3.4 Tissue Sectioning

1. Turn on the water bath and check that the temperature is 35–37 °C.
2. Place the blocks to be sectioned facedown on an ice block for 10 min.
3. Place a fresh blade on the microtome. Insert the block into the microtome chuck so the wax block faces the blade and is aligned in the vertical plane. Set the dial to cut 2–4 μm sections. The blade should angle ~10°. Face the block by cutting it down to the desired tissue plane and discard the paraffin ribbon. If the block is ribboning well, then cut another four sections and pick them up with forceps or a fine paintbrush and float them on the surface of the 37 °C water bath. Float the sections onto the surface of clean glass slides. If the block is not ribboning well, then place it back on the ice block to cool off and firm up the wax. If the specimens fragment when placed on the water bath, then they may be too hot.
4. Place the slides with paraffin sections in a 65 °C oven for 20 min (so the wax just starts to melt) to bond the tissue to the glass. Slides can be stored overnight at room temperature (*see* **Note 12**).

### 3.5 Beta Cell Identification (Cytoplasmic): Insulin Immunofluorescence

#### 3.5.1 Tissue Section Deparaffinization and Rehydration (in a Coplin Jar)

1. Wash the specimen in two changes of xylene for 7 min each wash.
2. Wash the specimen in two changes of absolute ethanol for 5 min each wash.
3. Wash the specimen in two changes of 95% ethanol for 5 min each wash.
4. Wash the specimen once in 70% ethanol for 10 min.
5. Rinse the specimen once in deionized water for 2 min.
6. Wash the specimen once in PBS for 5 min.

#### 3.5.2 Antigen Retrieval

1. Prepare citrate buffer, pour it in Coplin jars, allocate four slides per jar, and place them into the microwave. Use defrost function for two rounds of 7 min each.
2. Cool down slides by transferring them into Coplin jars with fresh washing buffer and incubate at room temperature for 20 min. Exercise care with hot Coplins and slides. Use forceps!

#### 3.5.3 Antibody Incubation and Mounting

1. Place slides in blocking solution for 15 min at room temperature.
2. Gently tap off excess liquid and carefully blot or aspirate around the section. Apply sufficient insulin antibody (1/100) to cover the specimen. Incubate o/n at 4 °C.
3. Warm buffers and slides up to room temperature for 1 h.
4. Wash the specimen in three changes of PBS for 5 min each wash.
5. Wipe around the specimen, and add sufficient Alexa Fluor® 555-labeled anti-rabbit IgG (1/400) antibody to cover the specimen. Incubate for 1 h at room temperature.
6. Wash the specimen in three changes of PBS for 5 min each wash.
7. Wipe around the specimen and add sufficient 4′-6-diamidino-2-phenylindole solution (DAPI) to cover the specimen. Incubate for 10 min at room temperature.
8. Wash the specimen in three changes of PBS for 5 min each wash.
9. Mount with mounting medium for immunofluorescence. Store the slides for 24 h at −80 °C prior visualization under the microscope.

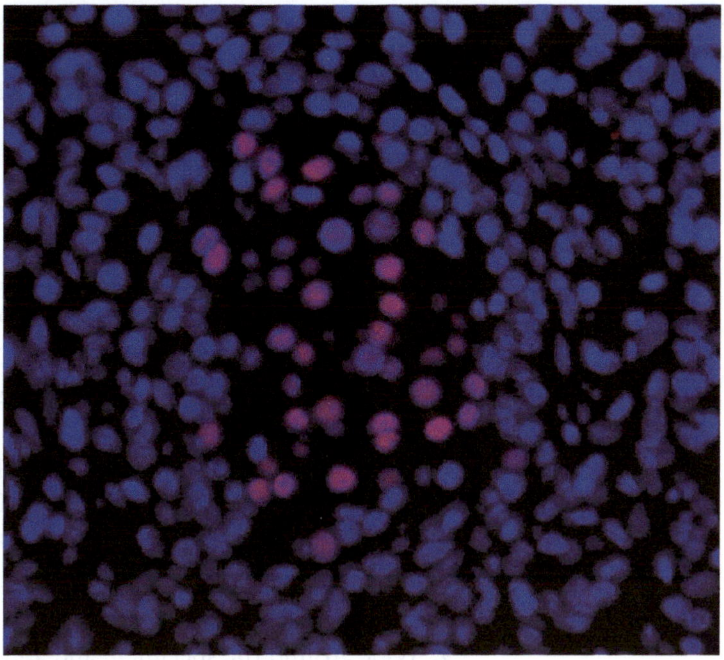

**Fig. 4** Beta cell identification (nuclear). NKX6.1 immunofluorescence. Representative fluorescence microscopy image of the islets of young adult rats stained for NKX6.1. This image was captured using a 20× microscope objective. NKX6.1 was visualized as described in Subheading 3.6

## *3.6 Beta Cell Identification (Nuclear): NKX6.1 Immunofluorescence (See Fig. 4)*

1. Deparaffinize and rehydrate tissue sections (*see* Subheading 3.5.1).
2. Carry out antigen retrieval (*see* **step 2** in Subheading 3.5.2).

### *3.6.1 Nuclear Antigen Retrieval*

1. Apply freshly diluted DNase I (10 μg/mL) to the specimen for 15 min at room temperature directly on the slide.
2. Wash the specimen in three changes of washing buffer for 5 min each wash.
3. Place slides in blocking solution for 15 min at room temperature.

### *3.6.2 Inhibition of Endogenous Biotin Activity*

1. Rinse the specimen in PBS for 5 min.
2. Apply streptavidin solution directly on the slide and incubate for 15 min at room temperature.
3. Rinse the specimen in PBS for 5 min.
4. Apply biotin solution directly on the slide and incubate for 15 min at room temperature.
5. Rinse the specimen in PBS for 5 min.

*3.6.3 Antibody Incubation and Mounting*

1. Gently tap off excess liquid and carefully blot or aspirate around the section.
2. Apply sufficient NKX6.1 antibody (1/100) to cover the specimen. Incubate o/n at 4 °C.
3. Warm buffers and slides up to room temperature for 1 h.
4. Wash the specimen in three changes of washing buffer for 5 min each wash.
5. Wipe around the specimen, and add sufficient biotinylated anti-mouse IgG (1/200) antibody to cover the specimen. Incubate for 1 h at room temperature.
6. Wash the specimen in three changes of washing buffer for 5 min each wash.
7. Wipe around the specimen, and add sufficient Alexa Fluor® 555-conjugated streptavidin (1/200) to cover the specimen. Incubate for 1 h at room temperature.
8. Wash the specimen in three changes of washing buffer for 5 min each wash.
9. Wipe around the specimen, and add sufficient 4′-6-diamidino-2-phenylindole solution (DAPI) to cover the specimen. Incubate for 10 min at room temperature.
10. Wash the specimen in three changes of washing buffer for 5 min each wash.
11. Mount with mounting medium for immunofluorescence. Store the slides for 24 h at −80 °C prior visualization under the microscope.

***3.7 Apoptosis: TUNEL Method (See Fig. 5)***

1. Deparaffinize the sections (as described in Subheading 3.5.1) in a Coplin jar.
2. Apply freshly diluted proteinase K (20 μg/mL) to the specimen for 15 min at room temperature in a Coplin jar or directly on the slide.
3. Wash the specimen in two changes of $ddH_2O$ in a Coplin jar for 2 min each wash.
4. Gently tap off excess liquid and carefully blot or aspirate around the section.
5. Immediately apply *equilibration buffer* directly on the specimen.
6. Incubate for at least 10 s at room temperature.
7. Gently tap off excess liquid and carefully blot or aspirate around the section.
8. Immediately pipette onto the section enough *working strength TdT enzyme*.

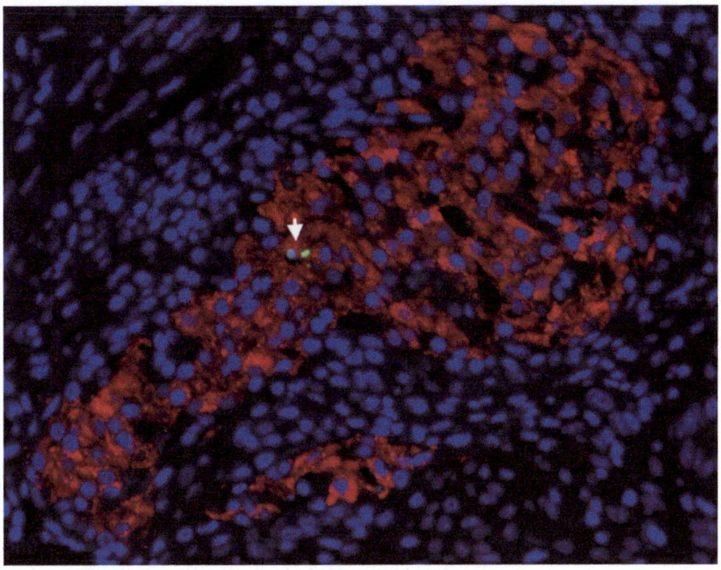

**Fig. 5** Beta cell apoptosis. TUNEL method. Representative fluorescence microscopy image of the islets of young adult rats stained for TUNEL (green) and insulin (red). This image was captured using a 20× microscope objective. TUNEL and insulin were visualized as described in Subheading 3.7

9. Incubate in a humidified chamber at 37 °C for 1 h.
10. Put the specimen in a Coplin jar containing *working strength stop/wash buffer*, agitate for 15 s, and incubate for 10 min at room temperature.
11. Warm up *anti-digoxigenin conjugate* to room temperature.
12. Wash the specimen in three changes of PBS for 1 min each wash.
13. Gently tap off excess liquid and carefully blot or aspirate around the section.
14. Apply room temperature fluorescein-conjugated *anti-digoxigenin conjugate* to the slide.
15. Incubate in a humidified chamber for 30 min at room temperature. (Protect from direct light.)
16. Wash the specimen in three changes of PBS in a Coplin jar for 3 min per wash at room temperature.
17. Perform insulin immunofluorescence starting at Subheading 3.5.3.
18. Visualize the slides under a fluorescence microscope (at a magnification of 200×, or 400× if needed) connected to a computer.
19. Screen the pancreas and capture all islets with the three channels (red, green, and UV).

20. By using an image editing software, overlay images acquired with the three channels.
21. Use merged images to count insulin-positive cells that are positive for TUNEL and divide by total number of beta cells (nucleus of insulin-positive cells) (*see* **Notes 13** and **14**).

### 3.8 Proliferation: Thymidine Analog Incorporation

#### 3.8.1 Beta Cell Proliferation: BrdU Incorporation (see Fig. 6)

1. 6 h before pancreas removal, administer BrdU solution at a dose of 100 mg/kg by intraperitoneal injection (*see* **Note 15**).
2. Remove and process the pancreas as described in Subheadings 3.1–3.4.
3. Deparaffinize tissues as described in Subheading 3.5.1.
4. Denature DNA by immersion of slides in 2 N HCl for 5 min at room temperature.
5. Wash the slides with 0.1 M Borax solution for 10 min at room temperature (*see* **Note 16**).
6. Wash the specimen in three changes of PBS for 5 min each wash.
7. Block the specimen in blocking solution for 15 min at room temperature.

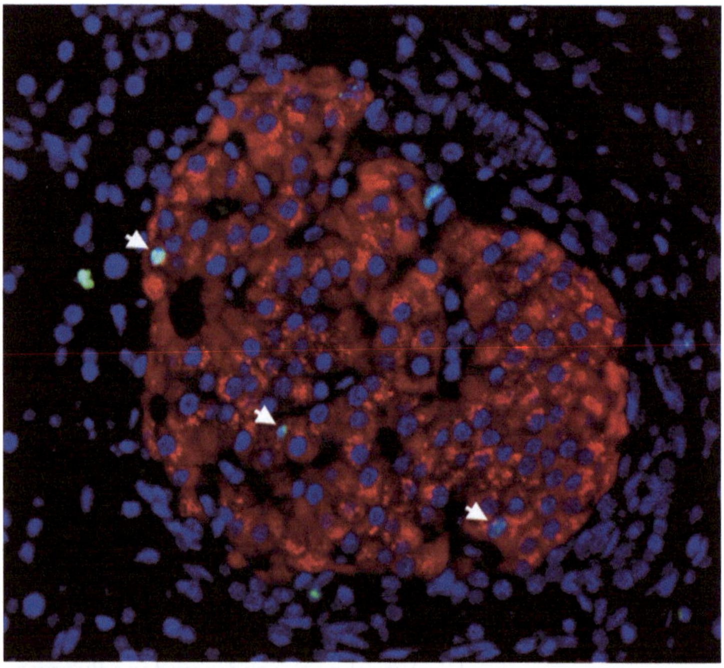

**Fig. 6** Beta cell proliferation. BrdU incorporation. Representative fluorescence microscopy image of the islets of 90% pancreatectomized young rats stained for BrdU (green) and insulin (red). 90% partial pancreatectomy induces beta cell replication. This image was captured using a 20× microscope objective. BrdU and insulin were visualized as described in Subheading 3.8.1

8. In the meantime, prepare enough cocktail of primary antibodies (rabbit anti-insulin (1/100) and mouse anti-BrdU (1/100)) in PBS (*see* **Note 17**).
9. Gently tap off excess liquid and carefully blot or aspirate around the section. Apply sufficient cocktail of primary antibodies to cover the specimen. Incubate o/n at 4 °C.
10. Warm buffers and slides up to room temperature for 1 h.
11. Wash specimens in three changes of PBS for 5 min each wash.
12. In the meantime, prepare enough cocktail of secondary antibodies (Alexa Fluor® 488 anti-mouse (1/400) and Alexa Fluor® 555 anti-rabbit (1/400)).
13. Wipe around the specimen and add sufficient cocktail of secondary antibodies to cover the specimen. Incubate for 1 h at room temperature protected from light (*see* **Note 18**).
14. Wash specimens in three changes of PBS for 5 min each wash.
15. Wipe around the specimen, and add sufficient DAPI solution to cover the specimen. Incubate for 10 min at room temperature.
16. Wash the specimen in three changes of PBS for 5 min each wash.
17. Mount with mounting medium for immunofluorescence. Store the slides for 24 h at −80 °C prior to visualization under the microscope.
18. Visualize the slides under a fluorescence microscope (at a magnification of 200×, or 400× if needed) connected to a computer.
19. Screen the pancreas and capture all islets with the three channels (green light (for red fluorescence), blue light (for green fluorescence), and UV light (for blue fluorescence)).
20. By using an image editing software, overlay images acquired with the three channels.
21. Use merged images to count insulin-positive cells that are positive for BrdU and divide by total number of beta cells (nucleus of insulin-positive cells) (*see* **Note 13**).

*3.8.2 Proliferation: EdU Incorporation (See Fig. 7)*

1. Inject EdU as described in Subheading 3.8.1.
2. Remove and process the pancreas as described in Subheadings 3.1–3.4.
3. Stain for insulin as outlined from Subheading 3.5.1 until Subheading 3.5.3.
4. After an overnight incubation with the insulin primary antibody at 4 °C, warm buffers and slides up to room temperature for 1 h.

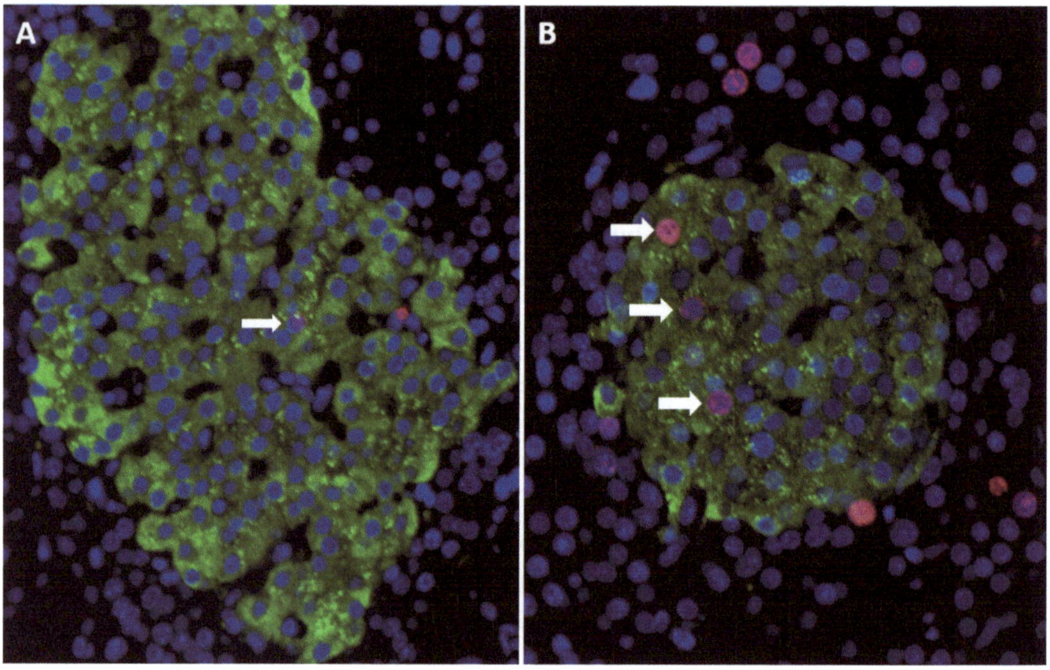


**Fig. 7** Beta cell proliferation. EdU incorporation. Representative fluorescence microscopy images of pancreatic islets stained for EdU (red) and insulin (green) in control rats (**a**) and after 90% partial pancreatectomy (**b**). 90% partial pancreatectomy induces beta cell replication. Images were captured using a 20× microscope objective. EdU and insulin were visualized as described in Subheading 3.8.2

5. Following manufacturer's instructions in the **Click-iT® EdU Alexa Fluor® 594 Imaging Kit**, prepare saponin solution in PBS from 10× concentrate provided with the kit.
6. Immerse slides in the saponin solution and incubate them for 10 min at room temperature.
7. In the meantime, prepare enough cocktail of the click reaction following manufacturer's instructions.
8. Wash specimens in three changes of PBS for 5 min each wash.
9. Wipe around the specimen and add sufficient click reaction cocktail to cover the specimen. Incubate for 30 min at room temperature protected from light.
10. Wash the specimen in three changes of PBS for 5 min each wash.
11. Gently tap off excess liquid and carefully blot or aspirate around the section. Apply sufficient Alexa Fluor® 488 anti-rabbit secondary antibody (1/400) to cover the specimen. Incubate for 1 h at room temperature.
12. Wash the specimen in three changes of PBS for 5 min each wash.

13. Wipe around the specimen and add sufficient DAPI solution to cover the specimen. Incubate for 10 min at room temperature.
14. Wash the specimen in three changes of PBS for 5 min each wash.
15. Mount with mounting medium for immunofluorescence. Store the slides for 24 h at −80 °C prior visualization under the microscope.
16. Visualize the slides under a fluorescence microscope (at a magnification of 200×, or 400× if needed) connected to a computer.
17. Screen the pancreas and capture all islets with the three channels (green light (for red fluorescence), blue light (for green fluorescence), and UV light (for blue fluorescence)).
18. By using an image editing software, overlay images acquired with the three channels.
19. Use merged images to count insulin-positive cells that are positive for EdU and divide by total number of beta cells (nucleus of insulin-positive cells) (*see* **Note 13**).

### 3.9 Proliferation: PHH3

1. Deparaffinize tissue sections as described in Subheading 3.5.1.
2. Carry out antigen retrieval as described in Subheading 3.5.2.
3. Apply freshly diluted trypsin (0.5 mg/mL) to the specimen and incubate for 15 min at room temperature.
4. Wash the specimen in three changes of washing solution for 5 min each wash.
5. Block the specimen in blocking solution for 15 min at room temperature.
6. In the meantime, prepare enough cocktail of primary antibodies (rabbit anti-insulin (1/100) and mouse anti-PHH3 (1/250)) in PBS (*see* **Note 17**).
7. Gently tap off excess liquid and carefully blot or aspirate around the section. Apply sufficient cocktail of primary antibodies to cover the specimen. Incubate o/n at 4 °C.
8. Warm buffers and slides up to room temperature for 1 h.
9. Wash specimens in three changes of washing buffer for 5 min each wash.
10. In the meantime, prepare enough cocktail of secondary antibodies (Alexa Fluor® 488 anti-mouse (1/400) and Alexa Fluor® 555 anti-rabbit (1/400)) in PBS.
11. Wipe around the specimen and add sufficient cocktail of secondary antibodies to cover the specimen. Incubate for 1 h at room temperature protected from light (*see* **Note 18**).

12. Wash specimens in three changes of washing buffer for 5 min each wash.
13. Wipe around the specimen, and add sufficient DAPI solution (3 nM) to cover the specimen. Incubate for 10 min at room temperature.
14. Wash the specimen in three changes of PBS for 5 min each wash.
15. Mount with mounting medium for immunofluorescence. Store the slides for 24 h at −80 °C prior visualization under the microscope.
16. Visualize the slides under a fluorescence microscope (at a magnification of 200×, or 400× if needed) connected to a computer.
17. Screen the pancreas and capture all islets with the three channels (green light (for red fluorescence), blue light (for green fluorescence), and UV light (for blue fluorescence)).
18. By using an image editing software, overlay images acquired with the three channels.
19. Use merged images to count insulin-positive cells that are positive for PHH3 and divide by total number of beta cells (nucleus of insulin-positive cells) (*see* **Note 13**).

### 3.10 Proliferation: Ki67

1. Deparaffinize tissue sections as described in Subheading 3.5.1.
2. Pour enough volume of citrate buffer in the pressure cooker. Place the pressure cooker on the hot plate and turn it on full power. Place the lid on top of the pressure cooker and do not secure it at this point. Wait until citrate buffer starts to boil and then transfer the slides placed in a plastic rack into the vessel.
3. Secure the pressure cooker lid following the manufacturer's instructions.
4. As soon as the cooker has reached full pressure, time 2 min.
5. After 2 min, activate the security valve and place the cooker in an empty sink. Cool down the pressure cooker with running tap water. Once depressurized, open the lid and run cold water into the cooker for 10 min to cool down slides.
6. Place slides in blocking solution for 30 min at room temperature.
7. In the meantime, prepare enough cocktail of primary antibodies (rabbit anti-insulin (1/100) and mouse anti-Ki67 (1/100)) in PBS (*see* **Note 17**).
8. Gently tap off excess liquid and carefully blot or aspirate around the section. Apply sufficient cocktail of primary antibodies to cover the specimen. Incubate o/n at 4 °C.

9. Warm buffers and slides up to room temperature for 1 h.
10. Wash specimens in three changes of washing buffer for 5 min each wash.
11. Wipe around the specimen and add sufficient cocktail of secondary antibodies (Alexa Fluor® 488 anti-mouse and Alexa Fluor® 555 anti-rabbit) to cover the specimen. Incubate for 1 h at room temperature protected from light (*see* **Note 18**).
12. Wash specimens in three changes of washing buffer for 5 min each wash.
13. Wipe around the specimen, and add sufficient DAPI solution (3 nM) to cover the specimen. Incubate for 10 min at room temperature.
14. Wash the specimen in three changes of PBS for 5 min each wash.
15. Mount with mounting medium for immunofluorescence. Store the slides for 24 h at −80 °C prior visualization under the microscope.
16. Visualize the slides under a fluorescence microscope (at a magnification of 200×, or 400× if needed) connected to a computer.
17. Screen the pancreas and capture all islets with the three channels (green light (for red fluorescence), blue light (for green fluorescence), and UV light (for blue fluorescence)).
18. By using an image editing software, overlay images acquired with the three channels.
19. Use merged images to count insulin-positive cells that are positive for Ki67 and divide by total number of beta cells (nucleus of insulin-positive cells) (*see* **Note 13**).

### 3.11 Beta Cell Mass: Computerized Image Processing

1. Weigh and process the pancreas as described in Subheadings 3.2–3.4.
2. Stain beta cells for insulin as described in Subheading 3.5.
3. Visualize the slides with the objective 20× under an epifluorescence microscope equipped with a high-resolution camera connected to a computer. Select the exposure time for each channel (insulin and DAPI) and save it. Ensure that the software is appropriately calibrated (*see* **Note 19**).
4. Systematically, capture consecutive fields along the *x*-axis acquiring the entire row and one out of every two fields along the *y*-axis.
5. In order to measure beta cell area, define the region of interest (ROI) based on wide-range tones of red positive pixels to select insulin-positive areas. Then, generate a macro to draw the ROI over the insulin-positive area on every single image. Calculate

the size of the area of the defined ROI in every image (*see* **Notes 20** and **21**).

6. In order to measure total pancreatic tissue area, use DAPI-labeled low-contrast images that will allow the identification of the tissue by the blue background.
7. Calculate the beta cell relative area by dividing beta cell area by total tissue area. Then estimate the beta cell mass by multiplying beta cell relative area by pancreas weight using the above formula.

### 3.12 Individual Beta Cell Size: Computerized Image Processing

1. Weigh and process the pancreas as described in Subheadings 3.2–3.4.
2. Identify beta cells by staining for insulin as described in Subheading 3.5.
3. Visualize the slides with the objective 20× under an epifluorescence microscope equipped with a high-resolution camera connected to a computer. Select the exposure time for each channel (insulin and DAPI) and save it. Ensure that the software is appropriately calibrated (*see* **Note 19**).
4. Systematically, capture consecutive fields along the $x$-axis acquiring the entire row and one out of every two fields along the $y$-axis.
5. In order to measure beta cell area, define the region of interest (ROI) based on wide-range tones of red positive pixels to select insulin-positive areas. Then, generate a macro to draw the ROI over the insulin-positive area on every single image. Calculate the size of the area of the defined ROI in every image (*see* **Notes 20** and **21**).
6. In order to measure total pancreatic tissue area, use DAPI-labeled low-contrast images that will allow the identification of the tissue by the blue background.
7. Calculate the beta cell relative area by dividing beta cell area by total tissue area.
8. Count all beta cell nuclei located in the selected beta cell area.
9. Obtain the individual cross-sectional beta cell area by dividing the total beta cell area by the total number of beta cell nuclei that it contains.

## 4 Notes

1. We use an insulin antibody from Santa Cruz Biotechnology, an NKX6.1 antibody from DSHB, an Alexa Fluor® 555-labeled anti-rabbit antibody from Invitrogen, a biotin-conjugated secondary antibody from Jackson, and an Alexa Fluor®

555-conjugated streptavidin antibody from Invitrogen. We found these antibodies to be useful for rat, mouse, and human pancreas. Numerous competitive reagents are available from other commercial sources.

2. We have found the Millipore kit to work very well in the pancreas. Numerous competitive reagents are available from other commercial sources.
3. This solution needs to be carefully heated (~40 °C) to dissolve.
4. We use clone MoBU-1, B35128, which does not cross-react with EdU which makes it valuable for dual-pulse labeling experiments.
5. We have found this antibody from Zymed to work very well in the human pancreas, but not in the rodent pancreas.
6. The grease-proof piece of paper preserves the gross tissue structure by providing an even base for the pancreas over the slots of the cassette.
7. Detach the pancreas from the intestines and stomach by gently pulling tissues apart with the two forceps.
8. The pancreas contains a high amount of digestive enzymes; thus, pancreas dissection must be carried out as quickly as possible in order to avoid potential autodigestion of the tissue.
9. Excision of the pancreas, placement in the cassette, and immersion in fixative must be done quickly if apoptosis is to be measured.
10. It is advisable to place the pancreas in the cassette in a way that maintains the spatial orientation of the pancreas, so that the position of the head, body, and tail can be identified. This will also facilitate the identification of the main duct and other ductal structures when the pancreas is sectioned.
11. Blocks can be left o/n in paraffin at 65 °C.
12. For beta cell mass quantification, eight to ten pancreas sections, 150 μm apart, must be analyzed.
13. A number of image editing and image analysis applications (commercial and free open-source) are currently available. GIMP and ImageJ are examples of open-source software.
14. We count all islets identified in the tissue section and a minimum of 1200 nuclei of beta cells per sample. If this number is not reached, a second section is counted.
15. The S phase duration of the cell cycle in beta cells is about 6 h [9]. Thus, a 6 h time period between BrdU injection and pancreas harvesting is recommended because it maximizes the number of proliferating beta cells that will be labeled.

16. HCl and Borax solutions are used to denature and stabilize DNA, respectively.
17. Double immunostaining requires the primary antibodies to be raised in different species.
18. In order to avoid cross-reactivity between antibodies in double immunostaining, ensure that secondary antibodies have been produced in the same species and are cross-absorbed against immunoglobulins from other species, particularly for the IgG species that will be combined.
19. Ensure that all images have been captured at the same magnification and size (pixels per inch, ppi).
20. ROIs must be manually verified and corrected if required.
21. Microscopy-derived images must be calibrated by the software. Note that calibration of images is critical, since the measurement given by the software will depend on it. Images with higher resolution would improperly show bigger beta cell area compared to those taken at lower resolution.

## Acknowledgments

This work has been supported by grants from the Catalan Diabetes Association (NT), University of Barcelona (NT), and Carlos III Health Institute (ISCIII) PI16/00462 co-funded by FEDER funds/European Regional Development Fund (ERDF) – "A Way to Build Europe" (EM) – and by CIBERDEM which is a project of ISCIII.

### References

1. Montanya E (2014) Insulin resistance compensation: not just a matter of β-Cells? Diabetes 63 (3):832–834. https://doi.org/10.2337/db13-1843
2. Tellez N, Joanny G, Escoriza J, Vilaseca M, Montanya E (2011) Gastrin treatment stimulates {beta}-cell regeneration and improves glucose tolerance in 95% pancreatectomized rats. Endocrinology 152(7):2580–2588
3. Montanya E, Nacher V, Biarnes M, Soler J (2000) Linear correlation between beta-cell mass and body weight throughout the lifespan in Lewis rats: role of beta-cell hyperplasia and hypertrophy. Diabetes 49 (8):1341–1346
4. Knip M (2017) Diabetes: loss of β-cell mass - an acute event before T1DM presentation? Nat Rev Endocrinol 13(5):253–254. https://doi.org/10.1038/nrendo.2017.33
5. Butler AE, Janson J, Bonner-Weir S, Ritzel R, Rizza RA, Butler PC (2003) Beta-cell deficit and increased beta-cell apoptosis in humans with type 2 diabetes. Diabetes 52(1):102–110
6. Eriksson O, Laughlin M, Brom M, Nuutila P, Roden M, Hwa A, Bonadonna R, Gotthardt M (2016) In vivo imaging of beta cells with radiotracers: state of the art, prospects and recommendations for development and use. Diabetologia 59(7):1340–1349. https://doi.org/10.1007/s00125-016-3959-7
7. Brom M, Woliner-van der Weg W, Joosten L, Frielink C, Bouckenooghe T, Rijken P, Andralojc K, Göke BJ, de Jong M, Eizirik DL, Béhé M, Lahoutte T, Oyen WJ, Tack CJ, Janssen M, Boerman OC, Gotthardt M (2014) Non-invasive quantification of the beta cell mass by SPECT with $^{111}$In-labelled exendin. Diabetologia 57(5):950–959. https://doi.org/10.1007/s00125-014-3166-3

8. Tellez N, Montanya E (2014) Gastrin induces ductal cell dedifferentiation and beta-cell neogenesis after 90% pancreatectomy. J Endocrinol 223(1):67–78. https://doi.org/10.1530/JOE-14-0222
9. Spijker HS, Ravelli RB, Mommaas-Kienhuis AM, van Apeldoorn AA, Engelse MA, Zaldumbide A, Bonner-Weir S, Rabelink TJ, Hoeben RC, Clevers H, Mummery CL, Carlotti F, de Koning EJ (2013) Conversion of mature human β-cells into glucagon-producing α-cells. Diabetes 62 (7):2471–2480. https://doi.org/10.2337/db12-1001
10. Téllez N, Vilaseca M, Martí Y, Pla A, Montanya E (2016) β-Cell dedifferentiation, reduced duct cell plasticity, and impaired β-cell mass regeneration in middle-aged rats. Am J Physiol Endocrinol Metab 311(3):E554–E563. https://doi.org/10.1152/ajpendo.00502.2015
11. Talchai C, Xuan S, Lin HV, Sussel L, Accili D (2012) Pancreatic β cell dedifferentiation as a mechanism of diabetic β cell failure. Cell 150 (6):1223–1234. https://doi.org/10.1016/j.cell.2012.07.029
12. Inada A, Nienaber C, Katsuta H, Fujitani Y, Levine J, Morita R, Sharma A, Bonner-Weir S (2008) Carbonic anhydrase II-positive pancreatic cells are progenitors for both endocrine and exocrine pancreas after birth. Proc Natl Acad Sci U S A 105(50):19915–19919
13. Thorel F, Nepote V, Avril I, Kohno K, Desgraz R, Chera S, Herrera PL (2010) Conversion of adult pancreatic alpha-cells to beta-cells after extreme beta-cell loss. Nature 464 (7292):1149–1154
14. Russ HA, Bar Y, Ravassard P, Efrat S (2008) In vitro proliferation of cells derived from adult human beta-cells revealed by cell-lineage tracing. Diabetes 57(6):1575–1583. https://doi.org/10.2337/db07-1283
15. Téllez N, Montolio M, Estil-les E, Escoriza J, Soler J, Montanya E (2007) Adenoviral overproduction of interleukin-1 receptor antagonist increases beta cell replication and mass in syngeneically transplanted islets, and improves metabolic outcome. Diabetologia 50 (3):602–611. https://doi.org/10.1007/s00125-006-0548-1
16. Estil les E, Téllez N, Escoriza J, Montanya E (2012) Increased β-cell replication and β-cell mass regeneration in syngeneically transplanted rat islets overexpressing insulin-like growth factor II. Cell Transplant 21(10):2119–2129. https://doi.org/10.3727/096368912X638955
17. Estil les E, Téllez N, Soler J, Montanya E (2009) High sensitivity of beta-cell replication to the inhibitory effects of interleukin-1beta: modulation by adenoviral overexpression of IGF2 in rat islets. J Endocrinol 203(1):55–63. https://doi.org/10.1677/JOE-09-0047
18. Téllez N, Montolio M, Biarnés M, Castaño E, Soler J, Montanya E (2005) Adenoviral overexpression of interleukin-1 receptor antagonist protein increases beta-cell replication in rat pancreatic islets. Gene Ther 12(2):120–128. https://doi.org/10.1038/sj.gt.3302351
19. Hija A, Salpeter S, Klochendler A, Grimsby J, Brandeis M, Glaser B, Dor Y (2014) G0-G1 transition and the restriction point in pancreatic β-cells in vivo. Diabetes 63(2):578–584. https://doi.org/10.2337/db12-1035
20. Diermeier-Daucher S, Clarke ST, Hill D, Vollmann-Zwerenz A, Bradford JA, Brockhoff G (2009) Cell type specific applicability of 5-ethynyl-2′-deoxyuridine (EdU) for dynamic proliferation assessment in flow cytometry. Cytometry A 75(6):535–546. https://doi.org/10.1002/cyto.a.20712
21. Salic A, Mitchison TJ (2008) A chemical method for fast and sensitive detection of DNA synthesis in vivo. Proc Natl Acad Sci U S A 105(7):2415–2420. https://doi.org/10.1073/pnas.0712168105
22. Carballar R, Canyelles ML, Fernández C, Martí Y, Bonnin S, Castaño E, Montanya E, Téllez N (2017) Purification of replicating pancreatic β-cells for gene expression studies. Sci Rep 7(1):17515. https://doi.org/10.1038/s41598-017-17776-2

# Index




Aileen J. F. King (ed.), *Animal Models of Diabetes: Methods and Protocols*, Methods in Molecular Biology, vol. 2128, https://doi.org/10.1007/978-1-0716-0385-7, © Springer Science+Business Media, LLC, part of Springer Nature 2020

MIX
Papier aus verantwortungsvollen Quellen
Paper from responsible sources
FSC® C105338

If you have any concerns about our products,
you can contact us on
**ProductSafety@springernature.com**

In case Publisher is established outside the EU,
the EU authorized representative is:
**Springer Nature Customer Service Center GmbH**
**Europaplatz 3, 69115 Heidelberg, Germany**

Printed by Libri Plureos GmbH
in Hamburg, Germany